T0132670

Experimental Endocrinology and Reproductive Biology

Experimental Endocrinology and Reproductive Biology

Editors

Chandana Haldar
Banaras Hindu University
Varanasi, India

Muniyandi Singaravel
Banaras Hindu University
Varanasi, India

S.R. Pandi-Perumal
Mount Sinai School of Medicine
New York, USA

Daniel P. Cardinali
University of Buenos Aires
Buenos Aires, Argentina

Science Publishers

Enfield (NH)   Jersey   Plymouth

SCIENCE PUBLISHERS
An imprint of Edenbridge Ltd., British Isles.
Post Office Box 699
Enfield, New Hampshire 03748
United States of America

Website: *http://www.scipub.net*

sales@scipub.net (marketing department)
editor@scipub.net (editorial department)
info@scipub.net (for all other enquiries)

Library of Congress Cataloging-in-Publication Data

Experimental endocrinology and reproductive biology/editors, Chandana Halda... [et al.]
    p. cm.
  Includes bibliographical references and index.
  ISBN 978-1-57808-518-7
    1. Endocrinology. 2. Reproduction--Endocrine aspects. I. Haldar, Chandana.

QP187.E97 2007
612.4--dc22

                                                                        2007040777

ISBN   978-1-57808-518-7

Published by Science Publishers, Enfield, NH, USA
An imprint of Edenbridge Ltd.
Printed in India

We dedicate this volume to all those who work on science and medicine

Foreword

This book deals with recent advances of experimental endocrinology and reproductive biology. Endocrinology is a field of study in which great advances have been made towards various directions in relation to other fields of biology such as developmental biology, environmental physiology, chronobiology, photobiology, reproductive biology, circulatory and digestive physiology, molecular biology, metabolic physiology, clinical and medical biology, etc. Comparative points of view have also accelerated the advancement of endocrinology. This book covers various topics of endocrinology from comparative, experimental, developmental, reproductive and clinical endocrine aspects. Another feature of this book is that more than half of the chapters were described in relation to the function of melatonin and the structure of the pineal organ. These trials of this book are reasonable and timely. Melatonin physiology has been reviewed from several points of view such as antioxidant and scavenger of hydroxyl radical, circadian clock and photoperiodic gonadal response including photoreceptor system, and development of vertebrates. I expect this book will give new insights and stimulate students and young scientists working on endocrinology and related fields of science.

Tadashi Oishi
President, Nara Saho College
Rokuyaon-cho 806, Nara 630-8566
Japan
Tel: +81-742-61-3858
Fax: +81-742-61-8054

Preface

Several recent discoveries both in laboratory and clinical settings have added to our understanding of experimental endocrinology and reproductive biology. These recent developments have also resulted in this field becoming increasingly interdisciplinary. A parallel development is that these fields are now reorganizing themselves at higher levels of complexity. The consequence of these developments is that it is becoming increasingly challenging for the researcher to assimilate, let alone master, the relevant findings in each of these fields.

To address this challenge the editors of the present volume have assembled chapters which summarize and review some of the latest discoveries concerning experimental endocrinology and reproductive biology. To this end contributions were sought from acknowledged experts, clinicians as well as basic researchers. The goal of this volume is therefore, to present the more recent developments in the fields of experimental endocrinology and to provide a context for considering them both in depth and from a multidisciplinary perspective.

This book is divided into 4 sections comprising of 17 chapters. SECTION I: Experimental Endocrinology; SECTION II: Reproductive Biology and Clinical Endocrinology; SECTION III: Developmental Endocrinology; and SECTION IV: Endocrine Physiology.

The volume begins with the new aspects of the protective actions and novel metabolites of melatonin. This chapter also addresses and summarizes the last three decades of progress in the field of gastrointestinal melatonin. The other three chapters deals with the studies on the sympathetic regulation of innate immunity, regulation and synthesis of maturation inducing hormone in fishes and the last chapter deals with the melatonin inhibition of gonadotropin-releasing hormone-induced calcium signaling and hormone secretion in neonatal pituitary gonadotrophs.

Section II contains five chapters: the roles of melatonin in photoperiodic gonadal response of birds, role of prenatal androgen excess on the development of Polycystic Ovarian Syndrome; the next three chapters address various aspects of the effects of antidiabetic drugs on reproductive system, life span and tumor development, obesity and male infertility. The final chapter deals with melatonin and induction and/or growth of experimental tumors.

The developmental endocrinology section contains four chapters dealing with: the role of melatonin in avian embryogenesis, the molecular aspects of early avian development, molecular aspects of avian development, the molecular development and regeneration aspects of the vertebrate retina.

The first chapter of the last section deals with the comparative aspects of the mammalian pineal gland ultrastructure. The other chapters are the ultimobranchial gland in poikilotherms morphological and functional aspects, and the role of melatonin and immunomodulation.

This volume is intended for endocrinologists and reproductive biologists (basic and clinical researchers), and generalists alike. It will also be useful for graduate students of biomedical subspecialties.

Chandana Haldar
Muniyandi Singaravel
S.R. Pandi-Perumal
Daniel P. Cardinali

Acknowledgements

The editors wish to express their sincere appreciation and gratitude to all our distinguished contributors for their scholarly contributions that facilitated the development of this volume. These outstanding authors who, regardless of how busy they were, managed to find time for this project. In a most diligent and thoughtful way they have brought a wide range of interests and disciplines to this volume.

We are most grateful to our families who provided love and support too valuable to measure. Their constant encouragement, understanding, and patience while the book was being developed are immensely appreciated. They were the source of inspiration for us, and we thank them for their continuing support, and for understanding the realities of academic life!

Chandana Haldar
Muniyandi Singaravel
S.R. Pandi-Perumal
Daniel P. Cardinali

Contents

Experimental Endocrinology

CHAPTER 1

Melatonin: New Aspects of its Protective Actions and Novel Metabolites

R. Hardeland[1],* and B. Poeggeler[2]

Abstract

Antioxidative protection by melatonin comprises various pleiotropic actions, not only direct radical scavenging, but also upregulations of antioxidant and downregulations of prooxidant enzymes and, presumably, effects on quinone reductase 2 mediated by melatonin binding. In addition, damage by free radicals is avoided by antiexcitotoxic effects of melatonin and its contribution to internal circadian timing. The circadian paradox of protection despite nocturnal phasing in both light- and dark-active animals seems to be solvable at the level of tissue melatonin, which oscillates with lower amplitudes. Moreover, formation of oxidative products in the kynuric pathway, having additional protective properties, may result in metabolite rhythms mainly determined by oxidant formation. A new model of radical avoidance aims to explain mitochondrial protection and decrease of mitochondrial oxidant formation by a cycle of electron exchange between melatonin and/or its metabolite AMK (N[1]-acetyl-5-methoxykynuramine) with the electron transport chain. Moreover, AMK does not only interact with reactive oxygen species, but also with the carbonate radical and reactive nitrogen species. Two novel metabolites formed from AMK by reactive nitrogens have been identified, i.e., 3-acetamidomethyl-6-methoxycinnolinone and 3-nitro-AMK.

INTRODUCTION

Melatonin is still good for surprises. This extremely pleiotropic molecule, which had first been identified as a mediator of dark signals released from

[1]Professor of Zoology, Institute of Zoology, Anthropology and Developmental Biology, University of Göttingen, Berliner Str. 28, D-37073 Göttingen, Germany.
[2]Institute of Zoology, Anthropology and Developmental Biology, University of Göttingen, Göttingen, Germany.
Corresponding author: Tel.: +49 551 395414 (office), +49 551 395410 (lab), Fax: +49 551 395438. E-mail: rhardel@gwdg.de

the pineal gland, was later shown to be produced in extrapineal sites (Huether, 1993; Hardeland, 1997; Bubenik, 2002) and in non-vertebrate organisms (Hardeland and Fuhrberg, 1996; Hardeland, 1999; Hardeland and Poeggeler, 2003), to modulate the immune system (Guerrero and Reiter, 2002; Haldar, 2002; Skwarlo-Sonta, 2002), and to participate in antioxidative defense mechanisms (Reiter, 1998; Reiter et al., 2002, 2003; Hardeland et al., 2003). In the beginning, antioxidative protection was sometimes regarded as a matter of radical scavening, although one problem was immediately evident, namely, the gap between low melatonin levels present in the circulation and much higher concentrations required in the majority of protection experiments, while other antioxidants were available in considerably higher amounts. Another fundamental difficulty relates to the circadian phasing of circulating melatonin, which always attains its maximum during night, in both dark- and light-active vertebrates. Since the formation of reactive oxygen and nitrogen species is associated with elevated metabolic activities in the tissues, high melatonin is associated with high rates of free radical generation in most tissues of dark-active species, but with low rates in the light-active ones, an apparent paradox (Hardeland et al., 2003). This problem cannot be resolved by assuming indirect antioxidant effects via signaling mechanisms. Moreover, the known upregulations of antioxidant and downregulations of prooxidant enzymes do not suffice for explaining the protective effects observed and are sometimes out of phase with circadian rhythms of oxidants formed (Hardeland et al., 2003). For example, downregulation of neuronal NO (nitric oxide) synthase by melatonin is well established and has frequently been confirmed; however, the physiological melatonin peak does not prevent a nocturnal peak in NO liberation from the rat cerebral cortex (Clément et al., 2003). With regard to melatonin's undoubtedly high protective potential, new explanations are awaited. This contribution aims to trace the newly emerging possibilities of additional protection mechanisms. Since some of them seem to involve metabolites of melatonin, recent findings on reactions of such products will also be presented, including the identification of substances previously unknown to both biologists and chemists.

TISSUE MELATONIN

The chronobiological paradox, as outlined above, which may at the first glance appear as a major problem, seems to be much less severe after a look at tissue melatonin. Concentrations of the indoleamine in the tissue should be decisive for cell protection. Additionally, the possibility of subcellular accumulation could be of importance. Tissue concentrations of melatonin are only partially known. Where determined, the amounts are much higher than in the circulation (Hardeland, 1997). An impressive example is the gastrointestinal system, which contains several hundred times more of melatonin than the pineal gland. Several sources contribute to this high amount, in particular, synthesis by enterochromaffin cells, uptake from the blood, and uptake from the lumen including enterohepatic cycling (Messner et al., 1998, 1999; Bubenik, 2001, 2002; Hardeland and Poeggeler, 2003). Another decisive difference in the levels of circulation concerns the amplitude of the gastrointestinal melatonin rhythm, which is much smaller (minimum/ maximum ratio rarely higher than 1:2) and sometimes even at the borderline of demonstrability (Bubenik, 2001, 2002; Hardeland et al., 2004a). In the rodent Harderian gland, the melatonin rhythm is also weakly expressed (Hardeland, 1997). It is, therefore, impossible to judge on melatonin's role in tissues by extrapolating from levels and rhythmicity in the blood plasma. If melatonin rhythmicity is weak or almost absent in a tissue, the phase of the melatonin maximum becomes less important. This may be especially relevant if the oxidative metabolites formed from melatonin by interactions with free radicals contribute to protection, what is assumed for good reasons. If melatonin concentrations are rhythmically varying in tissues by not more than a factor of 2 or even less, the amounts of oxidants produced may become rate limiting for the levels of oxidation products from melatonin, such as AFMK (N^1-acetyl-N^2-formyl-5-methoxykynuramine) and its deformylated metabolite AMK (N^1-acetyl-5-methoxykynuramine). Whether these products really oscillate in phase with radical formation remains to be demonstrated.

These assumptions should be seen on the background of numerous biological activities of kynuramines, representing a separate class of biogenic amines (Balzer and Hardeland, 1989). Protection by pharmacological levels of AFMK has been described, such as inhibition of 8-hydroxy-2-deoxyguanosine formation, reduction of lipid peroxidation, and rescue of hippocampal neurons from oxidotoxic cell death (Burkhardt et al., 2001; Tan et al., 2001). Its deformylated product AMK seems to be even more important: it is a radical scavenger of considerably higher reactivity than AFMK (Hardeland et al., 2004b; Ressmeyer et al., 2003; Behrends et al., 2004; Silva et al., 2004) and, as a cyclooxygenase inhibitor being much more potent than acetylsalicylic acid (Kelly et al., 1984), it may indirectly contribute to protection by an antiinflammatory action. For other mitochondrial effects subsequent paragraphs may be referred to.

The formation of kynuramines from melatonin seems to be of special significance in non-hepatic tissues. Extrahepatic P_{450} monooxygenase activities are too low for a quantitative turnover via this route, and the high amounts of gastrointestinal melatonin enter the circulation only to a limited extent, e.g., during a post-prandial response; most of it is either released to the lumen (Messner et al., 1998, 1999; Bubenik, 2001, 2002) or has to be metabolized in another pathway. According to recent estimations, about 30% of overall melatonin degradation is attributed to pyrrole-ring cleavage (Ferry et al., 2005), although the rate may be considerably higher in some tissues. AMK was identified as a main metabolite of melatonin in the central nervous system (Hirata et al., 1974), a finding known since long, which was, however, frequently ignored. After melatonin injection into the *cisterna magna*, AFMK and AMK were the only products found and no 6-hydroxymelatonin was detected. These results have recently gained high actuality, after melatonin was shown to be released directly to the brain, via the pineal recess, in much higher quantities than to the blood (Tricoire et al., 2002). The high turnover in the kynuric pathway of melatonin catabolism is particularly remarkable as it cannot be explained on the basis of enzymatic formation, neither by indoleamine 2,3-dioxygenase, which is upregulated by inflammatory signals in the microglia (Alberati-Giani et al., 1996; Lestage et al., 2002; Kwidzinski et al., 2003), nor by myeloperoxidase, which is associated with activated phagocytes (Silva et al., 2004; Ferry et al., 2005). To assume free radical reactions as the main cause of kynuric melatonin degradation in the brain is, therefore, highly suggestive.

INDIRECT ANTIOXIDANT EFFECTS OF MELATONIN, AN EXTENDED FIELD

The difficulty of explaining antioxidant effects of melatonin despite its low concentration in the blood had directed the attention to receptor-mediated mechanisms of upregulation of antioxidant and downregulation of prooxidant enzymes. In fact, stimulations of glutathione peroxidase, glutathione reductase, glucose-6-phosphate dehydrogenase as a source of reducing equivalents, γ-glutamylcysteine synthase as the rate-limiting enzyme of glutathione synthesis, Cu,Zn- and Mn-superoxide dismutases and, sometimes, catalase were described (Reiter et al., 2001, 2002, 2003). Additionally, dowregulations of lipoxygenases (Manev et al., 1998; Uz and Manev, 1998; Zhang et al., 1999) and of NO synthases (Pozo et al., 1994; Bettahi et al., 1996; Gilad et al., 1998; Reiter et al., 2001, 2002, 2003) were described.

Recently, many additional effects of indirect protection have been reported. In neuronal cell lines, glutathione peroxidase and superoxide dismutases (SODs) were not only shown to be enhanced at the mRNA level, but also the life-time of glutathione peroxidase and Cu,Zn-SOD mRNAs was extended by melatonin (Mayo et al., 2002). Indirect protection may be also assumed in the case of quinone reductase 2, a cytosolic enzyme which is implicated in detoxification of prooxidant quinones and which binds melatonin at upper physiological concentrations (Nosjean et al., 2000, 2001), so that it had originally been regarded as a melatonin receptor. However, the mechanistic consequences of melatonin binding to this putative antioxidant enzyme are not yet understood.

Melatonin also exerts indirect effects, especially in the central nervous system, by its antiexcitatory and antiexcitotoxic actions. Strong excitation is associated with high calcium influx and NO release and is, thus, usually associated with enhanced radical generation. Melatonin was shown to possess strong anticonvulsant properties and to efficiently counteract various excitotoxins (Lapin et al., 1998). Neuroprotection by melatonin against excitotoxicity was shown for glutamate (Espinar et al., 2000) and its agonists, ibotenate (Husson et al., 2002), kainic acid (Manev et al., 1996; Chung and

Han, 2003), domoic acid (Annanth et al., 2003), and, in particular, also for quinolinic acid (Southgate et al., 1998; Behan et al., 1999; Cabrera et al., 2000). When analyzed in detail, these effects turned out to be a superposition of antiexcitatory and direct antioxidant actions. However, the attenuation of excitatory responses already reduces radical formation via decreases of NO and Ca^{2+}-dependent mitochondrial effects and, therefore, represents indirect antioxidative protection.

The chronobiological role of melatonin as an endogenous coordinator of rhythmic time structures seems to represent another area of indirect protection. The importance of appropriate timing for maintaining low levels of oxidative damage has been overlooked for quite some time. However, temporal perturbations as occurring in short-period or arrhythmic circadian clock mutants were found to cause enhanced oxidative damage, effects which were observed in organisms as different as *Drosophila* (Coto-Montes and Hardeland, 1997, 1999) and the Syrian Hamster (Coto-Montes et al., 2001). Optimal circadian phasing of radical-generating and -detoxifying processes seems to be a strategy of minimizing damage by oxidants, too (Hardeland et al., 2003).

MITOCHONDRIAL EFFECTS

For several reasons, mitochondrial actions of melatonin are actually attracting particular attention: (i) Mitochondria represent a major source of free radicals and are thought to be involved in aging processes. (ii) The importance of mitochondrial diseases is increasingly perceived. (iii) Mitochondria play a key role in apoptosis. Mitochondrial protection by melatonin in a more general sense has been repeatedly described, including reduction of lipid peroxidation, attenuation of oxidative protein and DNA modifications, preservation of ultrastructure, and resistance against toxins (Wakatsuki et al., 2001; Reiter et al., 2003; Okatani et al., 2003; León et al., 2005). Melatonin was also shown to influence redox-active compounds in mitochondria, e.g., to decrease NO (Escames et al., 2003; Acuña-Castroviejo et al., 2003), to restore levels of reduced glutathione (Wakatsuki et al., 2001; Okatani et al., 2003) and coenzyme Q_{10} (Yalcin et al., 2004). More importantly, melatonin was found to enhance respiratory activity and ATP synthesis, mostly in conjunction with increased complex I and IV activities (Reiter et al., 2001; Martín et al., 2002; Escames et al., 2003; Acuña-Castroviejo et al., 2003; León et al., 2005). Effects on complex IV may be partially explained by stimulations of gene expression of complex IV components (Acuña-Castroviejo et al., 2003), but additional mechanisms are obviously involved, too.

In particular, melatonin was shown to antagonize critical mitochondrial parameters which are affected by excitatory actions of NO and damage by oxygen free radicals as well. In cells as different as cardiomyocytes, astrocytes and striatal neurons, calcium overload and the collapse of the mitochondrial membrane potential, as induced by H_2O_2, doxorubicin or oxygen/glucose deprivation, were prevented, and opening of the mitochondrial permeability transition pore (mtPTP) was inhibited (Xu and Ashraf, 2002; Andrabi et al., 2004; Jou et al., 2004). Interestingly, melatonin seemed to directly influence mtPTP currents, with an IC_{50} of 0.8 μM (Andrabi et al., 2004), a concentration which would require mitochondrial accumulation of melatonin; due to melatonin's amphiphilicity, this seems well possible.

Nevertheless, explanations should go beyond these findings. Recently a model of radical avoidance was proposed, based on mitochondrial effects in conjunction with comparable results on amphiphilic nitrones (Hardeland et al., 2003). In contrast to other models focussing on radical scavenging or oxidant destruction by protective enzymes, a mechanism for reducing radical formation was suggested. In fact, melatonin has been shown to diminish mitochondrial H_2O_2 (Poeggeler, 2004), which is formed from superoxide anions generated by single-electron donation to oxygen, mainly by the iron-sulfur cluster N2 in complex I. The explanation is based on single-electron exchange reactions of melatonin and/or its metabolite AMK with the respiratory chain. Since melatonin as well as AMK can reduce complexed iron, it is assumed that cytochrome c, normally mediating between complexes III and IV and acting as an electron acceptor from outside the respiratory chain, e.g., from $O_2 \bullet^-$ formed in complex I (Xu, 2004), may be also a site of electron donation by melatonin and/or AMK. Additionally, electrons may be transferred to the ubisemiquinone radical or to complex III. On the other hand, $O_2 \bullet^-$ formation at complex I is decreased by competing, electron-accepting melatonyl or AMK cation radicals formed by electron donation to the respiratory chain. Such a mechanism (Fig. 1) would only require very low, quasi-catalytic amounts of melatonin or AMK, in accordance with mitochondrial effects demonstrated at nanomolar concentrations of melatonin (Martín et al., 2002). As the recycled electrons do not get lost for the respiratory chain, this could also explain improvements in complex IV activity, oxygen consumption and ATP production. The prediction by the model of mitochondrial protection by AMK was recently confirmed (Acuña-Castroviejo et al., 2003): AMK was shown to exert effects on electron flux through the respiratory chain and ATP synthesis very similar to those observed with melatonin.

NEW METABOLITES FROM INTERACTIONS OF AMK AND REACTIVE NITROGEN SPECIES

Until recently, the kynuric pathway of melatonin catabolism was believed to end at AMK. After injections of melatonin, this compound was detected in the urine (Hirata et al., 1974) and, therefore, often believed to be a final, excreted end product. However, excretion after a pharmacological load of the precursor does not exclude a continual presence of AMK in tissues under physiological conditions. Since AMK was shown to be highly reactive (Ressmeyer et al., 2003), radical reactions of this kynuramine were also likely

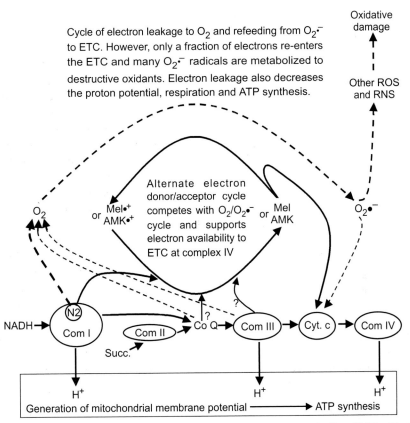

Fig. 1. A model of interactions of melatonin and/or its metabolite AMK with the mitochondrial respiratory chain (ETC). The model aims to explain reduction of electron leakage and oxidant formation as well as maintenance of mitochondrial membrane potential, support of respiration and ATP synthesis. Abbreviations: Com = complex; N2 = iron-sulfur cluster N2 within complex I; Co Q = coenzyme Q; Cyt. c = cytochrome c; RNS = reactive nitrogen species; ROS = reactive oxygen species; Succ. = succinate.

to occur in biological material. In fact, several oxidation products derived from AMK have recently been isolated (unpublished data). Moroever, products formed by interactions with reactive nitrogen species (RNS) were discovered (Guenther et al., 2005). Their physiological formation is very likely, although no information is available to date on their naturally occurring levels.

The first metabolite, 3-acetamidomethyl-6-methoxycinnolinone (AMMC) belongs to a class of substances, the cinnolines, which have not been known to be present in biological material. The compound is also chemically novel. It has been discovered when AMK was exposed to the air, in darkness, on the silica gel of a thin-layer chromatography plate. The small amounts of reactive nitrogens in the air are obviously sufficient for AMMC formation on the large surface of the silica gel, a confirmation of the high reactivity of AMK (Ressmeyer et al., 2003). It was also found that AMMC is produced by several chemically different NO donors, such as sodium nitroprussiate, S-nitroso-N-acetylpenicillamine (SNAP), and 3-(2-hydroxy-2-nitroso-1-propylhydrazino)-1-propanamine (PAPA NONOate); the reaction was partially inhibited by the NO scavenger N-acetylcysteine (Guenther et al., 2005). Since NO appears in three redox forms, the NO radical ($\bullet$NO), the nitrosonium cation (NO^+), and the nitroxyl anion (NO^-), the question was, of course, that of the preferential interaction partner of AMK. Several possibilities may exist. In the absence of proton-abstracting radicals, AMMC formation is most easily explained by the combination with NO^+ (Fig. 2A). As soon as electron abstraction by oxidants occurs, binding of $\bullet$NO seems also possible. Not only $\bullet$NO is a physiologically dangerous and potentially deleterious oxidant, but induction of cell damage, decrease of the mitochondrial potential and apoptosis was also described for NO^+ (Khan et al., 1997; Kayahara et al., 1998; Kim and Ponka, 2003), as well as S-nitrosation of thiol compounds (Hughes, 1999) and induction of heme oxygenase I (Naughton et al., 2002). The contribution of NO^+ to nitrosative stress is strongly supported by recent findings documenting its specific, also non-endothelial formation from various cellular sources. Firstly, NO^+ can be liberated from nitrosothiols such as S-nitrosocysteine (CysNO) and S-nitrosoglutathione (GSNO), however, it is also generated from copper- or iron-containing metalloproteins. This is particularly important in the case of deoxygenated hemoglobin. Contrary to the relatively stable $Hb(Fe^{II})\bullet NO$ complex, reduction of the $\bullet$NO metabolite nitrite by $Hb(Fe^{II})$ leads to unstable complexes, such as $Hb(Fe^{III})\bullet NO$ and $Hb(Fe^{II})NO^+$, which are in equilibrium and which can donate either $\bullet$NO or NO^+ (Nagababu et al., 2003), whereby the decomposition of $Hb(Fe^{III})\bullet NO$ to $Hb(Fe^{II})$ and NO^+ can be hardly distinguished from the decay of $Hb(Fe^{II})NO^+$. In any case, NO^+ is detected in arterioles, capillaries and venules even when NO synthase is experimentally inhibited and despite the rapid reaction of NO^+ with water (Kashiwagi et al., 2002). Therefore, physiological AMMC formation from AMK and NO^+ is very likely.

The second new compound is N^1-acetyl-5-methoxy-3-nitrokynuramine (AMNK = 3-nitro-AMK; Fig. 2B). It also appears on an air-exposed silica gel, but is readily formed by interaction with a physiologically important nitration mixture (Guenther et al., 2005). When AMK is incubated with hydrogen carbonate at physiological concentration, addition of peroxynitrite immediately leads to AMK nitration, whereas corresponding experiments in the absence of hydrogen carbonate do not cause AMNK formation at a substantial rate. These findings are easily explained by the formation of carbonate radicals and NO_2 from the peroxynitrite-CO_2 adduct ($ONOOCO_2^-$) (Fig. 2B). This nitration mixture seems to be physiologically more important than the alternate combination of $\bullet$OH and NO_2 from the decomposition of peroxynitrous acid (ONOOH), due to a pK of 6.8, competition with CO_2, formation of $CO_3\bullet^-$ from $\bullet$OH and HCO_3^-, and rapid disappearance of the

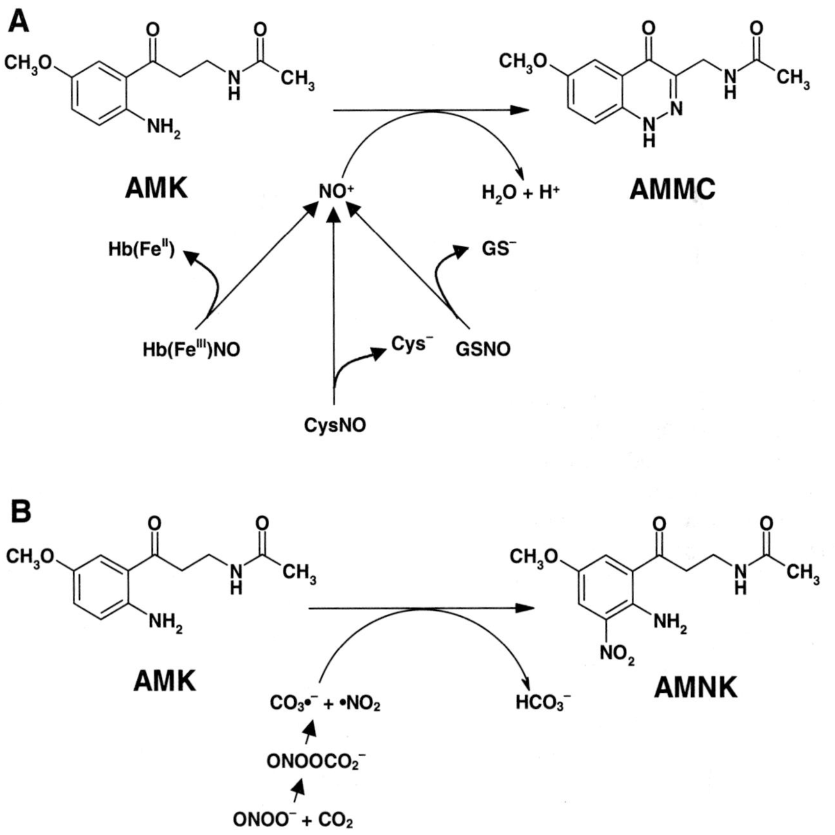

Fig. 2. Two novel melatonin metabolites deriving from interactions of AMK with reactive nitrogen species, and possible routes of formation. A: AMMC formation; B: AMNK formation. Abbreviations: AMK = N^1-acetyl-5-methoxykynuramine; AMMC = 3-acetamidomethyl-6-methoxycinnolinone; AMNK = N^1-acetyl-5-methoxy-3-nitro-kynuramine (= 3-nitro-AMK); CysNO = S-nitrosocysteine; GSNO = S-nitroso-glutathione; Hb = hemoglobin.

extremely shortlived •OH, compared to the resonance-stabilized CO_3•$^-$. These results also shed light on the frequently underrated relevance of HCO_3^- and CO_2 in the physiology and pathophysiology of radical reactions. According to these considerations, formation of AMNK in biological material seems very likely, although rates are not yet known.

CONCLUSION

Melatonin continues to be a fascinating molecule, particularly with regard to the steadily increasing wealth of information on its protective effects. The contribution of its metabolites seems to be an intriguing line of future investigation. By generating metabolites via the kynuric pathway, melatonin may act as a prodrug. Radical avoidance, rather than radical detoxification, may turn out to be decisive for the protective value of melatonin, and highest priority should be given to studies on mitochondrial effects exerted by this indoleamine and, at least, one of its metabolites, AMK.

REFERENCES

Acuña-Castroviejo, D., Escames, G., León, J., Carazo, A. and Khaldy, H. (2003). Mitochondrial regulation by melatonin and its metabolites. *Adv. Exp. Med. Biol.* **527**: 549-557.

Alberati-Giani, D., Ricciardi-Castagnoli, P., Köhler, C. and Cesura, A.M. (1996). Regulation of the kynurenine pathway by IFN-γ in murine cloned macrophages and microglial cells. *Adv. Exp. Med. Biol.* **398**: 171-175.

Ananth, C., Gopalakrishnakone, P. and Kaur, C. (2003). Protective role of melatonin in domoic acid-induced neuronal damage in the hippocampus of adult rats. *Hippocampus* **13**: 375-387.

Andrabi, S.A., Sayeed, I., Siemen, D., Wolf, G. and Horn, T.F. (2004). Direct inhibition of the mitochondrial permeability transition pore: a possible mechanism responsible for anti-apoptotic effects of melatonin. *FASEB J.* **18**: 869-871.

Balzer, I. and Hardeland, R. (1989). Action of kynuramine in a dinoflagellate: stimulation of bioluminescence in *Gonyaulax polyedra*. *Comp. Biochem. Physiol.* **94C**: 129-132.

Behan, W.M., McDonald, M., Darlington, L.G. and Stone, T.W. (1999). Oxidative stress as a mechanism for quinolinic acid-induced hippocampal damage: protection by melatonin and deprenyl. *Br. J. Pharmacol.* **128**: 1754-1760.

Behrends, A., Hardeland, R., Ness, H., Grube, S., Poeggeler, B. and Haldar, C. (2004). Photocatalytic actions of the pesticide metabolite 2-hydroxyquinoxaline: destruction of antioxidant vitamines and biogenic amines – implications of organic redox cycling. *Redox Rep.* **9**: 279-288.

Bettahi, I., Pozo, D., Osuna, C., Reiter, R.J., Acuña-Castroviejo, D. and Guerrero, J.M. (1996). Melatonin reduces nitric oxide synthase activity in rat hypothalamus. *J. Pineal Res.* **20**: 205-210.

Bubenik, G.A. (2001). Localization, physiological significance and possible clinical implication of gastrointestinal melatonin. *Biol. Signals Recept.* **10**: 350-366.

Bubenik, G.A. (2002). Gastrointestinal melatonin: localization, function, and clinical relevance. *Dig. Dis. Sci.* **47**: 2336-2348.

Burkhardt, S., Reiter, R.J., Tan, D.-X., Hardeland, R., Cabrera, J. and Karbownik, M. (2001). DNA oxidatively damaged by chromium(III) and H_2O_2 is protected by melatonin, N^1-acetyl-N^2-formyl-5-methoxykynuramine, resveratrol and uric acid. *Int. J. Biochem. Cell Biol.* **33**: 775-783.

Cabrera, J., Reiter, R.J., Tan, D.-X., Qi, W., Sainz, R.M., Mayo, J.C., Garcia, J.J., Kim, S.J. and El-Sokkary, G. (2000). Melatonin reduces oxidative neurotoxicity due to quinolinic acid: in vitro and in vivo findings. *Neuropharmacology* **39**: 507-514.

Chung, S.Y. and Han, S.H. (2003). Melatonin attenuates kainic acid-induced hippocampal neurodegeneration and oxidative stress through microglial inhibition. *J. Pineal Res.* **34**: 95-102.

Clément, P., Gharib, A., Cespuglio, R. and Sarda, N. (2003). Changes in sleep-wake cycle architecture and cortical nitric oxide release during ageing in the rat. *Neuroscience* **116**: 863-870.

Coto-Montes, A. and Hardeland, R. (1997). Diurnal time patterns of protein carbonyl in *Drosophila melanogaster*: Comparison of wild-type flies and clock mutants. In: *Biological Rhythms and Antioxidative Protection* (Hardeland, R., ed.), Cuvillier, Göttingen. 119-126.

Coto-Montes, A. and Hardeland, R. (1999). Diurnal rhythm of protein carbonyl as an indicator of oxidative damage in *Drosophila melanogaster*: Influence of clock gene alleles and deficiencies in the formation of free-radical scavengers. *Biol. Rhythm Res.* **30**: 383-391.

Coto-Montes, A., Tomás-Zapico, C., Rodríguez-Colunga, M.J., Tolivia-Cadrecha, D., Martínez-Fraga, J., Hardeland, R. and Tolivia, D. (2001). Effects of the circadian mutation 'tau' on the Harderian glands of Syrian hamsters. *J. Cell. Biochem.* **83**: 426-434.

Escames, G., León, J., Macías, M., Khaldy, H. and Acuña-Castroviejo, D. (2003). Melatonin counteracts lipopolysaccharide-induced expression and activity of mitochondrial nitric oxide synthase in rats. *FASEB J.* **17**: 932-934.

Espinar, A., García-Oliva, A., Isorna, E.M., Quesada, A., Prada, F.A. and Guerrero, J.M. (2000). Neuroprotection by melatonin from glutamate-induced excitotoxicity during development of the cerebellum in the chick embryo. *J. Pineal Res.* **28**: 81-88.

Ferry, G., Ubeaud, C., Lambert, P.H., Bertin, S., Cogé, F., Chomarat, P., Delagrange, P., Serkiz, B., Bouchet, J.P., Truscott, R.J. and Boutin, J.A. (2005). Molecular evidence that melatonin is enzymatically oxidized in a different manner than tryptophan. Investigation on both indoleamine-2,3-dioxygenase and myeloperoxidase. *Biochem. J.* **388**, Pt. **1**: 205-215.

Gilad, E., Wong, H.R., Zingarelli, B., Virag, L., O'Connor, M., Salzman, A.L. and Szabo, C. (1998). Melatonin inhibits expression of the inducible isoform of nitric oxide synthase in murine macrophages: role of inhibitions of NFκB activation. *FASEB J.* **12**: 685-693.

Guenther, A.L., Schmidt, S.I., Laatsch, H., Fotso, S., Ness, H., Ressmeyer, A.-R., Poeggeler, B. and Hardeland, R. (2005). Reactions of the melatonin metabolite AMK (N^1-acetyl-5-methoxykynuramine) with reactive nitrogen species: Formation of novel compounds, 3-acetamidomethyl-6-methoxycinnolinone and 3-nitro-AMK. *J. Pineal Res.* **39**: 251-260.

Guerrero, J.M. and Reiter, R.J. (2002). Melatonin–immune system relationships. *Curr. Top. Med. Chem.* **2**: 167-179.

Haldar, C. (2002). Apoptosis, cancer, immunity and melatonin. In: *Treatise on Pineal Gland and Melatonin* (Haldar, C., Singaravel, M. and Maitra, S.K., eds.). Science Publishers, Enfield, N.H., USA, 535-542.

Hardeland, R. and Fuhrberg, B. (1996). Ubiquitous melatonin – Presence and effects in unicells, plants and animals. *Trends Comp. Biochem. Physiol.* **2**: 25-45.

Hardeland, R. (1997). Melatonin: Multiple functions in signaling and protection. In: *Skin Cancer and UV Radiation* (Altmeyer, P., Hoffmann, K. and Stücker, M., eds.). Springer, Berlin – Heidelberg, 186-198.

Hardeland, R. (1999). Melatonin and 5-methoxytryptamine in non-metazoans. *Reprod. Nutr. Dev.* **39**: 399-408.

Hardeland, R. and Poeggeler, B. (2003). Non-vertebrate melatonin. *J. Pineal Res.* **34**: 233-241.

Hardeland, R., Coto-Montes, A. and Poeggeler, B. (2003). Circadian rhythms, oxidative stress and antioxidative defense mechanisms. *Chronobiol. Int.* **20**: 921-962.

Hardeland, R., Poeggeler, B., Huether, G., Cornélissen, G., Josza, R., Zeman, M., Balazova, K., Herichova, I., Schwartzkopff, O., Bubenik, G. and Halberg, F. (2004a). Chronomics support lead in phase of duodenal vs. pineal circadian rhythms of melatonin. In: *5th Workshop on Chronoastrobiology and Chronotherapy, 6th November, 2004* (Matsubayasi, K., ed.). Center for Southeast Asian Studies, Tokyo, Japan, 84-88.

Hardeland, R., Ressmeyer, A.-R., Zelosko, V., Burkhardt, S. and Poeggeler, B. (2004b). Metabolites of melatonin: Formation and properties of the methoxylated kynuramines AFMK and AMK. In: *Recent Advances in Endocrinology and Reproduction: Evolutionary, Biotechnological and Clinical Implications* (Haldar, C. and Singh, S.S., eds.). Banaras Hindu University, Varanasi, India, 21-38.

Hirata, F., Hayaishi, O., Tokuyama, T. and Senoh, S. (1974). *In vitro* and *in vivo* formation of two new metabolites of melatonin. *J. Biol. Chem.* **249**: 1311-1313.

Huether, G. (1993). The contribution of extrapineal sites of melatonin synthesis to circulating melatonin levels in higher vertebrates. *Experientia* **49**: 665-670.

Hughes, M.N. (1999). Relationships between nitric oxide, nitroxyl ion, nitrosonium cation and peroxynitrite. *Biochim. Biophys. Acta* **1411**: 263-272.

Husson, I., Mesples, B., Bac, P., Vamecq, J., Evrard, P. and Gressens, P. (2002). Melatoninergic neuroprotection of the murine periventricular white matter against neonatal excitotoxic challenge. *Ann. Neurol.* **51**: 82-92.

Jou, M.-J., Peng, T.-I., Reiter, R.J., Jou, S.-B., Wu, H.-Y. and Wen, S.-T. (2004). Visualization of the antioxidative effects of melatonin at the mitochondrial level during oxidative stress-induced apoptosis of rat brain astrocytes. *J. Pineal Res.* **37**: 55-70.

Kashiwagi, S., Kajimura, M., Yoshimura, Y. and Suematsu, M. (2002). Nonendothelial source of nitric oxide in arterioles but not in venules: alternative source revealed in vivo by diaminofluorescein microfluorography. *Circ. Res.* **91**: e55-e64.

Kayahara, M., Felderhoff, U., Pocock, J., Hughes, M.N. and Mehmet, H. (1998). Nitric oxide (NO•) and the nitrosonium cation (NO⁺) reduce mitochondrial membrane potential and trigger apoptosis in neuronal PC12 cells. *Biochem. Soc. Trans.* **26**: S340.

Kelly, R.W., Amato, F. and Seamark, R.F. (1984). N-Acetyl-5-methoxy kynurenamine, a brain metabolite of melatonin, is a potent inhibitor of prostaglandin biosynthesis. *Biochem. Biophys. Res. Commun.* **121**: 372-379.

Khan, S., Kayahara, M., Joashi, U., Mazarakis, N.D., Saraf, C., Edwards, A.D., Hughes, M.N. and Mehmet, H. (1997). Differential induction of apoptosis in Swiss 3T3 cells by nitric oxide and the nitrosonium cation. *J. Cell Sci.* **110**: 2315-2322.

Kim, S. and Ponka, P. (2003). Role of nitric oxide in cellular iron metabolism. *Biometals* **16**: 125-135.

Kwidzinski, E., Bunse, J., Kovac, A.D., Ullrich, O., Zipp, F., Nitsch, R. and Bechmann, U. (2003). IDO (indolamine 2,3-dioxygenase) expression and function in the CNS. *Adv. Exp. Med. Biol.* **527**: 113-118.

Lapin, I.P., Mirzaev, S.M., Ryzow, I.V. and Oxenkrug, G.F. (1998). Anticonvulsant activity of melatonin against seizures induced by quinolinate, kainate, glutamate, NMDA, and pentylenetetrazole in mice. *J. Pineal Res.* **24**: 215-218.

León, J., Acuña-Castroviejo, D., Escames, G., Tan, D.-X. and Reiter, R.J. (2005). Melatonin mitigates mitochondrial malfunction. *J. Pineal Res.* **38**: 1-9.

Lestage, J., Verrier, D., Palin, K. and Dantzer, R. (2002). The enzyme indoleamine 2,3-dioxygenase is induced in the mouse brain in response to peripheral administration of lipopolysaccharide and superantigen. *Brain Behav. Immun.* **16**: 596-601.

Manev, H., Uz, T., Kharlamov, A. and Joo, J.-Y. (1996). Increased brain damage after stroke or excitotoxic seizures in melatonin-deficient rats. *FASEB J.* **10**: 1546-1551.

Manev, H., Uz, T. and Qu, T. (1998). Early upregulation of hippocampal 5-lipoxygenase following systemic administration of kainate to rats. *Restor. Neurol. Neurosci.* **12**: 81-85.

Martín, M., Macías, M., León, J., Escames, G., Khaldy, H. and Acuña-Castroviejo, D. (2002). Melatonin increases the activity of the oxidative phosphorylation enzymes and the production of ATP in rat brain and liver mitochondria. *Int. J. Biochem. Cell Biol.* **34**: 348-357.

Mayo, J.C., Sainz, R.M., Antolin, I., Herrera, F., Martin, V. and Rodriguez, C. (2002). Melatonin regulation of antioxidant enzyme gene expression. *Cell. Mol. Life Sci.* **59**: 1706-1713.

Messner, M., Hardeland, R., Rodenbeck, A. and Huether, G. (1998). Tissue retention and subcellular distribution of continuously infused melatonin in rats under near physiological conditions. *J. Pineal Res.* **25**: 251-259.

Messner, M., Hardeland, R., Rodenbeck, A. and Huether, G. (1999). Effects of continuous melatonin infusions on steady-state plasma melatonin levels, metabolic fate and tissue retention in rats under near physiological conditions. *Adv. Exp. Med. Biol.* **467**: 303-313.

Nagababu, E., Ramasamy, S., Abernethy, D.R. and Rifkind, J.M. (2003). Active nitric oxide produced in the red cell under hypoxic conditions by deoxyhemoglobin-mediated nitrite reduction. *J. Biol. Chem.* **278**: 46349-46356.

Naughton, P., Foresti, R., Bains, S.K., Hoque, M., Green, C.J. and Motterlini, R. (2002). Induction of heme oxygenase 1 by nitrosative stress. A role for nitroxyl anion. *J. Biol. Chem.* **277**: 40666-40674.

Nosjean, O., Ferro, M., Cogé, F., Beauverger, P., Henlin, J.M., Lefoulon, F., Fauchère, J.L., Delagrange, P., Canet, E. and Boutin, J.A. (2000). Identification of the melatonin-binding site MT3 as the quinone reductase 2. *J. Biol. Chem.* **275**: 31311-31317.

Nosjean, O., Nicolas, J.P., Klupsch, F., Delagrange, P., Canet, E. and Boutin, J.A. (2001). Comparative parmacological studies of melatonin receptors: MT1, MT2 and MT3/QR2. Tissue distribution of MT3/QR2. *Biochem. Pharmacol.* **61**: 1369-1379.

Okatani, Y., Wakatsuki, A., Reiter, R.J., Enzan, H. and Miyahara, Y. (2003). Protective effect of melatonin against mitochondrial injury induced by ischemia and reperfusion of rat liver. *Eur. J. Pharmacol.* **469**: 145-152.

Poeggeler, B. (2004). Neuroprotection by indole and nitrone compounds acting as mitochondrial metabolism modifiers with potent antioxidant activity. *Neurobiol. Aging* **25**, Suppl. **2**: S587-S588.

Pozo, D., Reiter, R.J., Calvo, J.R. and Guerrero, J.M. (1994). Physiological concentrations of melatonin inhibit nitric oxide synthase in rat cerebellum. *Life Sci.* **55**: PL455-PL460.

Reiter, R.J. (1998). Oxidative damage in the central nervous system: protection by melatonin. *Prog. Neurobiol.* **56**: 359-384.

Reiter, R.J., Acuña-Castroviejo, D., Tan, D.-X. and Burkhardt, S. (2001). Free radical-mediated molecular damage. Mechanisms for the protective actions of melatonin in the central nervous system. *Ann. N.Y. Acad. Sci.* **939**: 200-215.

Reiter, R.J., Tan, D.-X. and Burkhardt, S. (2002). Reactive oxygen and nitrogen species and cellular and organismal decline: amelioration with melatonin. *Mech. Ageing Dev.* **123**: 1007-1019.

Reiter, R.J., Tan, D.-X., Mayo, J.C., Sainz, R.M., Leon, J. and Czarnocki, Z. (2003). Melatonin as an antioxidant: biochemical mechanisms and pathophysiological implications in humans. *Acta Biochim. Pol.* **50**: 1129-1146.

Ressmeyer, A.-R., Mayo, J.C., Zelosko, V., Sáinz, R.M., Tan, D.-X., Poeggeler, B., Antolín, I., Zsizsik, B.K., Reiter, R.J. and Hardeland, R. (2003). Antioxidant properties of the melatonin metabolite N^1-acetyl-5-methoxykynuramine (AMK): scavenging of free radicals and prevention of protein destruction. *Redox Rep.* **8**: 205-213.

Silva, S.O., Rodrigues, M.R., Carvalho, S.R.Q., Catalani, L.H., Campa, A. and Ximenes, V.F. (2004). Oxidation of melatonin and its catabolites, N1-acetyl-N2-formyl-5-methoxykynuramine and N1-acetyl-5-methoxykynuramine, by activated leukocytes. *J. Pineal Res.* **37**: 171-175.

Skwarlo-Sonta, K. (2002). Melatonin in immunity: comparative aspects. *Neuroendocrinol. Lett.* **23**, Suppl. **1**: 61-66.

Southgate, G.S., Daya, S. and Potgieter, B. (1998). Melatonin plays a protective role in quinolinic acid-induced neurotoxicity in the rat hippocampus. *J. Chem. Neuroanat.* **14**: 151-156.

Tan, D.-X., Manchester, L.C., Burkhardt, S., Sainz, R.M., Mayo, J.C., Kohen, R., Shohami, E., Huo, Y.-S., Hardeland, R. and Reiter, R.J. (2001). N^1-Acetyl-N^2-formyl-5-methoxykynuramine, a biogenic amine and melatonin metabolite, functions as a potent antioxidant. *FASEB J.* **15**: 2294-2296.

Tricoire, H., Locatelli, A., Chemineau, P. and Malpaux, B. (2002). Melatonin enters the cerebrospinal fluid through the pineal recess. *Endocrinology* **143**: 84-90.

Uz, T. and Manev, H. (1998). Circadian expression of pineal 5-lipoxygenase mRNA. *Neuroreport* **9**: 783-786.

Wakatsuki, A., Okatani, Y., Shinohara, K., Ikenoue, N., Kaneda, C. and Fukaya, T. (2001). Melatonin protects fetal rat brain against oxidative mitochondrial damage. *J. Pineal Res.* **30**: 22-28.

Xu, J.-X. (2004). Radical metabolism is partner to energy metabolism in mitochondria. *Ann. N.Y. Acad. Sci.* **1011**: 57-60.

Xu, M. and Ashraf, M. (2002). Melatonin protection against lethal myocyte injury induced by doxorubicin as reflected by effects on mitochondrial membrane potential. *J. Mol. Cell. Cardiol.* **34**: 75-79.

Yalcin, A., Kilinc, E., Kocturk, S., Resmi, H. and Sozmen, E.Y. (2004). Effect of melatonin cotreatment against kainic acid on coenzyme Q10, lipid peroxidation and Trx mRNA in rat hippocampus. *Int. J. Neurosci.* **114**: 1085-1097.

Zhang, H., Akbar, M. and Kim, H.Y. (1999). Melatonin: an endogenous negative modulator of 12-lipoxygenation in the rat pineal gland. *Biochem. J.* **344** (Pt. 2): 487-493.

CHAPTER 2

Gastrointestinal Melatonin– 30 Years of Research

G.A. Bubenik

Abstract

Gastrointestinal (GI) melatonin is produced in the mucosa of the vertebrate digestive tract and then transported via the hepatic portal vein into the general circulation. Various binding sites and two melatonin receptors, MT_1 and MT_2 were localized in the GIT (gastrointestinal tract). Besides hormonal action (such as the entraining of circadian activity), melatonin is also a very effective antioxidant and scavenger of hydroxyl radicals. This vitamin-like action of melatonin might provide protection to the gastric and intestinal mucosa from erosion and ulcer formation and promotes ulcer healing. Melatonin secreted into the lumen in response to food intake, may synchronize the digestive activity of its various segments. Melatonin influences intestinal motility, acting either directly on GI muscles or indirectly via the myenteric nervous system. Melatonin may also regulate the fecal water content by its action on the transmembrane transports of electrolytes. Melatonin influences the function of the digestive glands in the duodenum and the activity of the exocrine and endocrine pancreas. High concentrations of melatonin found in the liver and the bile may protect the mucosal membrane of the GI tract from oxidative stress. Finally, melatonin has been proposed as an effective and safe remedy for the treatment of gastric ulcers, ulcerative colitis, GI cancer and the irritable bowel syndrome.

INTRODUCTION

Melatonin (5-methoxy-N-acetyltryptamine) is a metabolite of the neurotransmitter serotonin (5-hydroxytryptamine) (Reiter, 1991). It was first identified in bovine pineal glands by Aaron Lerner and his coworkers (1958). Melatonin is a developmental relic present in bacteria, eukaryotic unicells, algae, plants, fungi and various invertebrates (Harderland and Poegeller, 2003). Melatonin exhibits multiple physiological effects; among others,

Department of Integrative Biology, University of Guelph, Guelph, Ontario, Canada, N1G 2W1.
Tel.: 519 763 2246, Fax: 519 767 1656. E-mail: gbubenik@uoguelph.ca

melatonin is effective as a human sleep inducer (in doses smaller than 1 mg) (Zhdanova et al., 1995). As a scavenger of free radicals, present in various foodstuffs, melatonin can be characterized as a vitamin (Tan et al., 2003), stronger than vitamins C and E (Poeggeler et al., 1993; Pieri et al., 1994; Hattori et al., 1995; Reiter, 2000). Until the mid 1970s, it was believed that pineal gland is the only source of this indole (Wurtman et al., 1963). However, in the early 1970s, using immunohistology, melatonin was identified in extrapineal tissues, such as the retina and the Harderian gland (Bubenik et al., 1974, 1976a,b, 1978). Around 30 years ago, melatonin was also identified in the GIT; first using paper chromatography (Raikhlin and Kvetnoy, 1974) and later immunohistology (Kvetnoy et al., 1975; Raikhlin et al., 1975; Raikhlin and Kvetnoy, 1976; Bubenik et al., 1977; Rakhlin et al., 1978; Bubenik, 1980a). After an initial interest in GIT melatonin, the research slowed down to a trickle. Before the development of reliable radioimmunoassay (RIA) (Kennaway et al., 1977; Brown et al., 1985), only a few immunohistological investigations of GIT melatonin were performed (Bubenik et al., 1977, 1980a, b; Holloway et al., 1980). In addition to melatonin localization, the effect of melatonin on peristalsis of the GIT tissues was also investigated (Bubenik, 1986; Harlow and Weekly, 1986). These *in vitro* studies were an extension of investigations first performed some 40 years ago by Quastel and Rahamimoff (1965) and continued 10 years later by Fioretti and his coworkers (1974a, b). The determination of melatonin concentrations in the GIT tissue by RIA (validated by high performance liquid chromatography) was first reported by Vakkuri et al. (1985a, b). After a slow start in the 1980s, the speed of research picked up in the 1990s and the number of publications related to GIT melatonin peaked recently in 2002-2003 period.

Localization of GIT Melatonin

Immunohistological evidence has indicated that melatonin is produced in the enteroendocrine (formerly known as enterochromaffine) cells of the GIT (Kvetnoy et al., 1975; Raikhlin et al., 1975; Raikhlin and Kvetnoy, 1976, 1978; Bubenik et al., 1977), which are the major source of serotonin in the body (Espamer and Asero, 1952). Melatonin has the same microscopic localization in Lieberkuhn's crypts as serotonin. In addition, melatonin distribution alongside the digestive tract corresponds to that of serotonin (Bubenik et al., 1977). Due to the large size of the gut, there is a substantial amount of melatonin in the GIT. Melatonin in the GIT was later measured by RIA, which was then validated by high performance liquid chromatography (Vakkuri et al., 1985a, b). In 1993, Huether calculated that GIT contains 400 times more melatonin, than is stored at any time in the entire pineal gland. Ontogenically, melatonin has been found in the duodenum of chick embryos (Herichova and Zeman, 1996) and in the entire digestive tract of bovine fetuses (Bubenik et al., 2000a). Phylogenically, melatonin was found in the GIT of all classes of vertebrates (Herichova and Zeman, 1996; Bubenik and Pang, 1997; Rajchard et al., 2000). Postnatally, high melatonin concentrations were reported a few days post partum in rat and mice brains and GIT with levels declining rapidly in the first few weeks of life (Bubenik, 1980a; Bubenik and Pang, 1994).

Is GIT Melatonin Synthesized in the Digestive Tract or is it of Pineal Origin?

Despite a great effort of two labs to detect the synthesis of melatonin in GIT tissues *in vitro* (M. Messner – personal communication; M. Zeman – personal communication), such attempts have not yet been successful. However, substantial evidence supports the hypothesis that melatonin is not only stored, but is also produced in all tissues of the GIT. This evidence includes:

(1) Two crucial enzymes involved in the synthesis of melatonin, the rate-limiting enzyme N-acetyltransferase (Hong and Pang, 1995) and the terminal enzyme, hydroxyindole-O-methyltransferase (Quay and Ma, 1976) were detected in the GIT. The presence of both enzymes in the gut was recently confirmed by transcription-polymerase chain reaction (Stefulj et al., 2001).

(2) Specific binding sites and receptors for melatonin were detected in the GIT (Bubenik et al., 1993; Lucchelli et al., 1997). These receptors may bind melatonin produced in the GIT and also melatonin which the pineal gland secretes into the general circulation during the night (Bubenik, 1980a).

(3) Parenteral or oral administration of melatonin or its precursor tryptophan (TRP) caused an accumulation of melatonin in the GIT (Kopin, 1961; Bubenik, 1980a; Huether et al., 1992; Bubenik et al., 1998a).

(4) Melatonin levels are higher in the portal vein than in the general circulation (Huether et al., 1992; Bubenik et al., 2000b). An oral administration of TRP increased plasma concentrations of melatonin more than intraperitoneally-applied TRP. Such an elevation of melatonin was not observed after the ligation of the portal vein.

(5) Significant melatonin concentrations were detected in rat, sheep, as well as pigeon plasma after pinealectomy (P_x) (Kennaway et al., 1977; Vakkuri et al., 1985a; Bubenik and Brown, 1997). While nighttime levels of melatonin in P_x animals were attenuated, daytime concentrations remained unchanged (Kennaway et al., 1977; Vaughan and Reiter, 1986). These extrapineal levels of melatonin persisted in plasma of fish which were not only pinealectomized, but in which both eyes were also removed (Kezuka et al., 1992).

(6) Administration of TRP increased blood levels of melatonin in intact as well as in P_x rats (Yaga et al., 1993).

(7) Cirrhosis, which impaired the function of the liver, the major site of melatonin degradation (Pardridge and Mietus, 1980), disrupted the diurnal rhythm and elevated melatonin concentrations in the general circulation (Steindl et al., 1995, 1997). As the pineal gland secretes only

small amounts of melatonin during the day (Reiter, 1991) and the contribution of the Harderian gland and the retina to the circulating levels of melatonin is negligible, it is now generally accepted that most of the daytime concentration of melatonin in plasma is of gastrointestinal origin (Bubenik, 2001, 2002). Finally, it must be pointed out, that not all melatonin in the GIT is of pineal or GIT origin. A small amount of melatonin might be derived from food sources, mostly plants (Hattori et al., 1995; Caniato et al., 2003).

Comparison between Gastrointestinal and Pineal Gland Melatonin

The major differences between pineal and GIT melatonin are:

(1) The concentrations of melatonin in the GIT tissues exceed the levels in plasma by 10-100x (Vakkuri et al. 1985a, b; Huether et al., 1992; Huether, 1994; Bubenik et al., 1996; Vician et al., 1999). Whereas the nighttime concentrations of melatonin in rat plasma were around 40 pg/ml, the corresponding levels of melatonin in the jejunum and ileum were more than 500 pg/ml (Bubenik and Brown, 1997). In the porcine colon mucosa, melatonin concentrations reached almost 900 pg/ml (Bubenik et al., 1999). Finally, in the hepatobiliary system the melatonin concentrations were 1,000 times higher than levels in the blood (Tan et al., 1999). Therefore, physiological concentrations of melatonin may vary among body fluids and tissues (Reiter, 2003a).

(2) The secretion of pineal melatonin is circadian (with peak levels achieved in the mid of scotophase) (Reiter, 1991, 1993); conversely, the release of GIT melatonin is either steady or episodic (Huether, 1993; Bubenik et al., 1996).

(3) Pineal-released melatonin acts in an endocrine fashion (Reiter, 1991, 1993), whereas GIT melatonin may act as an endocrine, paracrine or a luminal hormone (Lee et al., 1995; Chow et al., 1996; Bubenik et al., 1999; Sjoblom and Flemstron, 2004).

(4) Pineal-produced melatonin is immediately released (Reiter, 1991), but GIT produced melatonin is mostly released on demand (in response to food intake) (Huether, 1994; Bubenik et al., 1996, 2000b).

(5) Chronic underfeeding decreased levels of melatonin in the pineal gland (Ho et al., 1985; Chik et al., 1987), but increased its concentrations in the GIT (Bubenik et al., 1992; Huether, 1994).

(6) After food administration, the secretion of pineal melatonin did not change (Ho et al., 1985), but an increase of GIT melatonin was observed (Bubenik et al., 1996, 2000b).

Melatonin Binding and its Physiological Functions

Systemically or orally administered melatonin accumulated decisively in the GIT, particularly in the stomach and the colon (Kopin et al., 1961; Reppert and Klein, 1978; Bubenik, 1980a; de Boer, 1988; Bubenik et al., 1998a).

Immunohistological investigations revealed its location mostly in the enteroendocrine cells of the GIT mucosa (Raikhlin and Kvetnoy, 1976; Bubenik et al., 1977; Bubenik, 1980a; Lee et al., 1995; Bubenik et al., 1999). Maximum binding of melatonin was found in the mucosa and the intestinal villi (Lee and Pang, 1993; Lee et al., 1995). As less melatonin and fewer binding sites were detected in the submucosa and muscularis (Bubenik, 1980a; Lee and Pang, 1993; Pontoire et al., 1993; Lee et al., 1995; Chow et al., 1996), it was speculated, that although melatonin is produced in the enteroendocrine cells of the mucosa, it also acts as a paracrine hormone in other layers of the GIT (Chow et al., 1996). In that case, melatonin produced in the mucosa is transported into deeper layers of the GIT via blood vessels in the lamina propria and submucosa, and then acts in the muscularis, where a substantial amount of melatonin is also located (Bubenik, 1980a; Bubenik et al., 1999). At the subcellular level the strongest binding of melatonin was detected in the nuclear fraction, followed by the microsomial, mitochondrial and cytosolic fractions (Menendez-Pelaez et al., 1993; Chow et al., 1996). Melatonin can affect intestinal muscles either directly (Bubenik, 2001) or indirectly via myenteric nervous system, specifically by blocking nicotinic channels (Barajas-Lopez et al., 1996). The action of melatonin on smooth muscles is probably via ML_2 receptors (Lucchelli et al., 1997). As reported by Legris et al. (1982) melatonin located in the intestinal villi is involved in the transmembrane transport of electrolytes and ions. Melatonin also increased water content in the feces of intact mice (Bubenik and Pang, 1994) and the topical application of melatonin stimulated short-circuit current in human colonic cells (Chan et al., 1998).

Only few a years after the discovery of melatonin, Quastel and Rahamimoff (1965) reported that *in vitro* melatonin inhibits a spontaneous or serotonin-induced contraction of rat duodenum. Similar relaxation and restoration of motility was also later confirmed in the stomach, ileum and colon (Fioretti et al., 1974a, b; Bubenik, 1986; Harlow and Weekly, 1986). Modulation of intestinal motility was also reported by Delagrande et al. (2003). More recently, motor complexes in the duodenum and jejunum were modulated by endogenous as well as exogenous melatonin (Merle et al., 2000). Storr and co-workers (2000) reported that small conductance K^+ channels attenuated melatonin-induced relaxation of gastric muscles contracted by serotonin. Furthermore, Storr and co-workers (2002) demonstrated that the relaxing effect of melatonin in the GIT is via its effect on nitric oxide synthase. In addition to the inhibition of serotonin-induced effects *in vitro*, melatonin also appears to inhibit serotonin action *in vivo*. A facilitation of food transit time induced in mice by serotonin implants was partly restored by melatonin injection (Bubenik and Dhanvantari, 1989). Furthermore, studies using parachlorophenylalanine (PCPA) and serotonin implants (Bubenik and Pang, 1994; Bubenik et al., 1996; Bubenik et al., 1999), indicated that a mutual counterbalancing system exists between serotonin and melatonin.

It has been reported that melatonin influences mitotic activity in the GIT (Bindoni, 1971; Bogoyeva et al., 1998). The stimulatory effect of a pinealectomy may be mediated by the vagal or sympathetic nerves of the GIT (Callaghan, 1995). Conversely, an inhibition of epithelial cell proliferation after P_x was reported by Pawlikowski (1986). A single injection of melatonin increased the proliferation of epithelial cells in intact mice (Bogoeva et al., 1993). The opposing results of melatonin effects on mitosis of GIT cells mentioned previously might be due to the different doses used in those studies. Low doses of melatonin (1-10 µg/mouse) decreased mitotic activity in the jejunal epithelium, while high doses (100 µg/mouse) increased mitosis (Zerek-Melen et al., 1987). Similarly, small doses of melatonin (1-10 µg/kg) facilitated intestinal transit, but high doses (100-1000 µg/kg) reduced intestinal transit in rats (Drago et al., 2002). Opposing effects of low and high doses of melatonin were also reported by Kachi et al. (1999). High doses of melatonin induced intestinal elongation, whereas low doses induced intestinal contraction. Finally, low (0.5 mg), but not high (10 mg) doses of melatonin entrained a free-running blind person with a long circadian period (Lewy et al., 2002).

Melatonin, Food Intake and Digestion

Melatonin concentrations in the GIT and the peripheral circulation are influenced by food intake and digestion. Food consumption in mice was increased by melatonin implants (Bubenik and Pang, 1994). Melatonin injected intraperitoneally also increased short- and long- term food consumption in rats (Angers et al., 2000). Both intraperitoneal and oral administration of melatonin modified macronutrient selection in rats and sea bass (Angers et al., 2000; Rubio et al., 2004). Fasting elevated melatonin levels in the brain and most segments of the GIT (Bubenik et al., 1992; Huether, 1994). In addition, melatonin concentration in the peripheral blood of animals refed after fasting increased in several studies (DeBoer, 1988; Huether, 1994; Bubenik et al., 1996). The sudden pulse of melatonin released from the GIT might cause the shift of biological rhythms observed in animals after food intake (Zafar and Morgan, 1992; Rice et al., 1995). This shift is not mediated by the pineal gland, as P_x or the destruction of the suprachiasmatic nucleus (the area regulating the rhythm of the pineal gland) did not affect the shift (Krieger et al., 1977; Sanchez-Vasquez et al., 2000). Elevation of circulating melatonin levels might also be the cause of the postprandial sleepiness in humans (Bubenik et al., 2000; Bubenik, 2002). Finally, melatonin released into the lumen of the GIT upon food intake might synchronize the sequential digestive processes (Bubenik et al., 1996, 1999; Bubenik, 2002).

Melatonin in Digestive Glands and the Hepatobiliary System

Melatonin was first detected in the salivary gland of the rat palate by immunohistology (Bubenik, 1980b). In the saliva, melatonin might have a protective effect on periodontal conditions in human diabetics (Cutando et al., 2003). Melatonin is also involved in the function of glandular epithelium of the stomach and the intestines. In the duodenal lumen, melatonin is a strong stimulant of mucosal bicarbonate secretion (Sjoblom et al., 2001; Sjoblom and Flemstrom, 2003). Conversely, pretreatment with melatonin antagonist luzindol almost abolished this response (Sjoblom and Flemstrom, 2001). Furthermore, a release of melatonin from the duodenal mucosa mediates the duodenal secretory response to intracerebroventricular infusion of α_1-adrenoceptor agonist phenylnephrine (Sjoblom and Flemstrom, 2004). Finally, melatonin also induced intracellular calcium signaling in isolated human and rats, duodenal enterocytes, which strongly suggests that melatonin acts on the enterocyte membrane receptors (Sjoblom et al., 2003). Besides stimulation of bicarbonate secretion in the duodenum, which protects the gastric mucosa, melatonin also has a protective function in the exocrine pancreas. Melatonin precursor TRP protects the pancreas from the development of acute pancreatitis (Leja-Szpak et al., 2004) and the circadian rhythm of melatonin influences the variation in severity of caerulein-induced pancreatitis (Jaworek et al., 2002, 2004a). In addition, exogenous or TRP-derived melatonin reduced the damage to pancreas induced by caerulein or ischemia/reperfusion (Jaworek et al., 2003). Finally, melatonin stimulated secretion of pancreatic amylase under basal conditions or when pancreatic bile juices stimulated amylase secretion. Enzyme secretion was completely abolished by vagotomy, deactivation of sensory nerves or pretreatment with cholecystokinine (CCK_1) receptor antagonist (Jaworek et al., 2004b). Melatonin may also be involved in the function of the endocrine pancreas. Pinealectomy caused a severe hyperinsulinemia and the accumulation of triglycerides in the liver of type 2 diabetic rats (Nishida et al., 2003). Conversely, melatonin reduced hyperinsulinemia and improved the function of lipid metabolism in type 2 diabetic rats, probably by restoring insulin resistance (Nishida et al., 2002). Melatonin may be also involved in the secretion of CCK. Administration of melatonin alone caused a dose-dependant increase of CCK in plasma (Jaworek et al., 2004b).

Melatonin produced in the GIT is forwarded to the general circulation via the hepatic portal vein (Huether et al., 1992; Yaga et al., 1993; Bubenik et al., 1996; Bubenik et al., 2000). Melatonin was first reported in hepatic cells by Menendez-Pelaez et al. (1993). Later on, melatonin was detected in the liver in concentrations 15 times higher than found in blood (Messner et al., 2001). Degradation of melatonin in the liver is the main metabolic pathway for its deactivation (Kopin et al., 1960). However, a study of Huether and co-workers (1998) and Messner et al. (1998) indicated that below a certain threshold level (daytime concentrations in peripheral circulation) melatonin escapes liver deactivation. Above those levels, melatonin is quickly metabolized and then excreted via the bile. The discovery of melatonin receptor subtype MT_1 in the epithelium of the gallbladder indicates that melatonin could influence gallbladder contractions (Aust et al., 2004). The concentrations of melatonin in bile (ranging from 2,000 to 11,000 pg/ml), exceed the levels found in GIT tissues by 10-40 times (Tan et al., 1999). As a strong antioxidant, melatonin in bile might protect the liver tissues. For instance, melatonin was found to have a therapeutic effect on liver injuries caused by bile duct ligation (Ohta et al., 2003). Finally, it has also been hypothesized that high levels of melatonin in the bile might protect the GIT mucosa from oxidative stress (Tan et al., 1999). Indirectly, melatonin may promote the protection of GIT mucosa by activating the immune system. For example, in rats, melatonin increases the number and size of Payer's patches (the main immune system of the GIT) (Yanagisawa and Kachi, 1994).

Clinical Relevance of Melatonin in the Tubular GIT

A plethora of studies indicate that melatonin plays an important role in the gastrointestinal physiology and may be useful for the prevention or treatment of various gastrointestinal disorders of the stomach and the intestines (Bubenik et al., 1998b; Bubenik, 2001, 2002; Reiter et al., 2003). Therapeutic perspectives of melatonin were recently reviewed by Kvetnoy (2002) and Reiter (2003b), as well as by Delagrande and coworkers (2003). As a direct free radical scavenger and a powerful indirect antioxidant, melatonin was found to reduce toxicity and increase efficacy of numerous drugs used in human treatments (Reiter et al., 2002). Melatonin has a protective effect on the development of gastric ulcers (Cho et al., 1989; Khan et al., 1990; Brzozowski et al., 1997; Gruszka et al., 1997; Melchiorri et al., 1997; Komarov et al., 2000; Bandyopadhyay et al., 2000, 2004; Cabeza et al., 2001, 2002; Otsuka et al., 2001), probably via its antioxidative effects (Konturek et al., 1997a, b; Akbulut et al., 1997; Storr et al., 2002; Bilici et al., 2002; Reiter et al., 2003). Melatonin decreased stress-induced ulceration of mucosa in the stomach, ileum and the colon (Ercan et al., 2004) and also reduced symptoms of non-steroidal antiinflamatory drugs-induced gastric ulcers (Motilva et al., 2001). In pigs, orally applied melatonin reduced the incidence and the severity of gastric ulcers. The incidence of ulcers correlated with levels of melatonin in plasma and gastric tissues (Bubenik et al., 1998a). Results of several studies indicate that exogenous melatonin accelerates the healing of experimentally-induced gastric ulcers. This might be due to the enhancement of mucosal blood flow (Cho et al., 1989; Konturek et al., 1997b; Liaw et al., 2002). This effect is possibly mediated by cyclooxygenase-derived prostaglandins and nitric oxide. In the rat model of colitis, melatonin inhibited the activity of nitric oxide synthase and cyclooxgenase 2 (Dong et al., 2003). Another report suggests that gastrin and calcitonin-related peptide, released from sensory nerves, may be involved in the melatonin-induced healing of ulcers (Brzozowska et al., 2002). A circadian rhythm of melatonin is involved in the development of stress-induced mucosal lesions in rats (Kato et al., 2002). A diurnal rhythm in the protective effect of melatonin against gastric injury caused by ischemia-reperfusion was reported by Cabeza et al. (2002). Finally, melatonin prevented the development of gastric ulcers in rats during an artificially-shifted photoperiod (Komarov et al., 2000).

Melatonin also exhibited a protective effect in the intestines. For example, melatonin reduced the severity of methotrexate-induced enteritis in rats (Jahovic et al., 2004). Melatonin might also be useful for the treatment of inflammatory bowel diseases. Melatonin significantly reduced the symptoms of ulcerative colitis (such as diarrhea, loss of weight and the development of mucosal lesions) in the animal model (Pentney and Bubenik, 1995; Cuzzocrea et al., 2001) as well as in humans (Jan, 2003; Mann, 2003). Melatonin prevented intestinal injury in rats, caused by an ischemia-reperfusion (Kazez et al., 2000; Ustundag et al., 2000; Ozacmak et al., 2005) and also reduced bacterial translocation after intestinal ischemia-reperfusion (Sileri et al., 2004). In another study, melatonin reduced colon immunological injury in rats by regulating the activity of macrophages (Mei et al., 2002). The organoprotective effect of melatonin in the stomach and the pancreas was recently reviewed by Jaworek and co-workers (2005). In addition to the protective effect of melatonin, it has been speculated that because of the serotonin-inhibiting properties of melatonin, this indoleamine can be used as a treatment for infant colic (Weissbluth and Weisbluth, 1992). Melatonin is also a natural oncostatic agent (Maestroni, 1993; Fraschini et al., 1998; Schernhammer and Schulmeister, 2004), effective in prognosis (Vician et al., 1999) and *in vitro* as well as *in vivo* treatment of GIT cancer (Bartsch et al., 1997; Melen-Mucha, 1998; Farriol et al., 2000).

CONCLUSION

After more than 30 years, the field of GIT melatonin is now firmly established as a separate entity, different from the field of pineal melatonin. The multiple function of this indole acting in the GIT in the autocrine, paracrine, luminal or endocrine capacity, provides for a wide range of research studies in the basic as well as applied areas. There is a plethora of research and anecdotal evidence substantiating pilot clinical studies in the treatment of gastric ulcers, colitis, GIT cancer, children's colic and irritable bowel syndrome. The conflicting results of some melatonin studies, frequently related to its often opposite effects of high and low doses, requires further investigation as this may be an important point for possible clinical utilization.

REFERENCES

Akbulut, H., Akbulut, K.G. and Gonul, B. (1997). Malodialdehyde and glutathione in rat gastric mucosa and effects of exogenous melatonin. *Dig. Dis. Sci.* **42**: 1381-1388.

Angers, K., Haddad, N., Selmaoui, B. and Thibault, L. (2003). Effect of melatonin on total food intake and macronutrient choice in rats. *Physiol. Behav.* **80**: 9-11.

Aust, S., Thalhammer, T., Humpeler, S., Jaeger, W., Klimginger, M., Tucek, G., Obrist, P., Marktl, W., Penner, E. and Ekmekcioglu, C. (2004). The melatonin receptor subtype MT_1 is expressed in human gallbladder epithelia. *J. Pineal Res.* **36**: 43-48.

Bandyopadhyay, D., Biswas, D., Reiter, R.J. and Banerjee, R.K. (2000). Melatonin protects against stress-induced gastric lesions by scavenging hydroxyl radicals. *J. Pineal Res.* **29**: 143-151.

Bandyopadhyay, D., Ghosh, G., Bandyopathyay, A. and Reiter, R.J. (2004). Melatonin protects against piroxicam-induced gastric ulceration. *J. Pineal Res.* **36**: 195-203.

Barajas-Lopez, C., Perez, A.L., Espinosa-Luna, R., Reyes-Vasquez, C. and Prieto-Gomez , B. (1996). Melatonin modulates cholinergic transmission by blocking nicotinic channels in the guinea pig submucous plexus. *Eur. J. Pharmacol.* **312**: 319-325.

Bartsch, C., Kvetnoy, I., Kvetnaia, T., Bartsch, H., Molotkov, A., Franz, H., Raikhlin, N. and Mecke, D. (1997). Nocturnal urinary 6-sulfatoxymelatonin and proliferating cell nuclear antigen-immunopositive tumor cells show strong positive correlations in patients with gastrointestinal and lung cancer. *J. Pineal Res.* **23**: 9-96.

Bilici, D., Suleyman, H., Banoglu, Z.N., Kiziltunc, A., Ciftioglu, A. and Bilici, S. (2002). Melatonin prevents ethanol-induced gastric mucosal damage possibly due to its antioxidant effect. *Dig. Dis. Sci.* **47**: 856-861.

Bindoni, M. (1971). Relationship between the pineal gland and the mitotic activity in some tissues. *Arch. Sci. Biol.* **55**: 3-21.

Bogoeva, M., Mileva, M.S. and Tsanova, K.S. (1993). Effect of exogenous melatonin on the twenty-four-hour mitotic activity of some normal mouse tissues. *C. R. Acad. Bulgare. Sci.* **46**: 107-110.

Bogoeva, M., Mileva, M.S. and Gabev, E.E. (1998). Changes of the circadian rhythm of colonic mitotic activity after single melatonin application. *C. R. Acad. Bulgare. Sci.* **51**: 6-8.

Brown, G.M., Seggie, J. and Grota, L.J. (1985). Serum melatonin response to melatonin administration in the Syrian hamster. *Neuroendocrinol.* **41**: 31-35.

Brzozowska, I., Konturek, P.C., Brzozowski, T., Konturek, S.J., Kwiecien, S., Pajdo, R., Drozdowicz, D., Pawlik, M., Ptak, A. and Hahn, E.G. (2002). Role of prostaglandins, nitric oxide, sensory nerves and gastrin in acceleration of ulcer healing by melatonin and its precursor, L-tryptophan. *J. Pineal Res.* **32**: 149-162.

Brzozowski, T., Konturek, P.C., Konturek, S.J., Pajdo, R., Bielanski, W., Brzozowska, I., Stachura, J. and Hahn, E.G. (1997). The role of melatonin and L-tryptophan in prevention of acute gastric lesions induced by stress, ethanol, ischemia and aspirin. *J. Pineal Res.* **23**: 79-89.

Bubenik, G.A., Brown, G.M., Uhlir, I. and Grota, L.J. (1974). Immunohistological localization of N-acetylindolealkylamines in the pineal gland, retina and cerebellum. *Brain Res.* **81**: 233-242.

Bubenik, G.A., Brown, G.M. and Grota, L.J. (1976a). Differential localization of N-acetylated indolealkylamines in the CNS and the Harderian gland using immunohistology. *Brain Res.* **118**: 417-427.

Bubenik, G.A., Brown, G.M. and Grota, L.J. (1976b). Immunohistological localization of melatonin in the rat Harderian gland. *J. Histochem. Cytochem.* **24**: 1173-1177.

Bubenik, G.A., Brown, G.M. and Grota, L.J. (1977). Immunohistological localization of melatonin in the rat digestive tract. *Experientia* **33**: 662-663.

Bubenik, G.A., Purtil, R.A., Brown, G.M. and Grota, L.J. (1978). Melatonin in the retina and the Harderian gland. Ontogeny, diurnal variations and melatonin treatment. *Exp. Eye Res.* **27**: 323-333.

Bubenik, G.A. (1980a). Localization of melatonin in the digestive tract of the rat. Effect of maturation, diurnal variation, melatonin treatment and pinealectomy. *Horm. Res.* **12**: 313-323.

Bubenik, G.A. (1980b). Immunohistological localization of melatonin in the salivary gland of the rat. *Adv. Biosci.* **29**: 95-112.

Bubenik, G.A. (1986). Effect of serotonin, N-acetylserotonin and melatonin on spontaneous contractions of isolated rat intestine. *J. Pineal Res.* **3**: 41-54.

Bubenik, G.A. and Dhanvantari, S. (1989). The influence of melatonin on some parameters of gastrointestinal activity. *J. Pineal Res.* **7**: 333-344.

Bubenik, G.A., Ball, R.O. and Pang, S.F. (1992). The effect of food deprivation on brain and gastrointestinal tissue levels of tryptophan, serotonin, 5-hydroxyindoleacetic acid, and melatonin. *J. Pineal Res.* **12**: 7-16.

Bubenik, G.A., Pang, S.F., Niles, L.O. and Pentney, P.J. (1993). Diurnal variations and binding characteristics of melatonin in the mouse brain and the gastrointestinal tissues. *Comp. Biochem. Physiol.* **104**(A): 377-380.

Bubenik, G.A. and Pang, S.F. (1994). The role of serotonin and melatonin in the gastrointestinal physiology: ontogeny, regulation of food intake and mutual 5-HT, melatonin feedbacks. *J. Pineal Res.* **16**: 91-99.

Bubenik, G.A., Pang, S.F., Hacker, R.R. and Smith, P.S. (1996). Melatonin concentrations in serum, and tissues of porcine gastrointestinal tract and their relationship to the intake and passage of food. *J. Pineal Res.* **21**: 251-256.

Bubenik, G.A. and Pang, S.F. (1997). Melatonin levels in the gastrointestinal tissue of fish, amphibians and a reptile. *Gen. Comp. Endocrinol.* **106**: 415-419.

Bubenik, G.A. and Brown, G.M. (1997). Pinealectomy reduces melatonin levels in the serum but not in the gastrointestinal tract of the rat. *Biol. Signals* **6**: 40-44.

Bubenik, G.A., Ayles, H.L., Ball, R.O., Friendship, R.M. and Brown, G.M. (1998a). Relationship between melatonin levels in plasma and the incidence and severity of gastric ulcers in pigs. *J. Pineal Res.* **24**: 62-66.

Bubenik, G.A., Blask, D.E., Brown, G.M., Maestroni, G.J.M., Pang, S.F., Reiter, R.J., Viswanathan, M. and Zisapel, N. (1998b). Prospects of clinical utilization of melatonin. *Biol. Signals Recept.* **7**: 195-219.

Bubenik, G.A., Hacker, R.R., Brown, G.M. and Bartos, L. (1999). Melatonin concentrations in the luminal fluid, mucosa and muscularis of the bovine and porcine gastrointestinal tract. *J. Pineal Res.* **26**: 56-63.

Bubenik, G.A., Brown, G.M., Hacker, R.R. and Bartos, L. (2000a). Melatonin levels in the gastrointestinal tissues of fetal bovids. *Acta Vet.* **69**: 177-182.

Bubenik, G.A., Pang, S.F., Cockshut, J.R., Smith, P.S., Grovum, L.W., Friendship, R.M. and Hacker, R.R. (2000b). Circadian variation of portal, arterial and venous blood levels of melatonin in pigs and its relationship to food intake and sleep. *J. Pineal Res.* **28**: 9-15.

Bubenik, G.A. (2001). Localization, physiological significance and possible clinical implication of gastrointestinal melatonin. *Biol. Signals Recept.* **10**: 350-366.

Bubenik, G.A. (2002). Gastrointestinal melatonin: Localization, function and clinical relevance. *Dig. Dis. Sci.* **47**: 2336-2348.

Cabeza, J., Motilva, A. and Alarcon de la Lastra, C.A. (2001). Mechanism involved in the gastric protection of melatonin on ischemia-reperfusion-induced oxidative damage in rats. *Life Sci.* **68**: 1405-1415.

Cabeza, J., Alarcon de la Lastra, C.A., Martin, M.J., Herreira, J.M. and Motilva, V. (2002). Diurnal variation in the protective effect of melatonin against gastric injury caused by ischemia-reperfusion. *Biol. Rhythm Res.* **33**: 319-332.

Callaghan, B.D. (1995). The effect of pinealectomy and autonomic denervation on crypt cell proliferation in the rat small intestine. *J. Pineal Res.* **10**: 180-185.

Caniato, R., Filippini, R., Piovan, A., Puricelli, L., Borsarini, A. and Capelletti, E.M. (2003). Melatonin in plants. *Adv. Exp. Med. Biol.* **527**: 593-597.

Chan, H., Lui, K., Wong, W. and Poon, A. (1998). Effect of melatonin on chloride secretion by human colonic T$_{84}$ cells. *Life Sci.* **23**: 2151-2158.

Chik, C., Ho, A.K. and Brown, G.M. (1987). Effect of food restriction on 24-h serum and pineal melatonin. *Acta Endocrinol.* **115**: 507-513.

Cho, C.H., Pang, S.F., Chen, B.W. and Pfeiffer, C.J. (1989). Modulating action of melatonin on serotonin-induced aggravation of ethanol ulceration and changes of mucosal blood flow in rat stomach. *J. Pineal Res.* **6**: 89-97.

Chow, P.H., Lee, P.P.N., Poon, A.M.S., Shiu, S.Y.W. and Pang, S.F. (1996). The gastrointestinal system: A site of melatonin paracrine action. In: *Melatonin: A Universal Photoperiodic Signal With Diverse Action* (Tang, P.L., Pang, S.F. and Reiter, R.J., eds.). Karger, Basel, Switzerland, *Front. Horm. Res.* **21**: 123-132.

Cutando, A., Gomez-Moreno, G., Villalba, J., Ferrera, M.J., Escames, G. and Acuna-Castroviejo, D. (2003). Relationship between salivary melatonin levels and periodontal status in diabetic rats. *J. Pineal Res.* **35**: 239-244.

Cuzzocrea, S., Mazzon, E., Serraino, I., Lepore, V., Terranova, M., Ciccolo, A. and Caputi, A. (2001). Melatonin reduces dinitrobenzine sulfonic acid-induced colitis. *J. Pineal. Res.* **30**: 1-12.

DeBoer, H. (1988). The influence of photoperiod and melatonin on hormone levels and operand light demand in the pig. Ph.D. thesis. University of Guelph, Ontario, Canada.

Delagrande, P., Atkinson, J., Boutin, J.A., Casteilla, L., Lesieur, D., Misslin, R., Pellissier, S., Penicaud, L. and Renard, P. (2003). Therapeutic perspectives for melatonin agonists and antagonists. *J. Neuroendocrinol.* **15**: 442-448.

Dong, W-G., Mei, Q., Yu, J-P., Xu, J-M., Xiang, L. and Xu, U. (2003). Effects of melatonin on the expression of iNOS and COX-2 in rat model of colitis. *World J. Gastroenterol.* **9**: 1307-1311.

Drago, F., Macauda, S. and Salehi, S. (2002). Small doses of melatonin increase intestinal motility in rats. *Dig. Dis. Sci.* **47**: 1969-1974.

Ercan, F., Cetinel, S., Contuc, G., Cikler, E. and Sener, G. (2004). Role of melatonin in reducing water avoidance stress-induced degeneration of the gastrointestinal mucosa. *J. Pineal Res.* **37**: 113-121.

Espamer, V. and Asero, B. (1952). Identification of enteramine, the specific hormone of enterochromaffin cell system, as 5-hydroxytryptamine. *Nature* **169**: 800-801.

Farriol, M., Venerco, Y., Orta, X., Castellaos, J. and Segovia-Silvestre, T. (2000). *In vitro* effect of melatonin on cell proliferation in colon adenocarcinoma line. *J. Appl. Toxicol.* **21**: 21-24.

Fioretti, M.C., Menconi, E. and Ricardi, C. (1974a). Study on the type of antiserotonergic antagonism exerted in vitro on rat's stomach by pineal indole derivatives. *Il. Farmaco* (Ediziane Pratica) **29**: 401-412 (in Italian).

Fioretti, M.C., Menconi, E. and Ricardi, C. (1974b). Mechanism of the in vitro 5-hydroxytryptamine (5-HT) antagonism exerted by pineal indole derivatives. *Riv. Farmacol. Ter.* **5**: 43-49.

Fraschini, F., Demartini, G., Esposti, D. and Scaglione, F. (1998). Melatonin involvement in immunity and cancer. *Biol. Signals Recept.* **7**: 61-72.

Gruszka, A., Kunert-Radek, J., Pawlikowski, M. and Karasek, M. (1997). Melatonin, dehydroepiandrosterone and RZR/ROR-ligand CGP 52605 attenuate stress-induced gastric lesion formation in rats. *Neuroendocrinol. Lett.* **18**: 221-229.

Harderland, R. and Poegeller, B. (2003). Minireview: Non-vertebrate melatonin. *J. Pineal Res.* **34**: 233-241.

Harlow, H.J. and Weekly, B.L. (1986). Effect of melatonin on the force of spontaneous contraction of *in vitro* rat small and large intestine. *J. Pineal Res.* **3**: 277-284.

Hattori, A., Migitaka, H., Iigo, M., Yamamoto, K., Ohtaneki-Kaneko, T.M., Suzuki, T. and Reiter, R.J. (1995). Identification of melatonin in plants and its effect on plasma melatonin levels and binding to melatonin receptors in vertebrates. *Biochem. Mol. Biol. Ont.* **35**: 627-634.

Herichova, I. and Zeman, M. (1996). Perinatal development of melatonin production in gastrointestinal tract of domestic chicken. In: *Investigations of Perinatal Development in Birds* (Toenhardt, H. and Lewin, R., eds.), Free Univ., Berlin, 109-116.

Ho, T.K., Burns, T.G., Grota, L.J. and Brown, G.M. (1985). Scheduled feeding and 24-hour rhythms of N-acetylserotonin and melatonin in rats. *Endocrinol.* **116**: 1858-1862.

Holloway, W.R., Grota, L.J. and Brown, G.M. (1980). Determination of immunoreactive melatonin in the colon of rat by immunocytochemistry. *J. Histochem. Cytochem.* **28**: 255-262.

Hong, G.X. and Pang, S.F. (1995). N-acetyltransferase activity in the quail (*Coturnix coturnix jap*) duodenum. *Comp. Biochem. Physiol.* **112**(B): 251-255.

Huether, G., Poegeller, B., Reimer, R. and George, A. (1992). Effect of tryptophan administration on circulating melatonin levels in chicks and rats: evidence for stimulation of melatonin synthesis and release in the gastrointestinal tract. *Life Sci.* **51**: 945-953.

Huether, G. (1994). The contribution of extrapineal sites of melatonin to circulating melatonin levels in higher vertebrates. *Experientia* **49**: 665-670.

Huether, G. (1994). Melatonin synthesis in the gastrointestinal tract and the impact of nutritional factors on circulating melatonin. *Ann. N. Y. Acad. Sci.* **719**: 146-158.

Huether, G., Messner, M., Rodenback, A. and Hardeland, R. (1998). Effects of continuous infusion on steady state plasma melatonin levels under near physiological conditions. *J. Pineal Res.* **24**: 146-151.

Jahovic, N., Sener, G., Ersoy, Y., Arbak, S. and Yegen, B.C. (2004). Amelioration of methotrexate-induced enteritis by melatonin in rats. *Cell. Biochem. Function* **22**: 169-178.

Jan, J.E. (2003). Letter to editor. Re: Mann – Melatonin for ulcerative colitis. *Am. J. Gastroenterol.* **98**: 1446.

Jaworek, J., Konturek, S.J., Leja-Szpak, A., Nawrot, K., Bonior, J., Tomaszewska, R., Stachura, J. and Pawlik, W.W. (2002). Role of endogenous melatonin and its MT$_2$ receptor in the modulation of caerulein-induced pancreatitis in the rat. *J. Physiol. Pharmacol.* **53**: 791-804.

Jaworek, J., Leja-Szpak, A., Bonior, J., Nawrot, K., Tomaszewska, R., Stachura, J., Sendur, R., Pawlik, W., Brzozowski, T. and Konturek, S.J. (2003). Protective effect of melatonin and its precursor L-tryptophan on acute pancreatitis induced by caerulein overstimulation or ischemia/reperfusion. *J. Pineal Res.* **34**: 40-52.

Jaworek, J., Konturek, S., Tomaszewska, R., Leja-Szpak, A., Bonior, J., Nawrot, K., Palonek, M., Stachura, J. and Pawlik, W.W. (2004a). The circadian rhythm of melatonin modulates the severity of caerulein-induced pancreatitis in the rat. *J. Pineal Res.* **37**: 161-170.

Jaworek, J., Nawrot, K., Konturek, S.J., Leja-Szpak, T.P. and Pawlik, W.W. (2004b).

Melatonin and its precursor, L-tryptophan: influence on pancreatic amylase secretion *in vivo* and *in vitro*. *J. Pineal Res*. **36**: 155-164.

Jaworek, J., Brzozowski, T. and Konturek, S.J. (2005). Melatonin as an organoprotector in the stomach and the pancreas. *J. Pineal Res*. **38**: 73-83.

Kachi, T., Suzuki, T., Yanagisawa, M., Kimura, N. and Irie, T. (1999). Pineal-gut relations. *Hirosaki Med. J.* **51** (Suppl.): S209-S213.

Kato, K., Murai, I., Asai, S., Takahashi, Y., Nagata, T., Komura, S., Mizuno, S., Wasaki, A., Ishikawa, K. and Arakawa, Y. (2002). Circadian rhythm of melatonin and prostaglandins in modulatuion of stress-induced gastric mucosal lesions in rats. *Aliment. Pharmacol. Ther.* **16**: 29-34.

Kazez, A., Demirbag, M., Ustundag, B., Ozercan, H. and Saglam, M. (2000). The role of melatonin in the prevention of intestinal ischemia-reperfusion in rats. *J. Pediatric. Surg.* **35**: 1444-1448.

Kennaway, D.J., Firth, R.J., Philipou, G., Mathews, C.D. and Seamark, R.F. (1977). A specific radioimmunoassay for melatonin in biological tissues and fluids and its validation by gas chromatography-mass spectrometry. *Endocrinol.* **101**: 119-127.

Kezuka, H., Iigo, A., Furukawa, K., Aida, K. and Hanyu, I. (1992). Effects of photoperiod, pinealectomy and ophtalmectomy on circulating melatonin levels in the goldfish (*Carassius auratus*). *Zool. Sci.* **9**: 1147-1153.

Khan, R., Daya, S. and Potgieter, B. (1990). Evidence for the modulation of the stress response by the pineal gland. *Experientia* **46**: 860-862.

Komarov, F.I., Rapoport, S.I., Malinovskaja, N.K., Sudakov, K.B., Sosnovskij, A.S., Percov, S.S. and Wetterberg, L. (2000). Protective effect of melatonin in gastric ulcers in rats during an artificially-shifted photoperiod. *Vestnik Russ. Acad. Med. Nauk.* **8**: 21-25 (in Russian).

Konturek, P.C., Konturek, S.J., Brzozowski, T., Dembinski, A., Zembala, M., Mytar, B. and Hahn, E.G. (1997a). Gastroprotective activity of melatonin and its precursor, L-tryptophan, against stress-induced and ischemia-induced lesions is mediated by scavenge of oxygen free radicals. *Scand. J. Gastroenterol.* **32**: 433-438.

Konturek, P.C., Konturek, S.J., Majka, J., Zembala, H. and Hahn, E.G. (1997b). Melatonin affords protection against gastric lesions induced by ischemia-reperfusion due to its antioxidant and mucosal microcirculatory effect. *Eur. J. Pharmacol.* **122**: 73-77.

Kopin, I.J., Pare, C.M.B., Axelrod, J. and Weissbach, H. (1960). 6-hydroxylation, the major metabolic pathway for melatonin. *Biochem. Biophys. Acta* **40**: 370-377.

Kopin, I.J., Pare, C.M.B., Axelrod, J. and Weissbach, H. (1961). The fate of melatonin in animals. *J. Biol. Chem.* **236**: 3072-3075.

Krieger, D.T., Hauser, H. and Krey, L.C. (1977). Suprachiasmatic nucleus lesions do not abolish food-shifted circadian rhythmicity and temperature rhythmicity. *Science* **197**: 398-402.

Kvetnoy, I., Raikhlin, N.T. and Tolkachev, V.N. (1975). Chromatographical detection of melatonin (5-methoxy-N-acetylserotonin) and its biological precursors in enterochromaffine cells. *Dokl. Acad. Nauk. SSSR* **221**: 226-227 (in Russian).

Kvetnoy, I. (2002). Extrapineal melatonin in pathology: New perspectives for diagnosis, prognosis and treatment of illness. *Neuroendocrinol. Lett.* **23**(1): 92-96.

Lee, P.P.N. and Pang, S.F. (1993). Melatonin and its receptors in the gastrointestinal tract. *Biol. Signals* **2**: 181-193.

Lee, P.P.N., Shiu, S.Y.U., Choe, P.H. and Pang, S.F. (1995). Regional and diurnal studies on melatonin binding sites in the duck gastrointestinal tract. *Biol. Signals* **4**: 212-224.

Legris, G.J., Will, P.C. and Hopfer, U. (1982). Inhibition of amiloride-sensitive sodium conductance by indoleamines. *Proc. Natl. Acad. Sci.* **79**: 2040-2050.

Leja-Szpak, A., Jaworek, J., Tomaszewska, R., Nawrot, K., Bonior, J., Kot, M., Palonek, M., Stachura, J., Czuprina, A., Konturek, S.J. and Pawlik, W.W. (2004). Melatonin precursor, L-tryptophan protects the pancreas from development of acute pancreatitis through the central site action. *J. Physiol. Pharmacol.* **55**: 339-354.

Lerner, A.B., Case, J.D., Lee, T.H. and Mori, W. (1958). Isolation of melatonin, the pineal factor that lightens melanocytes. *J. Am. Chem. Soc.* **80**: 2587.

Lewy, A.J., Emens, J.S., Sack, R.L., Hasler, B.P. and Bernert, A. (2002). Low, but not high, doses of melatonin entrained a free-running blind person with a long circadian period. *Chronobiol. Int.* **19**: 649-658.

Liaw, S.J., Cheng, N.J., Ng, C. Cl, Ciu, D.F., Chen, M.F. and Chen, H.M. (2002). Beneficial role of melatonin on microcirculation in endotoxin-induced gastropathy in rats: Possible implication in nitrogen oxide reduction. *J. Formosan Med. Assoc.* **101**: 129-135.

Lucchelli, A., Santagostino-Barbone, M.G. and Tonini, M. (1997). Investigation into the contractile response of melatonin in the guinea pig isolated proximal colon: the role of 5-HT$_4$ and melatonin receptors. *Br. J. Pharmacol.* **121**: 1775-1781.

Maestroni, G.J.M. (1993). The immunoneuroendocrine role of melatonin. *J. Pineal Res.* **4**: 1-10.

Mann, S. (2003). Melatonin for ulcerative colitis? *Am. J. Gastroenterol.* **98**: 232-233.

Mei, Q., Yu, J.-M., We, W., Xiang, L. and Yue, L. (2002). Melatonin reduces colon immunological injury in rats by regulating activity of macrophages. *Acta Pharmacol. Sinica* **23**: 882-886.

Melchiorri, D.E., Sewerynek, E. , Reiter, R.J., Ortiz, G.G., Poegeller, B. and Nistico, G. (1997). Suppressive effect of melatonin administration on ethanol-induced gastroduodenal injury in rats *in vivo*. *Br. J. Pharmacol.* **121**: 264-270.

Melen-Mucha, G., Wynczyk, K. and Pawlikowski, M. (1998). Somatostatin analog ocreotide and melatonin inhibit bromodeoxyuridine incorporation into cell nuclei and enhance apoptosis in the transplantable murine colon 38 cancer. *Anticancer Res.* **18**: 3615-3620.

Menendez-Pelaez, A., Poeggeler, B., Reiter, R.J., Barlow-Walden, I., Pablos, M.I. and Tan, T-X. (1993). Nuclear localization of melatonin in the different mammalian tissues: Immunocytochemical and radioimmunoassay evidence. *J. Cell Biol.* **53**: 373-382.

Merle, A., Delagrande, P., Renard, P., Lesieur, D., Cuber, J.C., Roche, M. and Pellissier, S. (2000) Effect of melatonin on motility pattern of small intestine in rats and its inhibition by melatonin receptor antagonist S 22153. *J. Pineal Res.* **29**: 116-124.

Messner, M., Hardeland, R., Rodenback, A. and Huether, G. (1998). Tissue retention and subcellular distribution of continuously infused melatonin in rats under near physiological conditions. *J. Pineal Res.* **25**: 251-259.

Messner, M., Huether, G., Lorf, T., Ramdori, G. and Schwoerrer, H. (2001). Presence of melatonin in the human hepatobiliary tract. *Life Sci.* **69**: 543-551.

Motilva, V., Cabeza, J. and Alarcon de la Lastra, C. (2001). New issues about melatonin and its effect on the digestive system. *Curr. Pharmaceut. Design* **7**: 909-931.

Nishida, S., Segawa, T., Murai, I. and Nagakawa, S. (2002). Long-term melatonin administration reduces hyperinsulinemia and improves the altered fatty-acid compositions in type 2 diabetic rats via restoration of Δ-5 desaturase activity. *J. Pineal Res.* **32**: 26-33.

Nishida, S., Sato, R., Murai, I. and Nakagawa, S. (2003). Effect of pinealectomy on plasma levels of insulin and leptin and on hepatic lipids in type 2 diabetic rats. *J. Pineal Res.* **35**: 251-256.

Ohta, Y., Kongo, M. and Kishikawa, T. (2003). Melatonin exerts a therapeutic effect on cholestatic liver injury in rats induced by bile ligation. *J. Pineal Res.* **33**: 127-133.

Otsuka, M., Kato, K., Murai, I., Asai, S., Iwasaki, A. and Arakawa, Y. (2001). Roles of nocturnal melatonin and the pineal gland in modulation of water-immersion restrain stress-induced gastric mucosal lesions in rats. *J. Pineal Res.* **30**: 82-86.

Ozacmak, V.H., Sayan, H., Arslan, S.O., Altaner, S. and Aktas, R.G. (2005). Protective effect of melatonin on contractile activity and oxidative injury induced by ischemia and reperfusion of rat ileum. *Life Sci.* **76**: 1575-1588.

Pardridge, W.M. and Mietus, L.J. (1980). Transport of albumin-bound melatonin through the blood-brain barrier. *J. Neurochem.* **34**: 1761-1763.

Pawlikowski, M. (1986). The pineal gland and cell proliferation. *Adv. Pineal Res.* **1**: 27-30.

Pentney, P. and Bubenik, G.A. (1995). Melatonin reduces the severity of dextran-induced colitis in mice. *J. Pineal Res.* **19**: 31-39.

Pieri, C., Marra, M., Recchioni, R. and Marcheselli, F. (1994). Melatonin: a peroxyl radical scavenger more effective than vitamin E. *Life Sci.* **15**: 271-276.

Poegeller, B., Reiter, R.J., Tan, D-X., Chen, L-D. and Manchester, L.C. (1993). Melatonin, hydroxy-radical-mediated oxidative damage, and aging: a hypothesis. *J. Pineal Res.* **14**: 151-168.

Pontoire, C., Bernard, M., Silvain, C., Collin, J-P. and Voissin, P. (1993). Characterization of melatonin binding sites in chicken and human intestines. *Eur. J. Pharmacol.* **247**: 111-118.

Quastel, R. and Rahamimoff, R. (1965). Effect of melatonin on spontaneous contraction and response to 5-hydroxytryptamine of rat isolated duodenum. *Br. J. Pharmacol.* **24**: 455-461.

Quay, W.B. and Ma, Y.H. (1976). Demonstration of gastrointestinal hydroxyindole-O-methyl transferase. *IRCS Med. Sci.* **4**: 563.

Raikhlin, N.T. and Kvetnoy, I.M. (1974). Lightening effect of the extract of human appendix mucosa on frog skin melanophores. *Bull. Exp. Biol. Med.* **8**: 114-116 (in Russian).

Raikhlin, N.T., Kventoy, I.M. and Tolkachev, V.N. (1975). Melatonin may be synthesized in enterochromafinne cells. *Nature* **255**: 334-345.

Raikhlin, N.T. and Kvetnoy, I.M. (1976). Melatonin and enterochromaffine cells. *Acta Histochem.* **55**: 19-25.

Raikhlin, N.T., Kvetnoy, I.M., Kadagidze, Z.G. and Sokolov, A.V. (1978). Immuno-morphological studies on synthesis of melatonin in enterochromaffine cells. *Acta Histochem. Cytochem.* **11**: 75-77.

Rajchard, J., Hajek, I. and Sery, M. (2000). Melatonin level in guppy (*Poecilia reticulata*—Osteiichties, Poeciliidae). *Czech. J. Anim. Sci.* **45**: 105-111.

Reiter, R.J. (1991). Pineal melatonin: cell biology of its synthesis and its physiological interactions. *Endocrinol. Rev.* **12**: 151-180.

Reiter, R.J. (1993). The melatonin rhythm: both clock and calendar. *Experientia* **49**: 654-664.

Reiter, R.J. (2000). Melatonin: Lowering the high price of free radicals. *News Physiol. Sci.* **15**: 246-250.

Reiter, R.J., Tan, D-X., Sainz, R.M., Mayo, J.C. and Lopez-Burillo, S. (2002). Melatonin: reducing the toxicity and increasing efficacy of drugs. *J. Pharmacy and Pharmacol.* **54**: 1299-1321.

Reiter, R.J., Tan, D-X., Mayo, J.C., Sainz, R.M., Leon, J. and Bandyopadhyay, D. (2003). Neurally-mediated and neurally-independent beneficial actions of melatonin in the gastrointestinal tract. *J. Physiol. Pharmacol.* **54**(4): 113-125.

Reiter, R.J. (2003a). What constitutes a physiological concentration of melatonin? *J. Pineal Res.* **34**: 79-80.

Reiter, R.J. (2003b). Melatonin: clinical relevance. *Best practice & Res. Endocr. Metab.* **17**: 273-285.

Reppert, S.M. and Klein, D.C. (1978). Transport of maternal [^{3}H]melatonin to suckling rats and the fate of [^{3}H]melatonin in the neonatal rats. *Endocrinol.* **102**: 582-588.

Rice, J., Mayor, J., Tucker, H.A. and Bielski, R.J. (1995). Effect of light therapy on salivary melatonin in seasonal affective disorder. *Psychiatry Res.* **56**: 221-226.

Rubio, V.C., Sanchez-Vazquez, F.J. and Madrid, J.A. (2004). Oral administration of melatonin reduces food intake and modifies macronutrient selection in the European sea bass (*Dicentrachus labrax*, L.). *J. Pineal Res.* **37**: 42-47.

Sanchez-Vazquez, F.J., Iigo, M., Madrid, J.A., Zamora, S. and Tabata, M. (2000). Pinealectomy does not affect the entrainment of light nor the generation of the circadian-demand feeding rhythms of rainbow trout. *Physiol. Behav.* **69**: 455-461.

Schernhammer, E.S. and Schulmeister, K. (2004). Melatonin and cancer risk: does light at night compromise physiological cancer protection by lowering serum melatonin levels? *Br. J. Cancer* **90**: 941-943.

Sileri, P., Sica, G.S., Gentileschi, P., Venza, M., Benavoli, D., Jarzembowski, T., Manzelli, A. and Gasperi, A.L. (2004). Melatonin reduces bacterial translocation after intestinal ischemia-reperfusion injury. *Transplantation Proc.* **36**: 2944-2946.

Sjoblom, M. and Flemstrom, G. (2001). Central nervous stimuli increase duodenal bicarbonate secretion by release of mucosal melatonin. *J. Physiol. Pharmacol.* **52**: 671-678.

Sjoblom, M., Jedstedt, G. and Flemstrom, G. (2001). Peripheral melatonin mediates neural stimulation of duodenal mucosal bicarbonate secretion. *J. Clin. Invest.* **108**: 625-633.

Sjoblom, M. and Flemstrom, G. (2003). Melatonin in the duodenal lumen is a potent stimulant of mucosal bicarbonate secretion. *J. Pineal Res.* **34**: 288-293.

Sjoblom, M., Safsten, B. and Flemstrom, G. (2003). Melatonin-induced calcium signaling in clusters of human and rat duodenal enterocytes. *Am. J. Physiol. Gastrointest. Liver Physiol.* **284**: G1034-G1044.

Sjoblom, M. and Flemstrom, G. (2004). Central nervous α_1-adrenoceptor stimulation induces duodenal luminal release of melatonin. *J. Pineal Res.* **36**: 103-108.

Stefulj, J., Hoertner, M., Ghosh, M., Schauenstein, K., Rinner, I., Woelfeler, A., Semmier, J. and Liebman, M. (2001). Gene expression of the key enzymes of melatonin synthesis in extrapineal tissues of the rat. *J. Pineal Res.* **30**: 243-247.

Steindl, P.E., Finn, B., Bendok, B., Rothke, S., Zee, P.C. and Blei, A.C. (1995). Disruption of diurnal rhythm of plasma melatonin in cirrhosis. *Ann. Intern. Med.* **123**: 274-277.

Steindl, P.E., Ferenci, P. and Marktl, W. (1997). Impaired hepatic catabolism of melatonin in cirrhosis. *Ann. Intern. Med.* **127**: 494.

Storr, M., Schusdziarra, V. and Allescher, H.D. (2000). Inhibitory effect of melatonin on smooth muscle cells in rat gastric fundus: involvement of small conductance potassium channels. *Can. J. Physiol. Pharmacol.* **78**: 799-806.

Storr, M., Koppitz, P., Sibaev, A., Saur, D., Kurjak, M., Franck, H., Schusdziarra, V. and Allescher, H.-D. (2002). Melatonin reduces non-adrenergic, non-cholinergic relaxant neurotransmission by inhibition of nitric oxide synthase activity in the gastrointestinal tract of rodents in vitro. *J. Pineal Res.* **33**: 101-108.

Tan, D-X., Manchester, L.C., Reiter, R.J., Qi, W., Hanes, M. and Farley, N.J. (1999). High physiological levels of melatonin in the bile of mammals. *Life Sci.* **65**: 2523-2529.

Tan, D-X., Manchester, L.C., Lopez-Burillo, S.L., Mayo, J.C., Sainz, R.M. and Reiter, R.J. (2003). Melatonin: a hormone, a tissue factor, an autacoid, a paracoid, and an antioxidant vitamin. *J. Pineal Res.* **34**: 75-78.

Ustundag, B., Kazez, A., Demirgab, M., Canatan, H., Halifeoglu, I. and Ozercan, I. (2000). Protective effect of melatonin in experimental ischemia-reperfusion of rat small intestine. *Cell Physiol. Biochem.* **10**: 229-236.

Vakkuri, O., Rintamaki, H. and Leppaluoto, J. (1985a). Plasma and tissues concentrations of melatonin after midnight light exposure and pinealectomy in the pigeon. *J. Endocrinol.* **105**: 253-268.

Vakkuri, O., Rintamaki, H. and Leppaluoto, J. (1985b). Presence of immunoreactive melatonin in different tissues of the pigeon. *Gen. Comp. Endocrinol.* **58**: 69-75.

Vaughan, G.M. and Reiter, R.J. (1986). Pineal dependence of the Syrian hamster nocturnal serum melatonin surge. *J. Pineal Res.* **3**: 9-14.

Vician, M., Zeman, M., Herichova, I., Blazicek, P. and Matis, P. (1999). Melatonin content in plasma and large intestine of patients with colorectal carcinoma before and after surgery. *J. Pineal Res.* **29**: 56-63.

Weissbluth, L. and Weissbluth, L. (1992). Infant colic: the effect of serotonin and melatonin circadian rhythms on the intestinal smooth muscles. *Med. Hypoth.* **39**: 164-169.

Wurtman, R.J., Axelrod, J. and Phillips, L.S. (1963). Melatonin synthesis in the pineal gland: control by light. *Science* **142**: 1071-1073.

Yaga, H., Reiter, R.J. and Richardson, B.A. (1993). Tryptophan loading increases day time serum melatonin in intact and pinealectomized rats. *Life Sci.* **52**: 1231-1238.

Yanagisawa, M. and Kachi, T. (1994). Effects of pineal hormone on Payer's patches in the small intestine. *Acta Anat. Nippon* **69**: 522-527.

Zafar, N.P. and Morgan, E. (1992). Feeding entrains an endogenous rhythm of swimming activity in the blind Mexican cave fish. *Proc. 8th Meet. Europ. Soc. Chronobiol.* Noordwijkerhout, Netherlands, May 28-31. 165-166.

Zerek-Melen, G., Lewinski, A. and Kulak, J. (1987). The opposing effect of high and low doses of melatonin upon the mitotic activity of the mouse intestinal epithelium. *Endokrinol. Pol.* **37**: 317-323.

Zhdanova, I.V., Wurtman, R.J., Lynch, H.J., Yves, R.D., Dollins, A.B., Morabito, C., Matheson, J.K. and Schomer, D.L. (1995). Sleep-inducing effect of low doses of melatonin ingested in the evening. *Clin. Pharmacol. Ther.* **57**: 552-558.

CHAPTER 3

Studies on the Sympathetic Nervous Regulation of Innate Immunity

Georges J.M. Maestroni

Abstract

Dendritic cells (DCs) play a major role in innate immunity because of their ability to detect the presence of invading microorganisms and instruct the proper adaptive immune response. The identification of microbial pathogens occurs via the toll-like receptors (TLRs) gene family and the appropriate priming of T-helper (Th) cells via MHC (Major Histocompatibility Complex)-coupled antigen presentation and costimulatory molecules upregulation. The nature of the pathogen, the elicited inflammatory response and local microenvironmental factors, all contribute in the information conveyed by DCs to the adaptive immune system. The studies revealed that the sympathetic nervous system (SNS) influences such information by acting on specific adrenergic receptors (ARs) expressed in DCs. In particular, α1- and β2-ARs may modulate DCs migration and cytokine production, resulting in a modulation of the antigen-presenting capacity and Th priming. In vitro, ARs activation seems to modulate the DCs response to TLR agonists and thus contributes in shaping the immune response, which is more appropriate to clear the infection. However, the in vivo SNS influence on DCs function is not yet completely understood. With this objective, future studies should address the SNS influence on other cell players of innate immunity that may, however, have an impact on the inflammatory response and ultimately on the information that DCs convey to T cells.

INTRODUCTION

The innate immune system is at the intersection of several pathways that influence the balance between health and disease. Recent studies have shown that the innate immune system is endowed with a highly sophisticated ability to discriminate between indigenous and foreign pathogens. This discrimination relies on a family of receptors, known as TLRs, which plays a

Center for Experimental Pathology, Cantonal Institute of Pathology, PO Box, 6601 Locarno, Switzerland. Tel.: +41 91 816 07 91, Fax: 41 91 81607 99. E-mail: georges.maestroni@ti.ch crucial role in early host defense mechanisms. The TLRs-dependent activation of innate immunity is necessary for the induction of acquired immunity, in particular for Th1 priming. TLRs differ from each other in ligand specificities, expression pattern, and presumably in the target genes they can affect. To date, 11 TLRs are known. TLR4, the first TLR identified in humans and mice, recognizes lipopolysaccharide (LPS), a major component of gram-negative bacteria. In addition, TLR4 recognizes lipotechoic acid (LTA), the heat shock protein hsp60 and the fusion protein of the respiratory syncytial virus (Janeway and Medzhitov, 2002). TLR1 recognizes lipopeptides and soluble factors from *Neisseria Meningitis* (Akira and Takeda, 2004). TLR2 recognizes the largest number of ligands, including peptidoglicans (PGN), bacterial lipoproteins, and other pathogen-associated molecular patterns from a variety of pathogens (Janeway and Medzhitov, 2002). TLR3 is involved in recognition of viral double-stranded RNA (Alexopoulou et al., 2001). TLR5, TLR6, TLR7 and TLR8 recognize a variety of ligands from bacterial flagellin (TLR5) to synthetic compounds, fungi and viruses (Akira and Takeda, 2004). TLR9 recognizes unmethylated Cytosine-phospho-guanine (CpG) motifs present in bacterial and viral DNA (Janeway and Medzhitov, 2002) while the ligands recognized by TLR10 and TLR11 have not yet been determined.

The TLRs-mediated control of adaptive immune responses relies mainly on DC functions. DCs are a sparse population of antigen-presenting cells, irregular in shape and widely distributed in both lymphoid and non-lymphoid tissues (Shortman and Caux, 1997; Sallgaller and Lodge, 1998). After TLRs activation and antigen internalization, DCs leave the tissues interfacing with the external environment and enter the lymphatic vessels to reach the lymphoid organs and undergo maturation (Shortman and Caux, 1997; Sallgaller and Lodge, 1998; Weinlich et al., 1998). While still immature, the primary function of DCs is to capture and process antigens, then to present the antigenic peptides, and activate specific T cells (Shortman and Caux, 1997; Sallgaller and Lodge, 1998). Activation of naive T cells requires two signals. The first signal is delivered when the T cell receptor (TCR) engages the MHC/antigen complex, and the second costimulatory signal is delivered by costimulatory molecules on DCs (June et al., 1994). Activation of naive Th cells also results in their polarization towards the Th1 and/or Th2 type, which orchestrates the immune effector mechanism that is more appropriate for the invading pathogen. Th1 cells promote cellular immunity, protecting against intracellular infection and cancer, but carry the risk of organ-specific autoimmunity. Th2 cells promote humoral immunity, are highly effective against extracellular pathogens, and are involved in tolerance mechanisms and allergic diseases. Priming of Th1 cells is strictly dependent on cytokines such as IL-12 and IFN-γ, while that of Th2 cells is promoted by IL-4, IL-5 and IL-10 (Bancherau and Steinman, 1998; Schnare et al., 2001). Interestingly, DCs are uniquely able to either induce immune responses or to maintain the state of self-tolerance. Recent evidence has shown that the ability of DCs to induce tolerance in the steady state is critical to the prevention of the autoimmune response. Likewise, DCs have been shown to induce several types of regulatory T cells, depending on the maturation state of the DCs and the local microenvironment. DCs have been shown to have therapeutic value in models of allograft rejection and autoimmunity.

THE SYMPATHETIC NERVOUS SYSTEM AS REGULATOR OF DC FUNCTIONS

Effect of NE on Migration and Antigen Presenting Ability of DCs

The type of Th priming determines whether an infection is efficiently cleared, however, the decision making mechanisms linking the innate recognition of the pathogen and the type of Th priming are still poorly understood. Besides the type of invading pathogen and its route of entry into the organism, other local microenvironmental factors seem to play a role (Kalinski et al., 1999; Pulendran et al., 2001).

The sympathetic (noradrenergic) nervous system (SNS) which innervates all parts of the body, constitutes the largest and most versatile component of the autonomic nervous system. Nerve activity results in release of catecholamines which act on adrenoceptors (ARs). In the periphery, the sympathetic neurotransmitter norepinephrine (NE) is released nonsynaptically, i.e., from varicose axon terminals, without synaptic contacts. Thus, ARs on immune cells are targets of remote control, and NE may act as a modulator of the sympathetic-immune interface. The ARs mediate the functional effects of epinephrine and NE by coupling to several of the major signaling pathways modulated by G proteins. The AR family includes nine different gene products: three β (β1, β2, β3), three α2 (α2A, α2B, α2C) and three α1 (α1a, α1b, α1d) receptor subtypes.

In the studies carried out, it was found that immature bone marrow-derived murine DCs express the mRNA coding for the α1b-, β2-, β1-, α2A-, and α2C-ARs (Maestroni, 2000; Maestroni and Mazzola, 2003). Murine epidermal Langerhans cells (LC) mobilization was inhibited by local treatment with the specific α1-AR antagonist prazosin. Consistently, NE enhanced spontaneous emigration of DCs from ear skin explants, and prazosin inhibited this effect. In addition, local treatment with prazosin during sensitization with FITC inhibited the contact hypersensitivity response six days later. In vitro, bone marrow-derived immature, but not CD40-stimulated mature DCs migrated in response to NE, and this effect was neutralized by prazosin. NE seems, therefore, to exert both a chemotactic and chemokinetic activity on immature DCs influencing their antigen-presenting capacity (Maestroni, 2000). Furthermore, it was found that short-term exposure of bone marrow-derived DCs to NE at the beginning of LPS stimulation hampered IL-12 production and increased IL-10 release. The capacity of NE-exposed DCs to produce IL-12 upon CD40 cross-linking as well as to stimulate allogeneic Th lymphocytes was reduced. It is important to note that the ganglionic blocker pentolinium administered in mice before skin sensitization with FITC could increase the Th1-type response in the draining lymph nodes (Maestroni, 2002). More recently, it was detailed that the inhibition of IL-12 was due to activation of both β2-and α2A-ARs, while stimulation of IL-10 was a β2-AR phenomenon. IL-10, in turn, inhibited DCs migration in response to the homeostatic chemokines CCL21 and CCL19 reducing their Th1 priming ability (Maestroni and Mazzola, 2003). As has been shown that NE may enhance DCs migration via α1-ARs (Maestroni, 2000) and others have confirmed the expression of via α1-ARs in LC (Seiffert et al., 2002), the latter finding was seemingly in contrast with the previous study. A reasonable explanation is that, physiologically, the final NE effect on LC migration results from two opposing effects:

(i) chemotaxis/chemokinesis mediated by α1-ARs and,
(ii) inhibition mediated by β2-ARs (IL-10). The selective blockade of these two ARs results, in fact, in divergent effects on both LC migration and Th priming.

Other authors have also recently shown that epidermal Langerhans cells do express the mRNA for α1a-, β2-, β1-ARs and that catecholamines inhibit their antigen-presenting ability via β2-ARs (Seiffert et al., 2002).

The overall effect of the sympathetic neurotransmitter NE in innate immunity seems thus that of modulating DCs migration and Th1 priming. As such, the role of the DCs ARs would be to limit the inflammatory response to a given pathogen and to modulate the type and strength of the adaptive response. Consistently, recent reports have shown that NE depletion decreased the resistance to Pseudomonas aeruginosa and Listeria monocytogenes (Straub et al., 2000; Miura et al., 2001; Rice et al., 2001). Most interestingly, it was also found that a predominant Th1-type contact hypersensitivity response was induced by oxazolone, and it was not FITC that induces a prevailing Th2-type response, inhibiting the local NE turnover in the skin of mice during the first eight hours of sensitization. Oxazolone also induced higher expression of the inflammatory cytokines IL-1 and IL-6 mRNA in the skin. Furthermore, FITC and not oxazolone sensitization, in presence of the specific β2-AR antagonist ICI 118,551, enhanced the consequent response as well as the production of Th1 cytokines in draining lymph nodes; conversely Th2 cytokines were not affected. Thus, the extent of Th1 priming in the adaptive response to a sensitizing agent seems to depend also on its ability to modulate the local sympathetic nervous activity during the innate immune response (Maestroni, 2004).

Discrete Adrenergic Influence on TLR-dependent DCs Activation

The findings quoted above show that the extent of Th polarization in response to an antigen are influenced by the local sympathetic nervous activity in the early phase of dendritic cell stimulation, i.e., during the innate immune response. These results were in part obtained in DCs activated with LPS, a TLR4 agonist. It was then reasoned that if the local sympathetic nervous activity played a role in determining the type of innate and adaptive response to a pathogen, activation of different TLRs should result in different NE effects on cytokine production in DCs. When DCs were stimulated by TLR agonists, activation of TLR2 and TLR4 allowed NE to inhibit IL-12 production and to induce large amounts of IL-10, while upon activation of TLR3 and TLR9 the effect of NE was much smaller (Figs. 1A and 1B). In any case, TLR activation was accompanied by a rapid adrenergic desensitization, indicating that, to be effective, NE needs to be present in the very early phase of DC

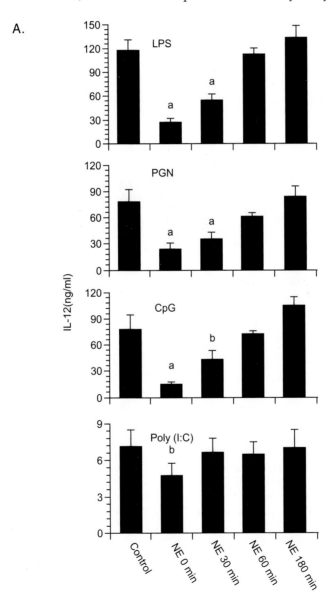

Fig. 1 Contd. ...

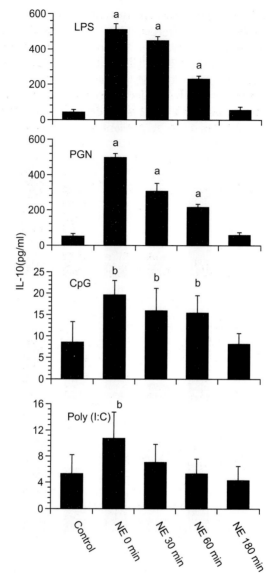

Fig. 1. Bone marrow derived myeloid DCs were cultured in presence of granulocyte-macrophage colony stimulating factor and purified by magnetic cell sorting. The cells (10^6/ml) were then incubated at 37°C in culture medium in presence of norepinephrine (NE, 10^{-6} M) added at the reported times and stimulated with *E. coli* lipopolysaccharide (LPS, TLR4 agonist), *S. aureus* peptidoglycan (PGN, TLR2 agonist), the oligonucleotide CpG (TLR9 agonist) and poly I:C (TLR3 agonist). Six hours later supernatants were collected and the concentration of IL-12 (Fig. 1A) and IL-10 (Fig. 1B) evaluated by ELISA.
a: p< 0.001; b: p<0.02 (ANOVA)

activation (Figs. 1A and 1B). The fact that activation of TLR2 and TLR4, but not TLR3 and TLR9, allows DCs to react to NE with inhibition of IL-12 and production of IL-10 has important implications. Firstly, it underlines the importance of adrenergic mechanisms in DCs function, and secondly, it could explain how bacterial components that activate TLR2 and TLR4 determine the appropriate Th response. IL-10 is a critical cytokine that blocks maturation of DCs, hampers IL-12 production, impairs the ability of DCs to generate Th1 responses, and is involved in tolerance induction (Wang et al., 1999; Akbari et al., 2001; Corinti et al., 2001; Demangel et al., 2002). Thus, again it seems that the effect of the sympathetic nervous regulation of DCs would be to limit the inflammatory response to a given pathogen and to modulate the type and strength of the adaptive response in order to render it more appropriate to clear the infection.

CONCLUSION

The reported findings indicate that the SNS may directly regulate DCs functions by activating various ARs expressed on their surface. The information conveyed by the ARs activation might be aimed at shaping the appropriate innate and adaptive response to an invading pathogen. However, a definite in vivo evidence of such SNS function is still elusive. This is probably due to the complexity of the innate immune response that requires the contribution of many cell types, which in turn, interact and condition DCs function. For example, CD4+, CD25+ T regulatory cells constitute a cell population that plays a crucial role in dampening exaggerated immune responses, as well as in the maintenance of immune tolerance to self or innocuous exogenous antigens. Part of their immunosuppressive effect depends on the ability to inhibit DCs maturation and antigen presentation (Misra et al., 2004). Recent reports have indicated that regulatory T cells might traffic to the skin in a manner very similar to that used by effector T cells (Colantonio et al., 2002). As such, CD4+, CD25+ T cells play a major role also in the immune response and are implicated in inflammatory disorders (McElwee et al., 2003; Graca et al., 2004; Ou et al., 2004; Saint-Mezard et al., 2004). Epidermal and mucosal keratinocytes are important albeit underappreciated players in the immune response. They produce large quantities and varieties of cytokines in response to infectious agents, kinetic and thermal trauma, and ultraviolet radiation (Kupper and Groves, 1995; Grone, 2002). These products have various effects on immune cells such as DCs, mast cells and macrophages, resulting in stimulation of the expression of other inducible mediators and costimulatory molecules. In fact, keratinocytes express TLRs (Medzhitov, 2001; Takeda et al., 2003). The resulting inflammatory reaction activates DCs, which emigrate to the draining lymph nodes carrying antigen for presentation to naive and memory T cells. In addition, inflammatory cytokines act on the local endothelial cells upregulating the expression of adhesion molecules and CCL21 production, which direct recruitment of additional innate immune cells and DCs emigration, respectively (Kupper and Fuhlbrigge, 2004; Martin-Fontecha et al., 2003).

In regard to the SNS influence, interestingly, CD4+, CD25+ T cells express dopaminergic receptors (Kipnis et al., 2004).

It has been also reported that keratinocytes express β2-ARs and may synthesize and release catecholamines. Activation of β2-ARs in keratinocytes regulates cell migration and differentiation, and is involved in wound healing processes (Schallreuter et al., 1995; Chen et al., 2002; Pullar et al., 2003). Catecholamine synthesis and release seems to be strictly related to β2-ARs expression, i.e., E and NE release seems to induce β2-ARs expression in keratinocytes (Schallreuter et al., 1995).

Thus, future studies should address the SNS influence on players of innate immunity other than DCs, but that may ultimately influence the inflammatory response, hence the ability of DCs to bridge the innate with the adaptive immune response. These studies could possibly provide important information for designing vaccines and/or boosting anti-infection and anti-tumor defense mechanisms.

REFERENCES

Akbari, O., DeKruyff, R.H. and Umetsu, D.T. (2001). Pulmonary dendritic cells producing IL-10 mediate tolerance induced by respiratory exposure to antigen. *Nat. Immunol.* **2**: 725-732.

Akira, S. and Takeda, K. (2004). Toll-like receptor signalling. *Nat. Rev. Immunol.* **4**: 499-511.

Alexopoulou, L., Holt, A.C., Medzhitov, R. and Flavell, R.A. (2001). Recognition of double-stranded RNA and activation of NF-kappaB by Toll-like receptor 3. *Nature* **413**: 732-738.

Bancherau, J. and Steinman, R.M. (1998). Dendritic cells and the control of immunity. *Nature* **392**: 245-252.

Chen, J., Hoffman, B.B. and Isseroff, R.R. (2002). Beta-adrenergic receptor activation inhibits keratinocyte migration via a cyclic adenosine monophosphate-independent mechanism. *J. Invest. Dermatol* **119**: 1261-1268.

Colantonio, L., Iellem, A., Sinigaglia, F. and D'Ambrosio, D. (2002). Skin-homing CLA+ T cells and regulatory CD25+ T cells represent major subsets of human peripheral blood memory T cells migrating in response to CCL1/I-309. *Eur. J. Immunol.* **32**: 3506-3514.

Corinti, S., Albanesi, C., La Sala, A., Pastore, S. and Girolomoni, A. (2001). Regulatory activity of autocrine IL-10 on dendritic cell functions. *J. Immunol.* **166**: 4312-4318.

Demangel, C., Bertolino, P. and Britton, W.J. (2002). Autocrine IL-10 impair dendritic cell (DC)-derived immune response to mycobacterial infection by suppressing DC trafficking to draining lymph nodes and local IL-12 production. *Eur. J. Immunol.* **32**: 994-1002.

Graca, L., Le Moine, A., Lin, C.Y., Fairchild, P.J., Cobbold, S.P. and Waldmann, H. (2004). Donor-specific transplantation tolerance: the paradoxical behavior of CD4+CD25+ T cells. *Proc. Natl. Acad. Sci. USA* **101**: 10122-10126.

Grone, A. (2002). Keratinocytes and cytokines. *Vet. Immunol. Immunopathol.* **88**: 1-12.

Janeway, A.C. and Medzhitov, R. (2002). Innate immune recognition. *Annu. Rev. Immunol.* **20**: 197-216.

June, C.H., Bluestone, J.A., Nadler, L.M. and Thompson, C.B. (1994). The B7 and CD28 receptor families. *Immunol. Today* **15**: 321-327.

Kalinski, P., Hilkens, C.M., Wierenga, E.A. and Kapsenberg, M.L. (1999). T-cell priming by type-1 and type-2 polarized dendritic cells: the concept of a third signal. *Immunol. Today* **20**: 561-567.

Kipnis, J., Cardon, M., Avidan, H., Lewitus, G.M., Mordechay, S., Rolls, A., Shani, Y. and Schwartz, M. (2004). Dopamine, through the extracellular signal-regulated kinase pathway, downregulates CD4+CD25+ regulatory T-cell activity: implications for neurodegeneration. *J. Neurosci.* **24**: 6133-6143.

Kupper, T.S. and Groves, R.W. (1995). The interleukin-1 axis and cutaneous inflammation. *J. Invest. Dermatol.* **105**: 62S-66S.

Kupper, T.S. and Fuhlbrigge, R.C. (2004). Immune surveillance in the skin: mechanisms and clinical consequences. *Nat. Rev. Immunol.* **4**: 211-222.

Maestroni, G.J. (2000). Dendritic cell migration controlled by alpha 1β-adrenergic receptors. *J. Immunol.* **165**: 6743-6747.

Maestroni, G.J. (2002). Short exposure of maturing, bone marrow-derived dendritic cells to norepinephrine: impact on kinetics of cytokine production and Th development. *J. Neuroimmunol.* **129**: 106-114.

Maestroni, G.J. and Mazzola, P. (2003). Langerhans cells beta 2-adrenoceptors: Role in migration, cytokine production, Th priming and contact hypersensitivity. *J. Neuroimmunol.* **144**: 91-99.

Maestroni, G.J.M. (2004). Modulation of skin norepinephrine turnover by allergen sensitization: Impact on contact hypersensitivity and Th priming. *J. Invest. Dermatol.* **122**: 119-124.

Martin-Fontecha, A., Sebastiani, S., Hopken, U.E., Uguccioni, M., Lipp, M., Lanzavecchia, A. and Sallusto, F. (2003). Regulation of dendritic cell migration to the draining lymph node: Impact on T lymphocyte traffic and priming. *J. Exp. Med.* **198**: 615-621.

McElwee, K.J., Freyschmidt-Paul, P., Zoller, M. and Hoffmann, R. (2003). *Alopecia areata* susceptibility in rodent models. *J. Invest. Dermatol. Symp. Proc.* **8**: 182-187.

Medzhitov, R. (2001). Toll-like receptors and innate immunity. *Nat. Rev. Immunol.* **1**: 135-145.

Misra, N., Bayry, J., Lacroix-Desmazes, S., Kazatchkine, M.D. and Kaveri, S.V. (2004). Cutting edge: human CD4+CD25+ T cells restrain the maturation and antigen-presenting function of dendritic cells. *J. Immunol.* **172**: 4676-4680.

Miura, K., Kudo, T., Matsuki, A., Sekikawa, K., Tagawa, Y., Iwakura, Y. and Nakane, A. (2001). Effect of 6-hydroxydopamine on host resistance against *Listeria monocytogenes* infection. *Infect. Immun.* **69**: 7234-7241.

Ou, L.S., Goleva, E., Hall, C. and Leung, D.Y. (2004). T regulatory cells in atopic dermatitis and subversion of their activity by superantigens. *J. Allergy Clin. Immunol.* **113**: 756-763.

Pulendran, B., Palucka, K. and Banchereau, J. (2001). Sensing pathogens and tuning immune responses. *Science* **293**: 253-256.

Pullar, C.E., Chen, J. and Isseroff, R.R. (2003). PP2A activation by beta2-adrenergic receptor agonists: novel regulatory mechanism of keratinocyte migration. *J. Biol. Chem.* **278**: 22555-22562.

Rice, P.A., Boehm, G.W., Moynihan, J.A., Bellinger, D.L. and Stevens, S.Y. (2001). Chemical sympathectomy increases the innate immune response and decreases the specific immune response in the spleen to infection with *Listeria monocytogenes*. *J. Neuroimmunol.* **114**: 19-27.

Saint-Mezard, P., Berard, F., Dubois, B., Kaiserlian, D. and Nicolas, J.F. (2004). The role of CD4+ and CD8+ T cells in contact hypersensitivity and allergic contact dermatitis. *Eur. J. Dermatol.* **14**: 131-138.

Sallgaller, M.L. and Lodge, P.A. (1998). Use of cellular and cytokine adjuvants in the immunotherapy of cancer. *J. Surg. Oncol.* **68**: 122-138.

Schallreuter, K.U., Lemke, K.R., Pittelkow, M.R., Wood, J.M., Korner, C. and Malik, R. (1995). Catecholamines in human keratinocyte differentiation. *J. Invest. Dermatol.* **104**: 953-957.

Schnare, M., Barton, G.M., Holt, A.C., Takeda, K., Akira, S. and Medzhitov, M. (2001). Toll-like receptors control activation of adaptive immune responses. *Nat. Immunol.* **2**: 947-950.

Seiffert, C., Hosoi, J., Torii, H., Ozawa, H., Ding, W., Campton, K., Wagner, J.A. and Granstein, R.D. (2002). Catecholamines inhibit the antigen-presenting capability of epidermal Langerhans cells. *J. Immunol.* **168**: 6128-6135.

Shortman, K. and Caux, C. (1997). Dendritic cell development: multiple pathways to nature's adjuvants. *Stem Cells* **15**: 409-419.

Straub, R.H., Linde, H.J., Mannel, D.N., Scholmerich, J. and Falk, W. (2000). A bacteria-induced switch of sympathetic effector mechanisms augments local inhibition of TNF-alpha and IL-6 secretion in the spleen. *FASEB J.* **14**: 1380-1388.

Takeda, K., Kaisho, T. and Akira, S. (2003). Toll-like receptors. *Annu. Rev. Immunol.* **21**: 335-376.

Wang, B., Zhuang, L., Fujisawa, H., Shinder, G.A., Feliciani, C., Shivji, G.M., Suzuki, H., Amerio, P., Toto, P. and Sauder, D.N. (1999). Enhanced epidermal Langerhans cell migration in IL-10 knockout mice. *J. Immunol.* **162**: 277-283.

Weinlich, G., Heine, M., Stössel, H., Zanella, M., Stoizner, P., Ortnet, U., Smolle, J., Koch, F., Sepp, N.T., Schuler, G. and Romani, N. (1998). Entry into lymphatics and maturation in situ of migrating murine cutaneous dendritic cells. *J. Invest. Dermatol.* **110**: 441-448.

CHAPTER 4

Regulation and Synthesis of Maturation Inducing Hormone in Fishes

R.M. Inbaraj[1,*] and Hanna Rosenfeld[2]

Abstract

The identification and the pathway involved in maturation inducing hormone (MIH) have been investigated by many authors since 1968. However, the consolidated idea in the production of MIH is yet to be evolved for the application and development of new techniques for the fish farmers. The glycoprotein hormones such as gonadotropin-I, gonadotropin-II, human chorionic gonadotropin and the steroid hormones of progestins are known to induce the maturation in male and female gonads. Various progestins were identified as MIH in promoting ovulation with the supportive enzymological studies. Time and again the views on the production and synthesis of the MIH are questioned based on its induction and the receptive factors. Although the synthesis and the action of the MIH are governed by endogenous factors, it is an indirect reciprocation of the exogenous factors. Within the species, males release the cues as pheromones for females and the females do likewise for males. The environmental cues predominantly govern the seasonal breeders, which also regulate the alteration of endogenous hormone release. Other steroids, produced as a consequence of necessity by the human body, form the by-products as endocrine disruptors, which also act as regulatory cues for maturation in fishes. This chapter fulfils the understanding of the total impact of the maturation inducing hormone synthesis in male and female gonads.

[1]Reader, Department of Zoology, Madras Christian College, Chennai 600 059, India.
[2]Scientist, Israel Oceanographic and Limnological Research, The National Center for Mericulture, P.O.B. 1212, Eilat 88112, Israel. Tel.: +9 7286361471. E-mail: hanna@ocean.org.il
*Corresponding author: Tel.: +91 9444850872, Fax: +91 542 2368174. E-mail: inbaraj@vsnl.com

INTRODUCTION

The reproduction of fishes culminates at the maturation of gonads. The various stages of development of testis and ovary were recorded in many species, and they correspondingly compared with the seasonal changes. In fact, the studies reveal that the development of gonads are controlled by endogenous hormones and by the exogenous environmental factors like temperature, photoperiod, humidity, rainfall and pH. Although the chromosomal variation initiates the early sexual differentiation, the various hormones can alter it while it undergoes development. The expression of sex dependent gene can be influenced by the down-regulating factors, which are presumed to be sex steroids in a few species. SR1 regulation is well understood in reptiles, birds and mammals but not much in fishes. The specific enzyme responsible gene controls the expression of sex steroid. Similarly, the growth and development of gonads are regulated by the various factors at different levels of the oogenesis and spermatogenesis process. Pituitary hormones play a key role in the regulation of gonadal maturation and ovulation in females (Masui and Clarke, 1979) and spermiation in males. The hormone secretion depends on the internal reciprocal mechanism, which directly depends on the receptivity of the gonads in both males and females (Goetz, 1983).

REGULATION OF GONADAL MATURATION

Gonadotropin-I (GtH-I) and gonadotropin-II (GtH-II) are released from the pituitary gland by the withdrawal of dopamine at SON (supra opticular nucleus) and PVN (paraventricular nucleus) centres, and upregulate the secretion of gonadotropin releasing hormone (GtH-RH). Although the FSH (follicle stimulating hormone) and LH (luteinizing hormone) are replaced with the term GtH-I and GtH-II in fishes, the hypothalamic secretion of the FSH-RH and LH-RH of fish considered as a single GtH-RH. Further study is required for classifying the expression at mRNA level for GtH-RH. The gonadotropins act at the follicular layer of the developing follicles and they promote the synthesis of maturation inducing steroids (MIS). The gonadotropin receptors were well documented in the ovarian follicles of different stages.

MATURATION INDUCING STEROIDS (MIS)

Several mediators of gonadotropin are needed for the adequate stimulation of oocyte maturation and ovulation. Among them, the oocyte stimulation by an ovarian steroid (MIS), which itself induces the maturation promoting factor (MPF) assumes great importance (Yamashita et al., 1992, Haider and Balamurugan, 1996). MIS is secreted in the ovary either by the thecal or granulosa layer or by both, depending on the species (Young et al., 1986). However in teleosts, MIS is found to be secreted in the granulosa layer (Young et al., 1986, Nagahama and Adachi, 1987). Studies conducted so far have revealed 17α, 20β-dihydroxy-4-pregnen-3-one (17,20β-P) as the MIS in most teleosts including salmonid and cyprinid fishes (Scott and Canario, 1987), and 17α,20β,21-trihydroxy-4-pregnen-3-one (17,20β,21-P) as the MIS in perciform fishes (Trant and Thomas, 1989).

The precursor 17α-hydroxy-4-pregnen-3-one (17-P) produced in the thecal layer is converted into estradiol-17β (E2). A shift in the steriodogenic pathway from E_2 (estradiol-17β) to the production of 17,20β-P occurs in ovarian follicle cells prior to oocyte maturation in salmonids and cyprinids (Nagahama, 1994), whereas the shift results in the production of 17,20β,21-P in perciformes. The conversion of 17-P into 17α,21-dihydroxy-4-pregnen-3-one (17,21-P) is constant, whereas the conversion of 17,21-P into 17,20β,21-P varies with time in *Repomucenus beniteguri* (Asahina et al., 1991). Large scale conversion of 17,21-P into 17,20β,21-P is reported in *Micropogonias undulatus* (Thomas, 1994). The conversion of the precursor pregnenolone (P_5) into various other metabolites is directed finally towards the synthesis of the steroids involved in the growth and maturation of oocytes and is aided by the presence of the enzyme 20β-hydroxysteroid dehydrogenase (20β-HSD), which enables the conversion of both 17-P into 17,20β-P, and 17,21-P into 17,20β,21-P (Trant and Thomas, 1989). In carps, 17,20β-P is identified as the MIS (Haider and Inbaraj, 1989; Yaron, 1995), though a few 5β-pregnane derivatives are also found to exhibit some activity. Besides these, certain gonadal factors are also known to induce the increase in 17,20β-P in the plasma to effect FOM (final oocyte maturation) (Scott and Liley, 1994).

The abundance of free 17,20β-P (< 10 ng/ml) is present only in the breeding period marked by the presence of maturing oocytes in the carps, *Catla catla, Labeo rohita* and *Cirrhinus mrigala*. The comparative analysis of the steroids revealed that, testosterone (T) is found in high amounts in the period starting September (degenerating follicles) to February (stage I oocytes) (southern part of India). T levels remained high at the start of vitellogenesis and decreased drastically with advancing vitellogenesis starting March and reached lowest levels just prior to ovulation (Fig. 2). Increased levels of E2 were observed in the month of April in all the three species indicating the occurrence of vitellogenesis, thereby suggesting the conversion of T to E2 in females with the advancing vitellogenesis.

In the Indian carps, 17,20β-P was found to be the most potent inducer of germinal vesicle breakdown (GVBD) and not the 17,20β,21-P *in vitro*. The synthesis of 17,20β-P was noticed in the medium when the matured oocytes were cultured with progesterone (P) and 17-P, but not with the P_5 and 17α-hydroxypregnenolone (17P_5), whereas the synthesis of 17,20β,21-P did not occur even in the presence of precursors (summarized in Fig. 1). The high level of 17,20β-P-sulphate (metabolic conjugation) in the blood plasma is another evidence for the 17,20β-P induction of the meiotic maturation. This would be the result of rapid metabolism of the MIS. Scott and Canario (1990, 1992) suggested the possible metabolites of 17,20β-P, found as the major

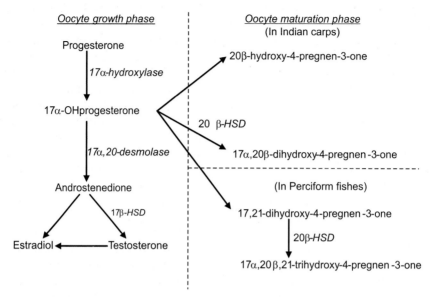

Fig. 1. Representation of the shift in the steroid pathway: A shift occurs in the steroidogenic pathway with 17α-hydroxyprogesterone getting converted largely to C_{18} and C_{19} steroids during the oocyte growth phase and early maturation phase and a shift occurs resulting in their conversion into C_{21} steroids during final maturation.

products of *in vitro* incubations of plaice ovaries, were sulphated 11-deoxycortisol and its 5β-reduced, 3α-hydroxylated (5β,3α) metabolite, 3α,17,21-P-5β. The 5β-reducing, 3α-hydroxylating, and sulphating enzymes occur in the plaice ovary suggesting the MIS (Inbaraj et al., 2001).

STEROID LEVELS IN THE BLOOD

Vitellogenic Phase

The prespawning period, characterized by the presence of previtellogenic and vitellogenic oocytes, exhibits a rise in the plasma E_2 and T levels. With the onset of vitellogenesis, E_2 levels increase reaching the highest levels during the final stages of vitellogenin accumulation. Increased plasma E_2 level was accompanied by oocyte growth, indicating a role for E_2 in regulating vitellogenesis (MacKenzie et al., 1989). Increase in E_2 levels during the vitellogenic phase is reported in striped bass (Berlinsky and Specker, 1991), channel cat fish (MacKenzie et al., 1989), rainbow trout *Oncorhynchus mykiss* (Scott et al., 1980) and Atlantic halibut *Hippoglossus hippoglossus* L. (Methven et al., 1992). In striped bass, plasma E_2 increased first and peaked prior to FOM. In coho salmon, spontaneous reproductive activity was characterized by a rapid decline in plasma E_2 10 days prior to ovulation. E_2 concentration also showed a steady decline prior to spawning in pike.

Increased levels of plasma T were reported in the early vitellogenic phase in striped bass (Berlinsky and Specker, 1991), coho salmon and lates. T levels peaked at the peripheral germinal vesicle (GV) stage in striped bass and after initiation of germinal vesicle migration in Atlantic salmon, remaining at high levels until ovulation; whereas in pike, T concentration did not show any marked difference when the germinal vesicle was peripheral, nor during germinal vesicle breakdown, but decreased rapidly at spawning.

As teleosts approach sexual maturity, the ovaries predominantly produce C_{21} steroids in place of C_{19} and C_{18} steroids (Young et al., 1983). In common carp (*C. carpio*), increase in gonadotropin and E_2 levels, and a sharp peak in 17,20β-P, occurred concomitantly with GVBD. 17,20β-P levels were very low in vitellogenic female amago salmon and sharply increased in mature and ovulated females (Foster et al., 1993). In cultures, *C. carpio* and *Chromis dispilus* showed a large increase in plasma 17,20β-P six days before ovulation. Terminal oocyte maturation *in vivo* is accelerated by 17,20β-P in trout, carp and coho salmon (Jalabert et al., 1978). In Atlantic salmon, 17,20β-P concentration increased progressively from 0.71 to 68 ng/ml from oocyte stage 1 (immature) to stage 6 (resorption), with the glucuronide predominating in stages 1-4 (mature) and the free steroid in stages 5-6 (ripe). Thus, very low levels of free 17,20β-P were present when migration of the germinal vesicle was initiated, but considerable free steroid was present during and after GVBD. Plasma levels of 17,20β-P peaked one week before the ovulation (Foster et al., 1993).

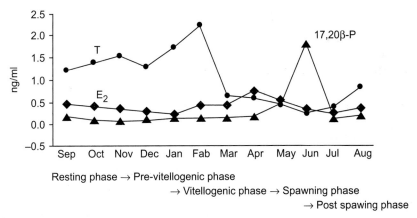

Resting phase → Pre-vitellogenic phase
→ Vitellogenic phase → Spawning phase
→ Post spawing phase

Fig. 2.   The reproductive cycle of Indian carp, *Catla catla* levels of steroids present in the blood plasma. 17,20β-P = 17α,20β-dihydroxy-4-pregnen-3-one(▲); T = Testo-sterone(●); E_2 = Estradiol-17β(◆).

Post-vitellogenic Phase

E_2 and T declined throughout post-vitellogenesis in striped bass, whereas in channel catfish, plasma E2 and T were lowest immediately after spawning, and increased the following season in association with the initiation of vitellogenesis (MacKenzie et al., 1989). T level decreased abruptly in grey mullets that spawned (Suzuki et al., 1991). In striped bass, 17,20β-P and 17,20β,21-P remained low and unchanged in nonmaturing females. In the channel catfish, $17P_5$ was most abundant when the ovary was regressed and during early vitellogenesis, and rapidly decreased prior to spawning. Plasma levels of 17,20β-P then abruptly declined immediately after in lake trout (Foster et al., 1993).

IN VITRO STUDIES

In rainbow trout, 17,20β-P and 17-P are effective mediators of oocyte maturation *in vitro* (Scott et al., 1983). In the rainbow trout, oocytes removed from the females at the end of the vitellogenic phase, in which germinal vesicles were still central, required 10-20 times more 17,20β-P to induce oocyte maturation *in vitro* than in oocytes with peripheral germinal vesicle. Other hormones also influence the sensitivity of oocytes to steroids, as in rainbow trout, where T and cortisol enhance and E_2 decreases the efficiency of 17,20β-P on oocyte maturation *in vitro* (Scott et al., 1983). In all the three Indian major carps, *in vitro* studies have revealed that 17,20β-P is the MIS (Haider and Inbaraj, 1989). 17,20β-P is also found to be the most potent steroid in inducing GVBD in brook trout, yellow perch, goldfish and pike. In rainbow trout and yellow perch, exposure to 17,20β-P for as short a period as one minute, can result in 100% GVBD. In striped bass (*Morone saxatilis*), isolated ovarian fragments exposed to hCG (human chorionic gonadotropin), dibutyryl cAMP (cyclic AMP), or forskolin produced significant amounts of 17,20β-P and the oocytes underwent germinal vesicle breakdown (GVBD). Slight increase in 17,20β,21-P production was also observed. Production of T and E_2 was relatively high at the beginning of *in vitro* treatment with hCG, but decreased as production of 17,20β-P increased and GVBD was initiated (Foster et al., 1993).

STEROID SYNTHESIS

During early maturational stages, progesterone was converted into 17-P, androstenedione and T. During late maturation, production of these steroids decreased markedly, and synthesis of 20β-hydroxy-4-pregnen-3-one (20β-P) and 17,20β-P were induced. 17-P was metabolized to androstenedione and T during the early stage; during final stage, the major resultant metabolite was 17,20β-P. The results suggest that the activities of 17α-hydroxylase and C17,20 lyase are inhibited and the activity of 20β-HSD is induced during the final stage of oocyte maturation and spawning (Suzuki et al., 1991). Inbaraj et al. (2001) have suggested that rapid conversion of free 17,20β-P to conjugated 17,20β-P occurs in the bloodstream under the influence of sulphating enzymes. The presence of conjugated steroids in most teleosts was also verified by Scott and Canario (1990). T is a precursor for both E2 and 11-keto-testosterone by two different biosynthetic pathways (Fostier et al., 1983). Striped bass follow the typical profile of changing plasma steroid levels seen in other teleosts during final maturation, with a clear shift from C_{18} and C_{19} steroids to C_{21} steroids. P was converted to 17-P, androstenedione and T during vitellogenesis and in the maturation phase 17-P was converted to 20α-hydroxy-4-pregnen-3-one (20α-P) and 17,20β-P in *Mugil cephalus* (Suzuki et al., 1991), converted to 17,21-P and 17,20β-P in *Tobinumeri dragonet* (Asahina et al., 1991). P_5 was entirely converted via $17P_5$ into 17-P, androstenodione and 17,20β,21-P in *C. carpio*. In rainbow trout, radiotracer studies demonstrated that the bioconversion of 17,20β-P into cortisol could fully account for the cortisol produced by the interrenal in response to 17,20β-P and demonstrated that rainbow trout interrenal cells contain an active 20βHSD. These data suggest that 17,20β-P may be a regulator of cortisol production during the periovulatory period in salmonid fishes.

In the Indian carps, *Catla catla, Labeo rohita* and *Cirrhinus mrigala* P_5 are converted into delta-4 steroids, but progesterone is detected only in *L. rohita*. However, in *C. catla* and *C. mrigala*, 17α-hydroxy pregnenolone could be the intermediary metabolite of 17α-hydroxyprogesterone. The 17,21-P production is found in the incubation of oocytes with progesterone and 17α-hydroxyprogesterone. In the pre-vitellogenic phase, progesterone was converted into 17-P, androstenedione and testosterone, when in maturation phase, synthesis of 20β-P and 17,20β-P was observed in *Mugil cephalus*. The pregnenolone and progesterone incubated oocytes gave the following pathway for the synthesis of maturation inducing steroids (Fig. 3).

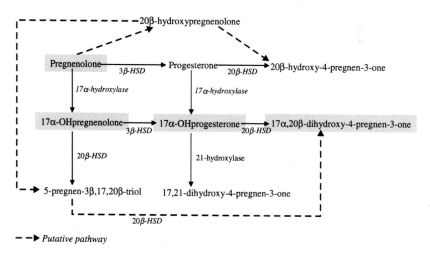

Fig. 3. The maturation inducing steroid synthesis pathway studied in Indian carps. The 20β-P acts as a major intermediary steroid for the production of 17,20β-P.

MIS RECEPTORS

Oocyte receptiveness to MIS is mediated by receptors. Both cytoplasmic/nuclear and membranous specific binding for 17,20β-P in salmoniformes (Maneckjee et al., 1989) and 17,20β,21-P in perciformes (Patino and Thomas, 1990) have been reported. In carps, 17,20β-P is identified as the MIS (Yaron 1995, Haider and Inbaraj, 1989). 17,20β-P acts through receptors on the plasma membrane of oocytes. A specific 17,20β-P receptor has been identified and characterized from defolliculated oocytes of several fishes. 17,20β-P receptor activity is found in both cytosol and zona radiata of brook trout oocytes. In goldfish, the receptors were identified as proteins present in the oocyte membrane (Nagahama, 1987). The receptor protein might be restricted to zona radiata in brook trout and rainbow trout (Maneckjee et al., 1989). The authenticity of the receptors was confirmed by high affinity specific binding using Radio-immunoassay (RIA) (Trant and Thomas, 1989, Nagahama, 1994). In addition, 17,20β,21-P receptors were noticed in the Atlantic croaker, sea trout and other perciform fishes in the oocyte membrane, where 17,20β,21-P is found as the MIS. Significant increases in 17,20β,21-P receptor concentrations were observed during final maturation in sea trout ovaries (Patino and Thomas, 1990). The receptor concentrations were found to increase gradually during ovarian recrudescence and subsequently increased rapidly during oocyte maturation.

MATURATION INDUCTION IN MALES

Testes show that: (i) the spermiation produce more testosterone (T), and especially 11-ketotestosterone (11-KT) *in vitro* than do testes in early spermatogenesis, and (ii) the production is more responsive in salmon gonadotropin. 11-KT was the major metabolite produced in the incubation of the gonads of spermiating males with 14C-labeled steroid precursors. *In vitro* 11-KT production was correlated with plasma levels of 11-KT and these levels were significantly higher than those in the early stages of males. The *in vitro* conversion of 17α-hydroxyprogesterone into 17α,20β-P was highest in winter during spawning in the protogynous fish, *Thalassoma duperrey* (Hourigan et al., 1991).

Developmental changes in the steroidogenic capacity of amago salmon (*Oncorhynchus rhodurus*) testicular fragments at six different stages during spermatogenesis and spermiation were studied (Sakai et al., 1989). Salmon gonadotropin (SGA)-induced testis revealed the production of 11-keto-testosterone and 17α,20β-dihydroxy-4-pregnen-3-one. Also, testes produced a considerable amount of 17α,20β-P in the presence of an exogenous precursor, 17 alpha-hydroxyprogesterone. These results indicate a shift in steroidogenesis from 11-KT to 17α,20β-P in the amago salmon testis immediately prior to or during the spermiation period. Studies on sea lamprey, *Petromyzon marinus* (Lowartz et al., 2003), when testis fragments incubated with tritium-labelled pregnenolone ([3H]P5), 17-hydroxyprogesterone

([3H]17OHP), or androstenedione ([3H]A), provided additional confirmation that the gonads synthesize a range of various steroids, and the metabolites found were consistent with the presence of a 15 alpha-hydroxylated (15α-OH) metabolic pathway common to testis and ovary. From the testis tissue, 15α-hydroxyprogesterone and 15α-hydroxytestosterone and small amounts of 15α-OHestrogen were identified. Although it was assumed that the E(2) was synthesized via the aromatization of T, [3H]T was not found as an intermediate metabolite. The study on sea lamprey 15α-OH compounds combined that only small amounts of E_2 or T were synthesized by the gonads at this stage of reproductive development. There was no direct evidence of progesterone (P(4)) synthesis from [3H]P(5), although the metabolites synthesized by both testis and ovary were indicative of a metabolic pathway that involved P(4) as an intermediate (Lowartz et al., 2003).

A novel complementary DNA (cDNA) was discovered in spotted sea trout, the ovaries and testis encoding a protein with seven transmembrane domains that has the characteristics of the membrane progestin receptor (mPR) mediating MIS, 17,20β,21–P induction of oocyte maturation. Preliminary results suggested the MIS also activates an mPR on the spermatozoa of spotted sea trout and a closely related species, Atlantic croaker, to stimulate sperm motility. The plasma membranes of croaker sperm demonstrate high affinity (Kd approximately 20 nM), limited capacity (Bmax 0.08 nM), specific and displaceable binding for progestins that is characteristic of mPRs. The MIS 17,20β,21-P displayed the greatest binding affinity for the sperm mPR among the steroids tested (Thomas et al., 2005).

MOLECULAR ASPECTS OF MATURATION

Steroid hormones are mainly synthesized in the adrenal and gonads. Their synthesis is stimulated by pituitary hormones through cAMP as an intracellular mediator. The first and rate-limiting step for steroid biosynthesis is catalyzed by CYP11A1 (Cytochrome P450SCC). Important regulatory elements for the control of the CYP11A1 gene expression have been characterized both *in vitro* and *in vivo*. The SF-1-binding sites are cis-acting elements controlling the basal and cAMP-stimulated gene expression (Hu et al., 2001). Steroidogenesis in the differentiating testis was demonstrated to be strongly altered by E_2 (feminizing effect) as estrogen treatment resulted in considerable decrease in 17-hydroxylase/lyase (P450c17), 3β-HSD and 11β-hydroxylase (P45011β) mRNAs in rainbow trout (Govoroun et al., 2001).

The level of cyclin B-associated cdc2 kinase (cyclin dependent cytokinase), a component of MPF, is known to be high during metaphase of the meiotic maturation of oocytes. The temporal relationship between GnRH-induced histone H1 kinase activity and the reinitiation of meiosis and steroidogenesis was established. Treatments with gonadotropin (GtH) or GnRH stimulated the histone H1 kinase activity which was observed to be maximum when follicles of goldfish tested for sGnRH (salmon GnRH)- and cGnRH-II(carp G_nRH)-induced histone H1 kinase activity (Pati et al., 2000).

The cloning and the characterization of zebra fish P450SCC and 3beta-HSD facilitate the study of steroidogenesis. P450SCC (SCC = side chain cleavage cytochrome) and 3beta-HSD cDNA were isolated from a zebra fish lambda gt10 cDNA library using trout SCC and 3β-HSD cDNA as the probes. The zebrafish SCC cDNA encodes a protein of 509 amino acids, which shares 78% similarities with the trout SCC and 58% with the human SCC. As for 3β-HSD, two forms of cDNA were isolated, termed HSD 5 and HSD 17, which encode proteins of 374 and 341 amino acids, respectively (Lai et al., 1998).

CONCLUSION

From the studies conducted so far, E_2 and 17,20β-P levels have shown a progressive and opposing shift in their levels from vitellogenesis through ovulation. E_2 was found in high concentrations at the end of vitellogenesis and progressively decreases during final maturation and ovulation, whereas 17,20β-P is relatively in fewer amounts during vitellogenesis and increases steadily towards the time of ovulation. T on the other hand steadily decreases during vitellogenesis, suggesting the process of their conversion into E_2, which promotes the vitellogenesis. *In vitro* studies have revealed that though 17-P and P enable GVBD, 17,20β-P achieves the highest percentage of GVBD, confirming their role in the FOM.

Though the MIS is identified and the seasonal cycle of steroid synthesis is understood to a certain level, the study on the synthesis of the enzymes responsible for the conversion of steroids through the metabolic pathway, as also the genes responsible for the synthesis of the enzymes, is lacking. A study on the receptors of the steroid hormones and the regulation of their activity in responding to the stimulus at the specific period needs to be understood. Contrary to popular belief that steroid receptors are found in the nuclear membrane, evidence in goldfish oocytes have shown that the steroids act at the plasma membrane suggesting the presence of MIS receptors on it (Nagahama, 1987). Plasma membrane receptor for 17,20β,21-P was reported by Patino and Thomas (1990) in the ovaries of spotted sea trout. Therefore, a thorough study of the receptors and their mechanism of action would largely prove beneficial for the improvement of the fishery yield.

ACKNOWLEDGEMENT

We appreciate the IFCPAR, New Delhi support 2003-3 for the research work.

REFERENCES

Asahina, K., Zhu, Y. and Higashi, T. (1991). Synthesis of 17,21-P and 17,20β,21-P in the ovaries of tobinumeri-dragonet, *Repomucemus beniteguri*, callionymidae teleostei. In: *Proc IV Inter. Symp. Rep. Phy. Fish* Norwich, UK. 80-82.

Berlinsky, D.L. and Specker, J. (1991). Changes in gonadal hormones during oocyte development in the striped bass, *Morone saxatilis*. *Fish Physiol Biochem.* 9: 51-62.

Foster, N.R., O'Connor, D.V. and Schreck, C.B. (1993). Gamete ripening and hormonal correlates in three strains of lake trout. *Transactions of the American Fisheries Society* **122**: 252-267.

Fostier, A., Jalabert, B., Billard, R., Breton, B. and Zohar, Y. (1983). The gonadal steroidogenesis. In: *Fish Physiology* (Hoar, W.S., Randall, D.J. and Donaldson, E.M., eds.), Academic Press, NY, **IXA**: 277-372.

Goetz, F.W. (1983). Hormonal control of oocyte final maturation and ovulation in fishes. In: *Fish Physiology* (Hoar, W.S., Randall, D.J. and Donaldson, E.M., eds.), Academic Press, NY, **IXA**: 117-170.

Govoroun, M., McMeel, O.M., Mecherouki, H., Smith, T.J. and Guiguen, Y. (2001). 17-β-estradiol treatment decreased steroidogenic enzyme messenger ribonucleic acid levels in the rainbow trout testis. *Endocrinol.* **142**: 1841-1848.

Haider, S. and Inbaraj, R.M. (1989). *In vitro* effectiveness of mammalian pituitary hormones on Germinal vesicle breakdown in oocytes of Indian major carps, *Labeo rohita, Cirrhinus mrigala, Catla catla* and *Cyprinus carpio. Indian J. Exp. Biol.* **27**: 653-655.

Haider, S. and Balamurugan, K. (1996). Identification and characterization of maturation promoting factor from catfish, *Clarias batrachus. Fish Physiol. Biochem.* **15**: 255-263.

Hourigan, T.F., Nakamura, M., Nagahama, Y., Yamauchi, K. and Grau, E.G. (1991). Histology, ultrastructure, and in vitro steroidogenesis of the testes of two male phenotypes of the protogynous fish, *Thalassoma duperrey* (Labridae). *Gen. Comp. Endocrinol.* **83**: 193-217.

Hu, M.C., Chiang, E.F., Tong, S.K., Lai, W., Hsu, N.C., Wang, L.C. and Chung, B.C. (2001). Regulation of steroidogenesis in transgenic mice and zebrafish. *Mol. Cell Endocrinol.* **22**: 9-14.

Inbaraj, R.M., Haider, S. and Baqri, S.S.R. (2001). Dynamics of 17α,20β-dihydroxy-4-pregnen-3-one in plasma and oocyte incubation media of catfish (*Clarias batrachus*) in response to salmon gonadotropin. *Current Sci.* **80**: 455-458.

Jalabert, B., Breton, B. and Fostier, A. (1978). Precocious induction of oocyte maturation and ovulation in rainbow trout (*Salmo gairdneri*): problems when using 17α-hydroxy-20ß-dihydroprogesterone. *Ann. Biol. Anim. Bioch. Biophys.* **18**: 977-984.

Lai, W.W., Hsiao, P.H., Guiguen, Y. and Chung, B.C. (1998). Cloning of zebrafish cDNA for 3beta-hydroxysteroid dehydrogenase and P450scc. *Endocr. Res.* **24**: 927-931.

Lowartz, S., Petkam, R., Renaud, R., Beamish, F.W., Kime, D.E., Raeside, J. and Leatherland, J.F. (2003). Blood steroid profile and in vitro steroidogenesis by ovarian follicles and testis fragments of adult sea lamprey, *Petromyzon marinus. Comp. Biochem. Physiol. A. Mol. Integr. Physiol.* **134**: 365-376.

MacKenzie, D.S., Thomas, P. and Farrer, S.M. (1989). Seasonal changes in thyroid and reproductive steroid hormones in female channel catfish (*Ictalurus punctatus*) in pond culture. *Aquaculture* **78**: 63-80.

Maneckjee, A., Weisbart, M. and Idler, D.R.(1989). The presence of 17α,20β-dihydroxy-4-pregnen-3-one receptor activity in the ovary of the brook trout *Salvelinus fontinalis*, during terminal stages of oocyte maturation. *Fish Physiol. Biochem.* **6**: 19-38.

Masui, Y. and Clarke, H.J. (1979). Oocyte maturation. *Int. Rev. Cytol.* 57: 185-282.

Methven, D.A., Crim, L.W., Norberg, B., Brown, J.A., Goff, G.P. and Huse, I. (1992). Seasonal reproduction and plasma levels of sex steroids and vitellogenin in Atlantic halibut (*Hippoglossus hippoglossus*). *Can. J. Fish. Aquat. Sci.* **49**: 754-759.

Nagahama, Y. (1987). Endocrine control of oocyte maturation. In: *Hormones and Reproduction in Fishes, Amphibians and Reptiles* (Norris, D.O. and Jones, R.E., eds.), Plenum, NY, 171-193.

Nagahama, Y. and Adachi, S. (1987). Identification of maturation inducing steroid in a teleost, the amago salmon (*Oncorhynchus rhodurus). Dev. Biol.* **109**: 428-435.

Nagahama, Y. (1994). Endocrine regulation of gametogenesis in fish. *Int. J. Dev. Biol.* **38**: 217-229.

Pati, D., Lohka, M.J. and Habibi, H.R. (2000). Time-related effect of GnRH on histone H1 kinase activity in the goldfish follicle-enclosed oocyte. *Can. J. Physiol. Pharmacol.* **78**: 1067-1071.

Patino, R. and Thomas, P. (1990). Effects of gonadotropin on ovarian intrafollicular processes during the development of oocyte maturational competence in a teleost, the Atlantic croaker: evidence for two distinct stages of gonadotropin control of final oocyte maturation. *Biol. Reprod.* **43**: 818-827.

Sakai, N., Ueda, H., Suzuki, N. and Nagahama, Y. (1989). Steroid production by amago salmon (*Oncorhynchus rhodurus*) testes at different development stages. *Gen. Comp. Endocrinol.* **75**: 231-240.

Scott, A.P., Bye, V.J., Baynes, S.M. and Springate, J.R.C. (1980). Seasonal variations in plasma-concentrations of 11-ketotestosterone and testosterone in male rainbow trout, *Salmo gairdneri* Richardson. *J. Fish Biol.* **17**: 495-505.

Scott, A.P., Sumpter, J.P. and Hardiman, P.A. (1983). Hormone changes during ovulation in the rainbow trout (*Salmo gairdneri* Richardson). *Gen. Comp. Endocrinol.* **49**: 128-134.

Scott, A.P. and Canario, A.V.M. (1987). Status of oocyte maturation-inducing steroids in teleosts. In: *Proc. III Inter. Symp. Rep. Phy. Fish* (Idler, D.R., Crim, L.W. and Walsh, J.M., eds.). Memorial University Press, St. John's, Newfoundland, 224-234.

Scott, A.P. and Canario, A.V.M. (1990). Plasma levels of ovarian steroids, including 17α,21-dihydroxy-4-pregnene-3,20-dione (11-deoxycortisol) and 3α,17α,21-trihydroxy-5β-pregnen-20-one in the ovaries of mature plaice (*Pleuronectes platessa*) induced to mature with human chorionic gonadotropin. *Gen. Comp. Endocrinol.* **78**: 286-298.

Scott, A.P. and Canario, A.V.M. (1992). 17α,20β-dihydroxy-4-pregnen-3-one 20-sulphate: A major metabolite of the teleost oocyte maturation-inducing steroid. *Gen. Comp. Endocrinol.* **85**: 91-100.

Scott, A.P. and Liley, N.R. (1994). Dynamics of excretion of 17α, 20β-dihydroxy-4-pregnen-3-one 20-sulphate, and of the glucuronides of testosterone and 17β-testradiol, by urine of reproductively mature male and female rainbow trout (*Oncorhynchus mykiss). J. Fish. Biol.* **44**: 117-129.

Suzuki, K., Asahina, K., Tamaru, C.S., Lee, C.S. and Inano, H. (1991). Biosynthesis of 17 alpha,20 beta-dihydroxy-4-pregnen-3-one in the ovaries of grey mullet (*Mugil cephalus*) during induced ovulation by carp pituitary homogenates and an LHRH analogue. *Gen. Comp. Endocrinol.* **84**: 215-221.

Thomas, P. (1994). Hormonal control of final oocytes maturation in Sciaenid fishes. In: *Perspectives in Comparative Endocrinology* (Dovey, K.G., Peter, R.E. and Tobe, S.S., eds.), National Research Council of Canada, Ottawa, Canada, 619-625.

Thomas, P., Tubbs, C., Detweiler, C., Das, S., Ford, L. and Breckenridge-Miller, D. (2005). Binding characteristics, hormonal regulation and identity of the sperm membrane progestin receptor in Atlantic croaker. *Steroids* **70**: 427-433.

Trant, J.M. and Thomas, P. (1989). Changes in ovarian steroidogenesis *in vitro* associated with final maturation of Atlantic croaker oocytes. *Gen. Comp. Endocrinol.* **75**: 405-412.

Young, G., Crim, L.M., Kagawa, H., Kambegawa, A. and Nagahama, Y. (1983). Plasma 17α,20β-dihydroxy-4-pregnen-3-one levels during sexual maturation of amago salmon (*Oncorhynchus rhodurus*): correlation with plasma goanadotropin and *in vitro* production by ovarian follicles. *Gen. Comp. Endocrinol.* **51**: 96-106.

Young, G., Adachi, S. and Nagahama, Y. (1986). Role of thecal granulose layers in gonadotropin-induced synthesis of a salmonid maturation-inducing substance (17,20β-P). *Dev. Biol.* **118**: 1-8.

Yamashita, M., Fukada, S., Yoshikuni, M., Bulet, P., Hirai, T., Yamaguchi, A., Lou, Y-H., Zhao, Z. and Nagahama, Y. (1992). Purification and characterization of maturation-promoting factor in fish. *Dev. Biol.* **149**: 8-15.

Yaron, Z. (1995). Endocrine control of gametogenesis and spawning induction in the carp. *Aquaculture* **129**: 49-73.

Melatonin Inhibition of Gonadotropin-releasing Hormone-induced Calcium Signaling and Hormone Secretion in Neonatal Pituitary Gonadotrophs

Hana Zemkova*, Ales Balik and Petr Mazna

Abstract

In gonadotrophs from neonatal rats, secretion of hormones LH (luteinizing hormone) and FSH (follicle stimulating hormone)) is stimulated by gonadotropin-releasing hormone (GnRH) and inhibited by melatonin, which is released from the pineal gland during the night. GnRH acts on Ca^{2+}-mobilizing GnRH receptors that are coupled to G_q protein, while melatonin receptors, MT_1 and MT_2, are negatively coupled to adenylyl cyclase by $G_{i/o}$. The effect of melatonin in neonatal gonadotrophs is due to inhibition of GnRH-induced calcium oscillations that precedes gonadotropin secretion. Single cell calcium and electrophysiological recordings revealed that melatonin blocks both extracellular calcium influx and IP_3-controlled calcium mobilization from intracellular pools. Calcium-induced calcium release from ryanodine-sensitive intracellular pools also seems to be involved. The mechanism by which melatonin receptor activation inhibits these GnRH-induced calcium signals is not yet well understood. The melatonin receptor molecule consists of a bundle of seven putative transmembrane (TM1-7) helices connected by three extracellular and three intracellular loops. Melatonin binds to amino acid residues within TM3, TM5, TM6 and TM7 in both receptor subtypes. The extracellular loops may participate in ligand recognition whereas intracellular loops are believed to directly activate G proteins. Melatonin-induced inhibition of adenylyl cyclase and cyclic adenosine monophosphate (cAMP)-activated/ protein kinase A pathway presumably decreases calcium influx through voltage-dependent calcium channels and the efficacy of its coupling to IP_3 or ryanodine receptors. The potent inhibition of GnRH-induced calcium signaling and

---

Department of Cellular and Molecular Neuroendocrinology: Institute of Physiology of the Academy of Sciences of the Czech Republic, Prague, Czech Republic.
* *Corresponding author:* Tel.: +420 2 4106 2574, Fax: +420 2 4106 2488.
E-mail: zemkova@biomed.cas.cz

gonadotropin secretion by melatonin provides an effective mechanism to preserve pulsatile GnRH release from hypothalamic neurons in immature animals, but to delay the prepubertal changes initiated by increase in gonadotropin plasma membrane levels. During the development, the tonic inhibitory effects of melatonin on GnRH action attenuate due to a progressive decrease in expression of functional MT receptors and developmental changes in GnRH signaling pathway. In adult animals, melatonin does not have an obvious direct effect on pituitary functions, whereas the connections between melatonin and hypothalamic functions, including melatonin effect on GnRH release, persist and are critically important in synchronization of external photoperiods and reproductive functions.

INTRODUCTION

Reproductive functions in vertebrates are controlled by releasing or inhibiting factors generated in neurons of hypothalamus. The key role is played by gonadotrophin releasing hormone (GnRH), also known as luteinizing hormone-releasing hormone (LHRH), which is synthesized by hypothalamic GnRH neurons, and secreted in a pulsatile manner into the hypophyseal portal system. A majority of wild species have seasonal reproduction in order to give birth at the optimal time of year, usually spring, allowing the newborn to grow and develop under favorable temperature and food availability conditions. The annual cycle of external light thus plays an important role in the reproduction and causes seasonal changes in reproductive behavior, weight gain, appetite, body fat storage, energy metabolism, and growth of fibers and horns or hibernation (Bartness et al., 1993). In this very complex process controlled by the suprachiasmatic nuclei (SCN) of hypothalamus, the production of hormone melatonin in the pineal gland plays a central role. Melatonin is necessary for entrainment of seasonal photoperiodic responses to the annual cycle of day length (Kennaway and Rowe, 1995). The activity of melatonin-producing enzyme arylalkylamine N-acetyltransferase (serotonin N-acetyltransferase, EC2.3.1.87) precisely reflects duration of the night (Klein et al., 1997). During the day, melatonin level is decreased due to inhibitory neuronal outputs from the SCN (Kalsbeek et al., 2000). In addition to controlling the seasonal biological rhythms in mammals, melatonin also participates in the coordination of circadian rhythms (Cassone, 1990; McArthur et al., 1991; Svobodova et al., 2003). Humans also secrete melatonin in a pattern that reflects the environmental light-dark cycle, but the seasonal melatonin information is not an integral part of their reproductive cycle.

Melatonin probably regulates reproduction at three levels: the hypothalamus, pituitary and gonads. Melatonin microimplants into the area of preoptic and mediobasal hypothalamus of mice (areas that contain GnRH neurons) elicit complete gonadal involution, whereas its injection in other areas was ineffective (Glass and Knotts, 1987). In accordance with this, it has been found that melatonin regulates GnRH gene transcription in the immortalized GnRH-secreting neurons in a cyclic manner (Roy et al., 2001;

Roy and Belsham, 2002). Melatonin also acts at the level of the gonads, where it modulates androgen production by Leydig cells (Valenti et al., 1997; Li et al., 1998). In prostate epithelial cells, melatonin suppresses cyclic guanosine monophosphate (cGMP) levels (Gilad et al., 1997). This chapter is focused on another aspect of melatonin effect in neuroendocrine axis: melatonin interaction with GnRH signaling in neonatal gonadotrophs. The possibility that melatonin directly modulates pituitary function was established in 1987 with the finding that melatonin binding sites are present in the pituitary (Reppert et al., 1988; Vanecek, 1988). The high expression level of melatonin receptors in the rat pituitary is limited to the first two postnatal weeks and is present namely in GnRH-sensitive cells. GnRH stimulates intracellular calcium increase, which underlies the secretion of luteinizing hormone (LH) and follicle-stimulating hormone (FSH). Both calcium increase and gonadotrophins secretion are effectively inhibited by melatonin (Martin and Klein, 1976; Vanecek and Klein, 1992a). This phenomenon is unique for neonatal pituitary cells and is absent in adult gonadotrophs. However, about 10% of neonatal level of melatonin receptors persist in adult gonadotrophs and their role is unknown (Vanecek, 1988). Interestingly, GnRH signaling in melatonin-sensitive neonatal gonadotrophs exhibits some specific features which are not found in gonadotrophs of adult rats. For example, GnRH-induced calcium increase is often non-oscillatory and more dependent on extracellular calcium than in adult gonadotorphs, and ryanodine receptor-controlled intracellular calcium stores play a role in calcium mobilization (Zemkova and Vanecek, 2000). It indicates that the neonatal cell-specific intracellular calcium handling mechanism might be a prerequisite for melatonin sensitivity in addition to a high level of melatonin receptors.

There are several reasons justifying the studies of melatonin effects in neonatal gonadotrophs. Firstly, it provides an explanation for the effective mechanism to down-regulate, but not to abolish, LH and FSH secretion during the early phase of development when reproductive functions have to be arrested. Secondly, the intracellular mechanism of melatonin action is largely unknown in most of the target cells expressing specific melatonin receptors, and the lack of specific antagonists limits investigations on their function. In neonatal gonadotrops, the effect of melatonin can be studied by both single cell calcium and electrophysiological measurements. Finally, these cells can be easily recognized in a mixed population of pituitary cells by typical calcium oscillations, which are not present in other cell types. Thus, neonatal gonadotrophs provide the best characterized model to study acute effect of melatonin at cellular level.

GnRH SIGNALING

Neuropeptide GnRH is synthesized by hypothalamic GnRH neurons and is secreted in a pulsatile manner into the hypophyseal portal system (Fig. 1). So far, sixteen forms of GnRH have been isolated from the brain of vertebrates. In the mammalian brain, two GnRH forms coexist. Surprisingly, the latter

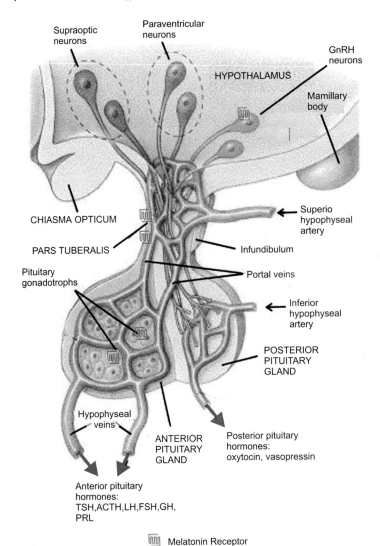

Supraoptic neurons

Paraventricular neurons

GnRH neurons

HYPOTHALAMUS

Mamillary body

Superio hypophyseal artery

CHIASMA OPTICUM

PARS TUBERALIS

Infundibulum

Pituitary gonadotrophs

Portal veins

Inferior hypophyseal artery

POSTERIOR PITUITARY GLAND

Hypophyseal veins

ANTERIOR PITUITARY GLAND

Posterior pituitary hormones: oxytocin, vasopressin

Anterior pituitary hormones: TSH,ACTH,LH,FSH,GH, PRL

Melatonin Receptor

Fig. 1. Pituitary gland with projection of GnRH neurons and localization of melatonin receptors in hypothalamo-pituitary reproductive neuroendocrine axis.

The hypothalamo-pituitary axis is an area where the CNS can affect the endocrine system. Neurons from supraoptic and paraventricular nuclei of hypothalamus send neuronal endings into the posterior part of the pituitary gland which represent a mechanism for the release of oxytocin and vasopressin. GnRH neurons, which axons end in the primary capillary plexus, produce a releasing factor GnRH through which the synthesis and release of LH and FSH from adenopituitary is controlled. Hormones of both anterior and posterior pituitary lobes are transported through the blood capillaries to the body circulation, and then to hypophyseal veins. Melatonin receptors have been identified in GnRH neurons, pars tuberalis and gonadotrophs of neonatal adenopituitary. Modified from:

http://www.biosbcc.net/barron/physiology/endo/hypopit.htm

identified type II GnRH is the most evolutionarily conserved form of GnRH (White et al., 1998) but its physiological role is unknown. In neonatal pituitary, both peptides stimulate oscillatory increase in calcium concentration and hormone secretion, the type II GnRH seems to be slightly less effective (Jindrichova and Zemkova, unpublished).

GnRH binds with high affinity to the plasma-membrane GnRH receptors expressed in gonadotroph cells (Perrin et al., 1989). GnRH receptor belongs to the rhodopsin-like family of seven transmembrane domain receptors (Reinhart et al., 1992; Tsutsumi et al., 1992). GnRH receptor is coupled to pertussis toxin-insensitive G_q/G_{11} proteins that stimulate phospholipase Cβ (PLCβ) pathway, leading to the generation of inositol (1,4,5) trisphosphate (IP_3) and diacylglycerol (DAG) production (Morgan et al., 1987; Chang et al., 1988) and activation of phospholipase D pathway (Zheng et al., 1994). However, the vast information compiled also indicates that GnRH receptor cross-couples to G_s and $G_{i/o}$-signaling pathway as well (Krsmanovic et al., 2003). The GnRH receptors found in mammals are unique, i.e., they lack C-terminal tails and they are, therefore, resistant to receptor desensitization and internalize slowly. In pituitary cells, sustained activation of GnRH receptors causes desensitization of gonadotrophin secretion because of down-regulation of IP_3 receptors and desensitization of Ca^{2+} mobilization (McArdle et al., 2002; Zemkova et al., 2004). The type II GnRH receptors, cloned so far only from the brain of Marmoset monkey (Millar et al., 2001), possess C-terminal tails and show rapid desensitization and internalization (McArdle et al., 2002).

GnRH-induced Calcium Oscillations

The intracellular signaling pathway of GnRH receptors is relatively well established in gonadotrophs from adult rats (Stojilkovic et al., 1994; Hille et al., 1995). GnRH-stimulated intracellular IP_3 binds to a specialized tetrameric IP_3 receptor-channel complex that spans the endoplasmic reticulum membrane (Berridge, 1993) and triggers oscillatory release of Ca^{2+} from the endoplasmic reticulum. Simultaneously generated intracellular DAG activates Ca^{2+}-dependent protein kinase-C, which in turn affects several other signaling pathways, including the extracellular Ca^{2+} entry via voltage-dependent calcium channels. Calcium influx is necessary for the long-lasting maintenance of oscillations in intracellular calcium concentrations ($[Ca^{2+}]_i$) (Shangold et al., 1988; Stojilkovic et al., 1991). The cross-coupling of GnRH receptors to G_s signaling pathway leads to an increase in intracellular levels of cAMP that may also participate in modulating voltage-gated Ca^{2+} influx and Ca^{2+} release from IP_3-sensitive intracellular pool. This pathway is yet to be understood clearly.

In neuroendocrine and endocrine cells, a prolonged rise in $[Ca^{2+}]_i$ is needed for two steps in exocytosis: priming the vesicles to the site of release, and fusion of vesicles with the plasma membrane (Stojilkovic, 2005). GnRH-stimulated single $[Ca^{2+}]$ spike in gonadotrophs lasts for several seconds, which is sufficient to trigger exocytosis. GnRH-stimulated increase in the $[Ca^{2+}]_i$ is oscillatory and gonadotropin release is also oscillatory (Hille et al., 1994). The recovery to the resting $[Ca^{2+}]_i$ levels is accomplished by Ca^{2+} sequestering into the endoplasmic reticulum and mitochondria and by efflux outside the cell by Ca^{2+} pump-ATPase and Na^+/Ca^{2+} exchange system in the plasma membrane (Zemkova et al., 2004). In addition to exocytosis, the GnRH-induced increase in $[Ca^{2+}]_i$ together with activated protein kinase C, and possibly protein kinase A, regulates ion channel activity and gene expression (Cesnjaj et al., 1994).

GnRH-induced Current and Voltage Oscillations

Gonaodotrophs express numerous plasma membrane ion channels, including TTX-sensitive and TTX-insensitive Na^+ channels, inactivating and non-inactivating voltage-gated Ca^{2+} channels, transient and delayed rectifying K^+ channels, and multiple Ca^{2+}-sensitive K^+ channel subtypes (Stojilkovic et al., 2005). The role of TTX-sensitive Na^+ channels in gonadotrophs is unknown. The inactivating Ca^{2+} current is mediated by the low voltage-activated or transient (T)-type Ca^{2+} channel, whereas the non-inactivating Ca^{2+} current is mediated by dihydropyridine-sensitive (L-type) (Stutzin et al., 1989) and dihydropyridine-insensitive, high-voltage activated Ca^{2+} channels (Stojilkovic and Catt, 1992; Kwiecien and Hammond, 1998; Stojilkovic et al., 2005). Gonadotrophs express slow-inactivating K^+ channels, transient K^+ channels (Van Goor et al., 2001) and apamin-sensitive small conductance Ca^{2+}-activated K^+ channels (SK channels) (Kukuljan et al., 1992), which are key determinants of cellular excitability. The large conductance Ca^{2+}-activated K^+ channels (BK channels) seem to be less important for gonadotrophs [Van Goor, 2001 #434] (Stojilkovic et al., 2005).

In electrophysiological experiments conducted, measurements of GnRH-stimulated Ca^{2+}-activated K^+ current mediated by SK channels could substitute for calcium measurements (Fig. 2A&B) (Kukuljan et al., 1992; Tse and Hille, 1992; Zemkova and Vanecek, 1997). The SK channels exhibit low sensitivity to membrane potential and their activation is determined only by intracellular calcium changes (Stojilkovic et al., 2005). In non-voltage clamped cells, the GnRH-induced $[Ca^{2+}]_i$ oscillations trigger oscillatory changes of membrane potential driven by rhythmic opening and closing of SK channels. Membrane potential oscillations and rhythmic depolarizations provide a mechanism for periodic firing of action potentials of higher frequency (Fig. 2C) and voltage-gated calcium entry, which sustains agonist-induced Ca^{2+} release and hormone secretion. The mechanism by which GnRH receptor activation leads to opening of calcium channels is not clear. L-type Ca^{2+} channels could be modulated directly by $\beta\gamma$ subunit released from activated G protein (Herlitze et al., 1996). Alternatively, GnRH could stimulate phosphorylation of Ca^{2+} channel protein by activation of protein kinase C (Bosma and Hille, 1992).

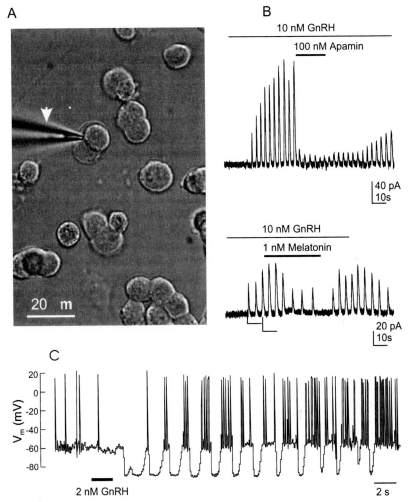

Fig. 2. Electrophysiological monitoring of GnRH-induced $[Ca^{2+}]_i$ oscillations.

[A] Patch-clamp electrode (arrow) and neonatal pituitary cells in culture. GnRH-induced calcium oscillations can be measured electrophysiologically using nystatin-perforated patch-clamp technique. For details, refer Zemkova and Vanecek, 1997.

[B] GnRH-induced current oscillations. Inhibitory effect of apamin, a specific blocker of SK-type channels (upper part), and attenuation of GnRH-induced current oscillations by melatonin (lower part). The time of GnRH, apamin and melatonin application is indicated by horizontal bars. The cell was voltage-clamped to −40 mV (Zemkova and Vanecek, 1997).

[C] Electrical excitability of gonadotrophs. When the Ca^{2+}-mobilizing pathway is activated by GnRH, spontaneous firing of action potentials is transiently abolished by Ca^{2+}-induced hyperpolarization. This is followed by recovery of membrane potential and action potential firing of higher frequency, but regular hyperpolarizing waves again interrupt such firing. The membrane potential oscillations were recorded in current-clamp measurement (Stojilkovic et al., 2005).

Critical Features of Intracellular GnRH Signaling in Neonatal Gonadotrophs

Agonist-induced calcium signaling in immature gonadotrophs exhibit some additional characteristics not present in gonadotrophs from adult animals. The GnRH-induced $[Ca^{2+}]_i$ oscillations of neonatal gonadotrophs rapidly disappear in the absence of extracellular calcium (Tomic et al., 1994; Zemkova et al., 2004), whereas they last for ten minutes in adult gonadotrophs (Kukuljan et al., 1992; Tse and Hille, 1992). The short-lasting Ca^{2+} oscillations in neonatal gonadotrophs indicated that their intracellular stores contain less amounts of Ca^{2+} and that their refilling is highly dependent on calcium influx (Zemkova et al., 2004). When stimulated with GnRH, neonatal gonadotrophs quite often exhibit non-oscillatory responses. More importantly, the coupling of voltage-dependent calcium influx with Ca^{2+} release from ryanodine-sensitive intracellular stores plays a critical role in initiation of GnRH-induced $[Ca^{2+}]_i$ oscillations in neonatal gonadotrophs (Zemkova and Vanecek, 2000). In cells with blocked voltage-gated calcium influx (by removal of extracellular calcium or by the addition of nifedipine, a blocker of L-type voltage-dependent Ca^{2+} channels), the latency preceding the GnRH-induced response is prolonged more than three times. Ryanodine has a similar effect in concentrations that block Ca^{2+} release from intracellular pools. Provided that both IP_3 and ryanodine receptors co-exist and interact in one cell, Ca^{2+} released from the ryanodine receptor-controlled stores significantly contributes to the rise in $[Ca^{2+}]_i$ (Bezprozvanny et al., 1991; Larini et al., 1995). In neonatal gonadotrophs, ryanodine-sensitive calcium-induced calcium release is very small (Zemkova and Vanecek, 2000), but is sufficient to amplify voltage-dependent calcium influx to the level needed to trigger IP_3-controlled Ca^{2+} release (Fig. 3). The ryanodine receptor was not found in gonadotrophs from adult animals (Stojilkovic et al., 1992) but it has been found in immortalized GH_3 pituitary cells (Kramer et al., 1994). Thus, the expression and functional coupling of ryanodine-sensitive channels in neonatal gonadotrophs coincides with the expression and function of melatonin receptors in these cells.

MELATONIN RECEPTORS

Localization of Melatonin Receptors in Reproductive Neuroendocrine Axis

Melatonin receptors involved in the control of reproduction are expressed in the SCN, and 1 GnRH-secreting neurons localized in hypothalamus. High density of melatonin receptors has also been found in pars tuberalis and pars distalis regions of pituitary (Fig. 1). Melatonin receptors in the SCN are responsible for seasonal changes in physiological and behavioral functions and the phase-shifting effects on the circadian rhythms (Armstrong et al., 1986; McArthur et al., 1991; Liu et al., 1997). Hypothalamic GnRH neurons

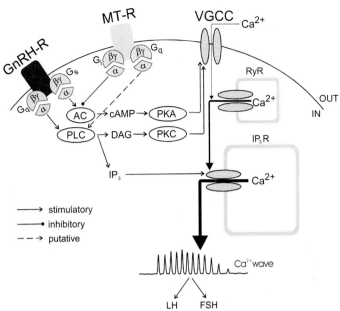

Fig. 3. Model of melatonin interaction with GnRH signaling in neonatal gonadotrophs Stimulation of gonadotropin-releasing hormone receptors (GnRH-R) causes activation of phospholipase C (PLC) through G_q-dependent signaling pathway, leading to the production of diacylglycerol (DAG) and inositol (1,4,5)-triphosphate (IP_3). DAG together with Ca^{2+} stimulates protein kinase C (PKC). The cross-coupling of GnRH signaling pathway with G_s accounts for stimulation of adenylyl cyclase (AC), leading to increase in cAMP production and activation of protein kinase A (PKA). Both PKC and PKA could be involved in control of voltage-gated calcium channels (VGCC). The influx of Ca^2 triggers Ca^{2+} release from ryanodine-sensitive pool (RyR). Calcium and IP_3 coordinately regulate the opening of IP_3 receptor-channels expressed in endoplasmic reticulum (IP_3R), leading to a massive Ca^{2+} release from this pool, which occurs in an oscillatory manner. The rise in Ca^{2+} is sufficient to trigger the release of luteinizing hormone (LH) and follicle-stimulating hormone (FSH). Melatonin receptors (MT-R) signal through G_i pathway to inhibit AC. In some cells, melatonin also stimulates PLC, but only at high concentrations and $\beta\gamma$ dimer of G_i or α subunit of G_q accounts for this stimulation. PKA phosphorylates IP_3 receptor-channel, an action that facilitates IP_3/Ca^{2+}/controlled calcium release. Such organization of calcium signaling provides an effective mechanism for amplification of signals from Ca^2 channels, through RyR store to IP_3R store (indicated by thickness of arrows), whereas down-regulation of AC activity by MT-R provides an effective mechanism for inhibition of this cascade.

(Roy et al., 2001; Roy and Belsham, 2002) and pituitary cells (Vanecek and Klein, 1992a; Zemkova and Vanecek, 1997) are the main sites of the reproductive actions of melatonin, to which melatonin transduces information about daily and seasonal photoperiods. The acute action of melatonin in pituitary, which is a subject of this chapter, is thus only a part of a more complex mechanism by which this hormone controls reproductive functions.

The radioligand ^{125}I-melatonin was the first to be used to localize melatonin binding sites and melatonin receptors in the central and peripheral tissues. Autoradiographic studies indicated that melatonin receptors are expressed in the brain, with high density of receptors in hypothalamic SCN (Vanecek et al., 1987; Reppert et al., 1988; Duncan et al., 1989; Siuciak et al., 1990), as well as outside the brain, including the retina (Dubocovich, 1985), anterior pituitary (Reppert et al., 1988; Vanecek, 1988; Williams and Morgan, 1988), some arteries (Viswanathan et al., 1990), and in cells of the immune system (Lopez-Gonzalez et al., 1992). The pars tuberalis of anterior pituitary contains the highest concentration of melatonin receptors of all mammalian tissues, whereas in pars distalis melatonin binding is present only in a gonadotroph fraction of secretory cells (Williams et al., 1991, 1997; Skinner and Robinson, 1995). Melatonin receptors are expressed more widely in fetal and newborn animals, both in the CNS (Williams et al., 1991) and pituitary (Vanecek 1988; Laitinen et al., 1992), compared to more restricted distribution in adult animals (Davis, 1997). In the pituitary, the melatonin receptor number gradually declines over the perinatal period (Fig. 4A & B).

Melatonin Receptor Subtypes

Two high-affinity melatonin receptor subtypes, called MT_1 and MT_2, have been identified in mammals by cloning (Reppert et al., 1994, 1995a; Roca et al., 1996; Gauer et al., 1998). Both receptors have also been found in the pituitary of neonatal rats (Fig. 4C). The mammalian MT_1 subtype seems to be widely expressed and functionally a more important subtype, both in the SCN and pituitary (Reppert et al., 1994; Liu et al., 1997), whereas the expression of MT_2 subtype is less abundant (Weaver et al., 1996; Liu et al., 1997). Analysis of MT_1 mRNA expression and melatonin receptor density in Syrian hamster revealed that decrease in the development of melatonin receptor number in the pars tuberalis and SCN could not be attributed to inhibition of the mRNA expression, but rather is related to a post-transcriptional blockade of the MT_1 receptor expression (Gauer et al., 1998). Significant circadian variations in pars tuberalis MT_1 receptor density, which has been found in Syrian hamster, is directly regulated by the daily variations of melatonin itself (Gauer et al., 1993; Recio et al., 1998).

The mammalian MT_2 melatonin receptor has been found in the retina and brain (Reppert et al., 1995a). The role of MT_2 melatonin receptor is less clear than that of MT_1 and it has been suggested to mediate the melatonin inhibition of dopamine release in the retina (Reppert et al., 1995a; Liu et al., 1997; Dubocovich et al., 1998). In MT_1 receptor-deficient mice, the MT_2 melatonin receptor substitutes the role for MT_1 in the phase-shifting response (Liu et al., 1997). Other forms of melatonin receptor have been also found in the mammalian brain. This includes the expression of low-affinity melatonin

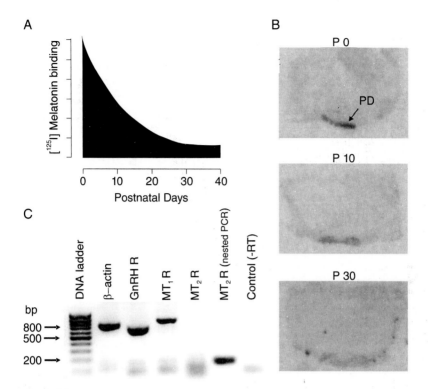

Fig. 4. Transient expression of melatonin receptors in anterior pituitary.

[A] Developmental decrease in radioligand ^{125}I-melatonin binding in pituitary of male rats (schematic representation). For experimental information about measurement, refer Vanecek, 1988.

[B] Developmental loss of MT_1 mRNA expression in pars distalis (PD) of pituitary within the first four weeks of postnatal life. Autoradiograph of coronal brain sections of the rat at three different postnatal days: P0 (magnification 5×), P10 (magnification 4.5×), and P30 (magnification 3.25×). Note the greatly reduced intensity of signal: high expression occurs at P0, evident decrease at P10 and strong decrease at P30 (Balik, unpublished observations).

[C] Agarose gel represents RT-PCR analysis of transcripts for MT_1 and MT_2 receptors in the pituitary from neonatal rats. In samples of total RNA isolated from pituitary cells of 6- to 10-day-old rats, the mRNAs for β-actin, GnRH receptor (GnRH R) and MT_1 melatonin receptor (MT_1 R) were easily detected, whereas two-round nested-PCRs had been applied to detect the message for the MT_2 melatonin receptor (MT_2 R). These results indicate that MT_1 receptor provides the major pathway, but MT_2 melatonin receptor could also be involved in melatonin inhibition of gonadotropin secretion (Balik et al., 2004).

receptor, called MT_3 receptor, identified in the central and peripheral hamster tissues (Paul et al., 1999; Nosjean et al., 2000). The functional role of this putative melatonin receptors remains to be established.

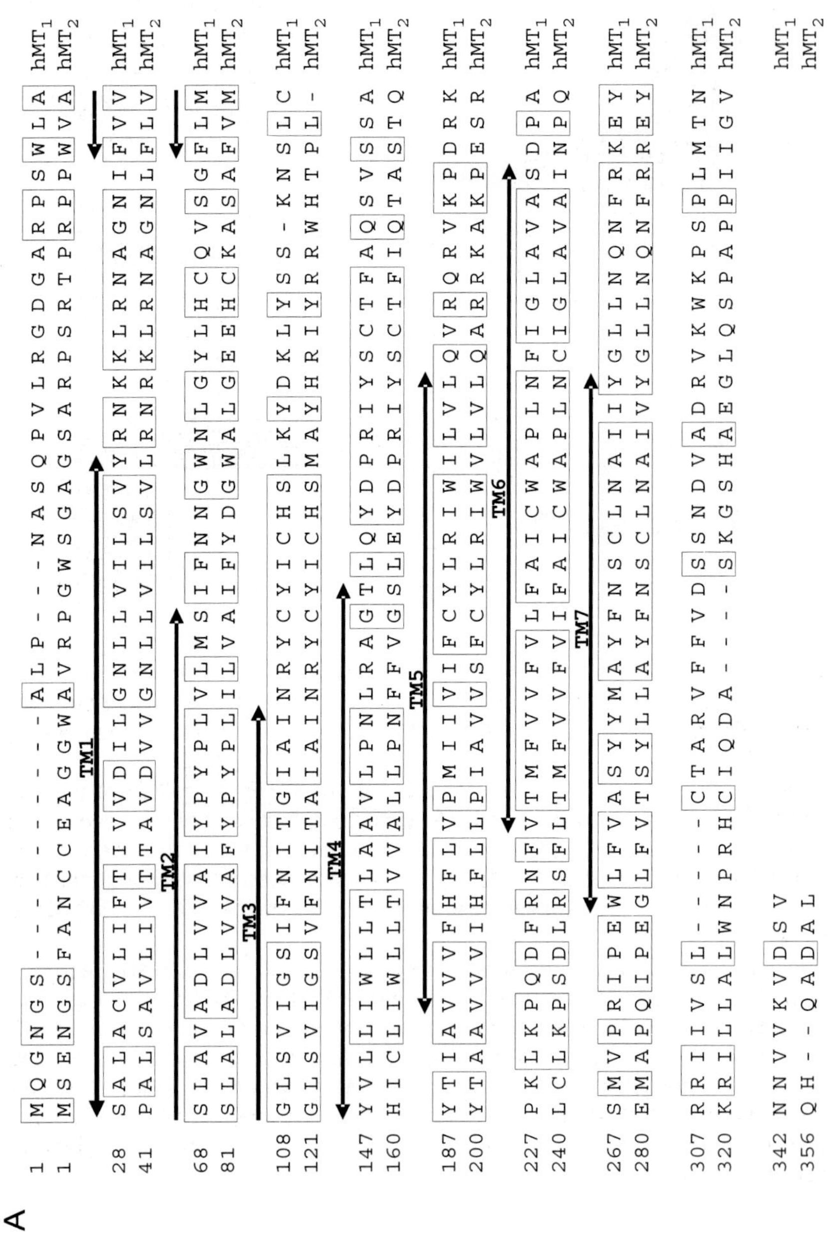

A

```
  1  M Q G N G S - - - - - - - - A L P - - N A S Q P V L R G D G A R P S W L A   hMT1
  1  M S E N G S F A N C C E A G G W A V R P G W S G A G S A R P S R T P R P P W V A   hMT2
                        TM1

 28  S A L A C V L I F T I V V D I L G N L L V I L S V Y R N K K L R N A G N I F V V   hMT1
 41  P A L S A V L I V T T A V D V V G N L L V I L S V L R N R K L R N A G N L F L V   hMT2
                        TM2

 68  S L A V A D L V V A I Y P Y P L V L M S I F N N G W N L G Y L H C Q V S G F L M   hMT1
 81  S L A L A D L V V A F Y P Y P L I L V A I F Y D G W A L G E E H C K A S A F V M   hMT2
              TM3

108  G L S V I G S I F N I T G I A I N R Y C Y I C H S L K Y D K L Y S S - K N S L C   hMT1
121  G L S V I G S V F N I T A I A I N R Y C Y I C H S M A Y H R I Y R R W H T P L -   hMT2
                  TM4

147  Y V L L I W L L T L A A V L P N L R A G T L Q Y D P R I Y S C T F A Q S V S S A   hMT1
160  H I C L I W L L T V V A L L P N F F V G S L E Y D P R I Y S C T F I Q T A S T Q   hMT2
                      TM5

187  Y T I A V V V F H F L V P M I I V I F C Y L R I W I L V L Q V R Q R V K P D R K   hMT1
200  Y T A A V V V I H F L L P I A V V S F C Y L R I W V L V L Q A R R K A K P E S R   hMT2
                          TM6

227  P K L K P Q D F R N F V T M F V V F V L F A I C W A P L N F I G L A V A S D P A   hMT1
240  L C L K P S D L R S F L T M F V V F V I F A I C W A P L N C I G L A V A I N P Q   hMT2
                  TM7

267  S M V P R I P E W L F V A S Y Y M A Y F N S C L N A I I Y G L L N Q N F R K E Y   hMT1
280  E M A P Q I P E G L F V T S Y L L A Y F N S C L N A I V Y G L L N Q N F R R E Y   hMT2

307  R R I I V S L - - - C T A R V F F V D S S N D V A D R V K W K P S P L M T N   hMT1
320  K R I L L A L W N P R H C I Q D A - - - S K G S H A E G L Q S P A P P I I G V   hMT2

342  N N V V K V D S V   hMT1
356  Q H - - Q A D A L   hMT2
```

Fig. 5 Contd. ...

Fig. 5 Contd. ...

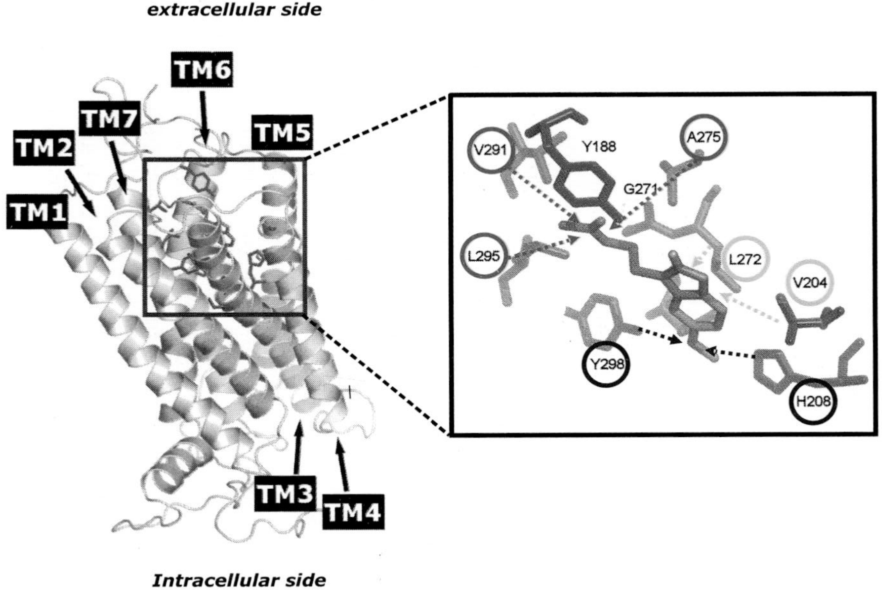

extracellular side

Intracellular side

Fig. 5. The primary sequence of human melatonin receptors and the model of melatonin binding site.

[A] Alignment of the human MT$_1$ (hMT$_1$) and MT$_2$ (hMT$_2$) melatonin receptors. The receptor molecule contains seven transmembrane domains (TM1-TM7) that are indicated by a horizontal bar above the relevant amino acid sequence. The amino acids conserved within both melatonin receptors are enclosed in boxes.

[B] Three-dimensional model of the hMT$_2$ melatonin receptor with presumptive position of 2-iodomelatonin molecule used in binding experiments (green). Amino acid residues in transmembrane domains 3, 5, 6 and 7 of the hMT2 receptor, which are involved in melatonin-receptor interaction, are shown in orange. TM domains are shown as white ribbons. Inset: enhanced view of residues important for ligand binding to the hMT2 receptor. Color scheme is as above. This model suggests that residues H208 and Y298 could specifically interact with 5-methoxy group of the 2-iodomelatonin (blue dotted arrows), while residues V204 and L272 may participate in hydrophobic interactions with the indole group of the ligand (yellow dotted arrows). Residues A275, V291 and L295 may interact with the N-acetyl chain of docked radioligand (*red* dotted arrows). Based on the hMT2 model and binding data, it is assumed that the side chain of residue Y188 in second extracellular loop could specifically interact with the N-acetyl group of 2-iodomelatonin (Mazna et al., 2004, 2005).

Signaling Pathways

Both MT_1 and MT_2 melatonin receptors (Fig. 5A) belong to the class of G protein-coupled receptor superfamily that signals through pertussiss toxin-sensitive pathways (White et al., 1987; Carlson et al., 1989; Morgan et al., 1994). Melatonin receptors exhibit high affinity for melatonin; MT_1 binds melatonin with K_d of <200 pM, whereas MT_2 binds this agonist with K_d between 1 and 10 nM (Duncan et al., 1989). The pharmacology of MT_1 and MT_2 receptors is not much advanced (Sugden et al., 1998; Spadoni et al., 1999). Luzindole and 4P-PDOT are the only known effective antagonists that are used in studies for investigating the function of melatonin receptors. Luzindole is a MT_1/MT_2 receptor ligand that has slightly higher (15-26 times) selectivity for the MT_2 receptor subtype. 4P-PDOT is more than 300 times more selective for MT_2 receptor than the MT_1 receptor (Dubocovich and Markowska, 2005).

The principal effect of melatonin observed in all cells expressing specific high-affinity melatonin MT_1 receptor is inhibition of the activity of adenylyl cyclase, leading to a decrease in intracellular levels of cAMP (White et al., 1987; Carlson et al., 1989; Reppert et al., 1994, 1995b; Liu et al. 1997; Teh and Sugden, 1999) and inhibition of cellular processes regulated by this intracellular messenger (Barrett et al., 1998; Ross et al., 1998). This effect of melatonin is probably mediated via α-subunit of PTX-sensitive G_i proteins (Fig. 3), as indicated in amphibian melanophores (White et al., 1987) or pars tuberalis cells from Djungarian hamsters (Carlson et al., 1989), and confirmed in cells expressing recombinant melatonin receptor (Brydon et al., 1999). In neonatal gonadotrophs, melatonin also has an inhibitory effect on GnRH-induced cAMP production (Vanecek and Vollrath, 1989), but its main effect is the inhibition of GnRH-stimulated increase in $[Ca^{2+}]_i$ (Vanecek and Klein, 1992; Zemkova and Vanecek, 1997). This phenomenon, unique for neonatal pituitary gonadotrophs, is not yet completely understood (Morgan et al., 1994; Reppert, 1997; Vanecek, 1998; Hazlerigg, 2001; Balik et al., 2004). Melatonin seems to inhibit both GnRH-stimulated calcium release from intracellular stores and extracellular calcium entry (Vanecek and Klein, 1992; Zemkova and Vanecek, 1997, 2000). Activation of the MT_2 receptors leads to inhibition of soluble guanylyl cyclase activity (Petit et al., 1999) and decrease in the intracellular levels of cGMP (Vanecek and Vollrath, 1989), but to what extent this is a receptor-specific function is not clear.

In transfected cell lines expressing MT_1 receptors, inhibition of adenylyl cyclase pathway is occasionally accompanied by simultaneous facilitation of phospholipase C activity and production of IP_3 and diacylglycerol (Godson and Reppert, 1997; Brydon et al., 1999). This dual effect of melatonin is most probably due to the melatonin receptor coupling simultaneously with two G proteins, $G_{i/o}$ and G_q, indicating that the effect of melatonin might depend on the cell type and the G proteins they contain.

Ligand Binding to Melatonin Receptors

The cloning of genes encoding the melatonin receptors and subsequent analysis of their protein products confirmed that these receptors have all the major structural characteristics found in other G protein-coupled receptors (Fig. 5B). These receptors consists of a bundle of seven putative transmembrane helices (TM) that form a ligand binding site, connected by three extracellular loops and three intracellular loops (Baldwin, 1994). Extracellular loops may participate in ligand recognition by the receptor, whereas intracellular loops are believed to directly activate G proteins (Wess, 1997; Ballesteros et al., 2001). However, melatonin receptors also have some unique features and contain several highly conserved sequence motifs not found in other G protein-coupled receptors (Barrett et al., 2003). Initial models of the melatonin binding site have been generated by using the known MT_1 and the third melatonin receptor primary structures fitted into a 3D structure of bacteriorhodopsin, or into a low-resolution structure of bovine rhodopsin (Grol and Jansen, 1996; Navajas et al., 1996). Recently, more accurate models based on high-resolution structures of bovine rhodopsin (Palczewski et al., 2000) have been published (Uchikawa et al., 2002; Mazna et al. 2004, 2005; Rivara et al., 2005). Subsequently, the validity of these models has been tested in studies using site-directed mutagenesis. Mutational analysis conducted on MT_1 (Conway et al., 1997, 2001; Gubitz and Reppert, 2000) and MT_2 (Gerdin et al., 2003; Mazna et al., 2004, 2005) melatonin receptors showed that similar amino acid residues within TM3, TM5, TM6 and TM7 play important roles in ligand binding to both receptor subtypes (Fig. 5B). Intracellular loops of G protein-coupled receptors are believed to directly activate G proteins. In case of the MT_1 melatonin receptor, single point mutations in TM3 almost eliminated the receptor's ability to inhibit coupling of adenylyl cyclase activity to voltage-dependent calcium channels (Nelson et al., 2001).

Melatonin Interaction with GnRH in Neonatal Pituitary Gonadotrophs

In neonatal pituitary pars distalis of the rat, melatonin at picomolar to low nanomolar concentration range inhibits or modulates GnRH-induced $[Ca^{2+}]_i$ increase in 40-70% of gonadotrophs (Vanecek and Klein, 1992a, b, 1993; Zemkova and Vanecek, 1997, 2000). This modulatory effect of melatonin was mimicked by Ca^{2+}-depeleted media or nifedipine indicating that melatonin inhibits extracellular Ca^{2+} entry (Vanecek and Klein, 1992, 1993; Zemkova and Vanecek, 1997). Provided that GnRH-stimulates Ca^{2+} entry by phosphorylation of Ca^{2+} channels using cAMP/PKA pathway (Stojilkovic and Catt, 1995; Grosse et al., 2000), melatonin-induced inhibition of Ca^{2+} entry could be a consequence of adenylyl cyclase inhibition (Vanecek and Vollrath, 1989). However, melatonin also inhibits GnRH-induced Ca^{2+} release from intracellular Ca^{2+} stores because the inhibitory effect of melatonin, when applied during ongoing oscillations, was observed in cells bathed in Ca^{2+}-deficient medium (Slanar et al., 1997; Zemkova and Vanecek, 1997, 2000). The mechanism by which melatonin inhibits IP_3-dependent calcium signaling is largely unknown. Melatonin has no effect on Ca^{2+} oscillations evoked by intracellular injection of IP_3 (Zemkova and Vanecek, 2000, 2001) indicating that melatonin inhibits the GnRH-stimulated signaling pathway upstream of IP_3 receptor activation. Melatonin has been shown to inhibit the GnRH-induced increase in DAG (Vanecek and Vollrath, 1990), which might be due to inhibition of phospholipase C activity. However, this is an unlikely explanation, because potentiation of phospholipase C has been found in cells transiently expressing melatonin receptor (Godson and Reppert, 1997; Brydon et al., 1999). As the majority of GnRH-stimulated DAG production comes from phospholipase D pathway (Zheng et al., 1994), it is likely that melatonin inhibits this pathway. An alternative explanation for the observed effects on GnRH-induced Ca^{2+} mobilization is inhibition of adenylyl cyclase pathway. Cross-talk between the cAMP and phosphoinositide signaling pathway is well documented. For example, in hepatocytes the frequency of agonist-stimulated calcium oscillations linked to IP_3 production is increased by activation of receptors positively coupled to adenylyl cyclase (Walaas et al., 1986; Furuichi et al., 1989; Mignery et al., 1990). As both melatonin receptor subtypes expressed in neonatal gonadotrophs are negatively coupled to adenylyl cyclase or guanylyl cyclase, it is reasonable to speculate that sensitivity of IP_3 receptors for IP_3 is lowered in cells with activated melatonin receptors.

DEVELOPMENTAL CHANGES IN ANTERIOR PITUITARY

In the anterior pituitary gland, melatonin mediates the effect of photoperiod acting primarily at two secretory cell types: lactotrophs that secrete prolactin, and gonadotrophs that secrete two gonadotropins, LH and FSH. The mechanisms of melatonin actions in lactotrophs are unknown at large (Morgan, 2000), whereas its actions in gonadotrophs are well documented (as mentioned above). However, these effects were observed only in neonatal gonadotrophs, as within the pituitary, the melatonin receptor persists until adulthood only in the pars tuberalis (Morgan and Williams, 1996).

In contrast to the physiological relevance of transient expression of melatonin receptors in gonadotrophs, the mechanism by which the developmental down-regulation of melatonin receptor expression is mediated is not clear at the present time. During embryonic development, anterior pituitary cells express mRNAs for several hormones in various combinations. For example, in mice at embryonic day 16, plurihormonal population accounts for more than 60% of cells, at postnatal day 1, it is approximately 35%; and at postnatal day 38, it is only about 25% (Seuntjens et al., 2002). Thus, it appears that developmental disappearance of melatonin receptors in anterior pituitary cells parallels the development loss of plurihormonal population of cells. One possibility is that melatonin-sensitive cells are not yet fully differentiated and melatonin receptors disappear as a consequence of programmed cell differentiation (Hazlerigg, 2001). Interestingly, melatonin receptors are primarily located in thyrotrophs, and gonadotrophs are the secondary population of cells to which the expression of melatonin receptors is restricted during the embryonal development (Johnston et al., 2006).

CONCLUSION

It appears contradictory to have a very sophisticated mechanism operative for pulsatile GnRH release in hypothalamus, and at the same time also to have an effective system to block the action of GnRH in the target pituitary cells, as is demonstrated in neonatal animals. However, such a dual control of gonadotroph function is physiologically justified. It provides an effective mechanism to down-regulate, but not to abolish, LH and FSH secretion. During development, low levels of plasma gonadotropins are needed for keeping operative gonadal steroidogenesis, which is critical for numerous cellular pathways. Thus, GnRH release is required in prepubertal period, but its action should be attenuated in order to keep plasma gonadotropin levels below the threshold for initiation of peripubertal changes.

The studies on neonatal gonadotrophs revealed two important aspects of melatonin actions: (1) a gradual loss of melatonin receptors, which temporally coincides with the loss of multi-hormonal cells, and (2) the coupling of GnRH receptors to ryanodine receptors, which also appears to be limited for prepubertal period. These observations raised the questions about the mechanism of gonadotroph differentiation and the physiological relevance of the delay in this process in order to provide an effective control of GnRH action by melatonin in neonatal cells. Also, it appears that both subtypes of high-affinity melatonin receptors, MT_1 and MT_2, mediate the effect of melatonin on calcium signaling in neonatal gonadotrophs. Which type of melatonin receptor subtype is responsible for inhibitory effect of melatonin remains to be established.

REFERENCES

Armstrong, S.M., Cassone, V.M., Chesworth, M.J., Redman, J.R. and Short, R.V. (1986). Synchronization of mammalian circadian rhythms by melatonin. *J. Neural. Transm.* Suppl. **21**: 375-394.

Baldwin, J.M. (1994). Structure and function of receptors coupled to G proteins. *Curr. Opin. Cell Biol.* **6**: 180-190.

Balik, A., Kretschmannova, K., Mazna, P., Svobodova, I. and Zemkova, H. (2004). Melatonin action in neonatal gonadotrophs. *Physiol. Res.* **53**(1): S153-S166.

Ballesteros, J.A., Shi, L. and Javitch, J.A. (2001). Structural mimicry in G protein-coupled receptors: implications of the high-resolution structure of rhodopsin for structure-function analysis of rhodopsin-like receptors. *Mol. Pharmacol.* **60**: 1-19.

Barrett, P., Davidson, G., Hazlerigg, D.G., Morris, M.A., Ross, A.W. and Morgan, P.J. (1998). Mel1a melatonin receptor expression is regulated by protein kinase C and an additional pathway addressed by the protein kinase C inhibitor Ro 31-8220 in ovine pars tuberalis cells. *Endocrinol.* **139**: 163-171.

Barrett, P., Conway, S. and Morgan, P.J. (2003). Digging deep-structure-function relationships in the melatonin receptor family. *J. Pineal Res.* **35**: 221-230.

Bartness, T.J., Powers, J.B., Hastings, M.H., Bittman, E.L. and Goldman, B.D. (1993). The timed infusion paradigm for melatonin delivery: what has it taught us about the melatonin signal, its reception, and the photoperiodic control of seasonal responses? *J. Pineal Res.* **15**: 161-190.

Berridge, M.J. (1993). Inositol trisphosphate and calcium signalling. *Nature* **361**: 315-325.

Bezprozvanny, I., Watras, J. and Ehrlich, B.E. (1991). Bell-shaped calcium-response curves of Ins(1,4,5)P3- and calcium-gated channels from endoplasmic reticulum of cerebellum. *Nature* **351**: 751-754.

Bosma, M.M. and Hille, B. (1992). Electrophysiological properties of a cell line of the gonadotrope lineage. *Endocrinol.* **130**: 3411-3420.

Brydon, L., Roka, F., Petit, L., de Coppet, P., Tissot, M., Barrett, P., Morgan, P.J., Nanoff, C., Strosberg, A.D. and Jockers, R. (1999). Dual signaling of human Mel1a melatonin receptors via G(i2), G(i3), and G(q/11) proteins. *Mol. Endocrinol.* **13**: 2025-2038.

Carlson, L.L., Weaver, D.R. and Reppert, S.M. (1989). Melatonin signal transduction in hamster brain: inhibition of adenylyl cyclase by a pertussis toxin-sensitive G protein. *Endocrinol.* **125**: 2670-2676.

Cassone, V.M. (1990). Effects of melatonin on vertebrate circadian systems. *Trends Neurosci.* **13**: 457-464.

Cesnjaj, M., Catt, K.J. and Stojilkovic, S.S. (1994) Coordinate actions of calcium and protein kinase-C in the expression of primary response genes in pituitary gonadotrophs. *Endocrinol.* **135**: 692-701.

Chang, J.P., Morgan, R.O. and Catt, K.J. (1988). Dependence of secretory responses to gonadotropin-releasing hormone on diacylglycerol metabolism. Studies with a diacylglycerol lipase inhibitor, RHC 80267. *J. Biol. Chem.* **263**: 18614-18620.

Conway, S., Canning, S.J., Barrett, P., Guardiola-Lemaitre, B., Delagrange, P. and Morgan, P.J. (1997). The roles of valine 208 and histidine 211 in ligand binding and receptor function of the ovine Mel1a beta melatonin receptor. *Biochem. Biophys. Res. Commun.* **239**: 418-423.

Conway, S., Mowat, E.S., Drew, J.E., Barrett, P., Delagrange, P. and Morgan, P.J. (2001). Serine residues 110 and 114 are required for agonist binding but not antagonist binding to the melatonin MT(1) receptor. *Biochem. Biophys. Res. Commun.* **282**: 1229-1236.

Davis, F.C. (1997). Melatonin: role in development. *J. Biol. Rhythms.* **12**: 498-508.

Dubocovich, M.L. (1985). Characterization of a retinal melatonin receptor. *J. Pharmacol. Exp. Ther.* **234**: 395-401.

Dubocovich, M.L., Yun, K., Al-Ghoul, W.M., Benloucif, S. and Masana, M.I. (1998). Selective MT2 melatonin receptor antagonists block melatonin-mediated phase advances of circadian rhythms. *FASEB J.* **12**: 1211-1220.

Dubocovich, M.L. and Markowska, M. (2005). Functional MT1 and MT2 melatonin receptors in mammals. *Endocrinol.* **27**: 101-110.

Duncan, M.J., Takahashi, J.S. and Dubocovich, M.L. (1989). Characteristics and autoradiographic localization of 2-[^{125}I]iodomelatonin binding sites in Djungarian hamster brain. *Endocrinol.* **125**: 1011-1018.

Furuichi, T., Yoshikawa, S., Miyawaki, A., Wada, K., Maeda, N. and Mikoshiba, K. (1989). Primary structure and functional expression of the inositol 1,4,5-trisphosphate-binding protein P400. *Nature* **342**: 32-38.

Gauer, F., Masson-Pevet, M. and Pevet, P. (1993). Melatonin receptor density is regulated in rat pars tuberalis and suprachiasmatic nuclei by melatonin itself. *Brain Res.* **602**: 153-156.

Gauer, F., Schuster, C., Poirel, V.J., Pevet, P. and Masson-Pevet, M. (1998) Cloning experiments and developmental expression of both melatonin receptor Mel1A mRNA and melatonin binding sites in the Syrian hamster suprachiasmatic nuclei. *Brain Res. Mol. Brain Res.* **60**: 193-202.

Gerdin, M.J., Mseeh, F. and Dubocovich, M.L. (2003). Mutagenesis studies of the human MT2 melatonin receptor. *Biochem. Pharmacol.* **66**: 315-320.

Gilad, E., Matzkin, H. and Zisapel, N. (1997). Inactivation of melatonin receptors by protein kinase C in human prostate epithelial cells. *Endocrinol.* **138**: 4255-4261.

Glass, J.D. and Knotts, L.K. (1987). A brain site for the antigonadal action of melatonin in the white-footed mouse (*Peromyscus leucopus*): involvement of the immuno-reactive GnRH neuronal system. *Neuroendocrinol.* **46**: 48-55.

Godson, C. and Reppert, S.M. (1997). The Mel1a melatonin receptor is coupled to parallel signal transduction pathways. *Endocrinol.* **138**: 397-404.

Grol, C.J. and Jansen, J.M. (1996). The high affinity melatonin binding site probed with conformationally restricted ligands-II. Homology modeling of the receptor. *Bioorg. Med. Chem.* **4**: 1333-1339.

Grosse, R., Schmid, A., Schoneberg, T., Herrlich, A., Muhn, P., Schultz, G. and Gudermann, T. (2000). Gonadotropin-releasing hormone receptor initiates multiple signaling pathways by exclusively coupling to G(q/11) proteins. *J. Biol. Chem.* **275**: 9193-9200.

Gubitz, A.K. and Reppert, S.M. (2000). Chimeric and point-mutated receptors reveal that a single glycine residue in transmembrane domain 6 is critical for high affinity melatonin binding. *Endocrinol.* **141**: 1236-1244.

Hazlerigg, D.G. (2001). What is the role of melatonin within the anterior pituitary? *J. Endocrinol.* **170**: 493-501.

Herlitze, S., Garcia, D.E., Mackie, K., Hille, B., Scheuer, T. and Catterall, W.A. (1996). Modulation of Ca^{2+} channels by G-protein beta gamma subunits. *Nature* **380**: 258-262.

Hille, B., Tse, A., Tse, F.W. and Almers, W. (1994). Calcium oscillations and exocytosis in pituitary gonadotropes. *Ann. NY Acad. Sci.* **710**: 261-270.

Hille, B., Tse, A., Tse, F.W. and Bosma, M.M. (1995). Signaling mechanisms during the response of pituitary gonadotropes to GnRH. *Recent Prog. Horm. Res.* **50**: 75-95.

Johnston, J.D., Klosen, P., Barrett, P. and Hazlerigg, D.G. (2006). Regulation of MT1 melatonin receptor expression in the foetal rat pituitary. *Neuroendocrinol.* **18**: 50-56.

Kalsbeek, A., Garidou, M.L., Palm, I.F., Van Der Vliet, J., Simonneaux, V., Pevet, P. and Buijs, R.M. (2000). Melatonin sees the light: blocking GABA-ergic transmission in the paraventricular nucleus induces daytime secretion of melatonin. *Eur. J. Neurosci.* **12**: 3146-3154.

Kennaway, D.J. and Rowe, S.A. (1995). Melatonin binding sites and their role in seasonal reproduction. *J. Reprod. Fertil.* **49**: 423-435.

Klein, D.C., Coon, S.L., Roseboom, P.H., Weller, J.L., Bernard, M., Gastel, J.A., Zatz, M., Iuvone, P.M., Rodriguez, I.R., Begay, V., Falcon, J., Cahill, G.M., Cassone, V.M. and Baler, R. (1997). The melatonin rhythm-generating enzyme: molecular regulation of serotonin N-acetyltransferase in the pineal gland. *Recent. Prog. Horm. Res.* **52**: 307-357.

Kramer, R.H., Mokkapatti, R. and Levitan, E.S. (1994). Effects of caffeine on intracellular calcium, calcium current and calcium-dependent potassium current in anterior pituitary GH3 cells. *Pflugers. Arch.* **426**: 12-20.

Krsmanovic, L.Z., Mores, N., Navarro, C.E., Arora, K.K. and Catt, K.J. (2003). An agonist-induced switch in G protein coupling of the gonadotropin-releasing hormone receptor regulates pulsatile neuropeptide secretion. *Proc. Natl. Acad. Sci. USA* **100**: 2969-2974.

Kukuljan, M., Stojilkovic, S.S., Rojas, E. and Catt, K.J. (1992). Apamin-sensitive potassium channels mediate agonist-induced oscillations of membrane potential in pituitary gonadotrophs. *FEBS Lett.* **301**: 19-22.

Kwiecien, R. and Hammond, C. (1998). Differential management of Ca^{2+} oscillations by anterior pituitary cells: a comparative overview. *Neuroendocrinol.* **68**: 135-151.

Laitinen, J.T., Viswanathan, M., Vakkuri, O. and Saavedra, J.M. (1992). Differential regulation of the rat melatonin receptors: selective age-associated decline and lack of melatonin-induced changes. *Endocrinol.* **130**: 2139-2144.

Larini, F., Menegazzi, P., Baricordi, O., Zorzato, F. and Treves, S. (1995). A ryanodine receptor-like Ca^{2+} channel is expressed in nonexcitable cells. *Mol. Pharmacol.* **47**: 21-28.

Li, L., Xu., J.N., Wong, Y.H., Wong, J.T., Pang, S.F. and Shiu, S.Y. (1998). Molecular and cellular analyses of melatonin receptor-mediated cAMP signaling in rat corpus epididymis. *J. Pineal Res.* **25**: 219-228.

Liu, C., Weaver, D.R., Jin, X., Shearman, L.P., Pieschl, R.L., Gribkoff, V.K. and Reppert, S.M. (1997). Molecular dissection of two distinct actions of melatonin on the suprachiasmatic circadian clock. *Neuron.* **19**: 91-102.

Lopez-Gonzalez, M.A., Calvo, J.R., Osuna, C. and Guerrero, J.M. (1992). Interaction of melatonin with human lymphocytes: evidence for binding sites coupled to potentiation of cyclic AMP stimulated by vasoactive intestinal peptide and activation of cyclic GMP. *J. Pineal Res.* **12**: 97-104.

Martin, J.E. and Klein, D.C. (1976). Melatonin inhibition of the neonatal pituitary response to luteinizing hormone-releasing factor. *Science* **191**: 301-302.

Mazna, P., Obsilova, V., Jelinkova, I., Balik, A., Berka, K., Sovova, Z., Ettrich, R., Svoboda, P., Obsil, T. and Teisinger, J. (2004). Molecular modeling of human MT2 melatonin receptor: the role of Val204, Leu272 and Tyr298 in ligand binding. *J. Neurochem.* **91**: 836-842.

Mazna, P., Berka, K., Jelinkova, I., Balik, A., Svoboda, P., Obsilova, V., Obsil, T. and Teisinger, J. (2005) Ligand binding to the human MT2 melatonin receptor: The role of residues in transmembrane domains 3, 6, and 7. *Biochem. Biophys. Res. Commun.* **332**: 726-734.

McArdle, C.A., Franklin, J., Green, L. and Hislop, J.N. (2002). Signalling, cycling and desensitisation of gonadotrophin-releasing hormone receptors. *Endocrinol.* **173**: 1-11.

McArthur, A.J., Gillette, M.U. and Prosser, R.A. (1991). Melatonin directly resets the rat suprachiasmatic circadian clock in vitro. *Brain Research* **565**: 158-161.

Mignery, G.A., Newton, C.L., Archer, B.T. and 3rd, Sudhof, T.C. (1990). Structure and expression of the rat inositol 1,4,5-trisphosphate receptor. *J. Biol. Chem.* **265**: 12679-12685.

Millar, R., Lowe, S., Conklin, D., Pawson, A., Maudsley, S., Troskie, B., Ott, T., Millar, M., Lincoln, G., Sellar, R., Faurholm, B., Scobie, G., Kuestner, R., Terasawa, E. and Katz, A. (2001). A novel mammalian receptor for the evolutionarily conserved type II GnRH. *Proc. Natl. Acad. Sci. USA* **98**: 9636-9641.

Morgan, R.O., Chang, J.P. and Catt, K.J. (1987). Novel aspects of gonadotropin-releasing hormone action on inositol polyphosphate metabolism in cultured pituitary gonadotrophs. *J. Biol. Chem.* **262**: 1166-1171.

Morgan, P.J., Barrett, P., Howell, H.E. and Helliwell, R. (1994). Melatonin receptors: localization, molecular pharmacology and physiological significance. *Neurochem. Inter.* **24**: 101-146.

Morgan, P.J. and Williams, L.M. (1996). The pars tuberalis of the pituitary: a gateway for neuroendocrine output. *Rev. Reprod.* **1**: 153-161.

Morgan, P.J. (2000). The pars tuberalis: the missing link in the photoperiodic regulation of prolactin secretion? *J. Neuroendocrinol.* **12**: 287-295.

Navajas, C., Kokkola, T., Poso, A., Honka, N., Gynther, J. and Laitinen, J.T. (1996). A rhodopsin-based model for melatonin recognition at its G protein-coupled receptor. *Eur. J. Pharmacol.* **304**: 173-183.

Nelson, C.S., Ikeda, M., Gompf, H.S., Robinson, M.L., Fuchs, N.K., Yoshioka, T., Neve, K.A. and Allen, C.N. (2001). Regulation of melatonin 1a receptor signaling and trafficking by asparagine-124. *Mol. Endocrinol.* **15**: 1306-1317.

Nosjean, O., Ferro, M., Coge, F., Beauverger, P., Henlin, J.M., Lefoulon, F., Fauchere, J.L., Delagrange, P., Canet, E. and Boutin, J.A. (2000). Identification of the melatonin-binding site MT3 as the quinone reductase 2. *J. Biol. Chem.* **275**: 31311-31317.

Palczewski, K., Kumasaka, T., Hori, T., Behnke, C.A., Motoshima, H., Fox, B.A., Le Trong, I., Teller, D.C., Okada, T., Stenkamp, R.E., Yamamoto, M. and Miyano, M. (2000). Crystal structure of rhodopsin: A G protein-coupled receptor. *Science* **289**: 739-745.

Paul, P., Lahaye, C., Delagrange, P., Nicolas, J.P., Canet, E. and Boutin, J.A. (1999). Characterization of 2-[125I]iodomelatonin binding sites in Syrian hamster peripheral organs. *J. Pharmacol. Exp. Ther.* **290**: 334-340.

Perrin, M.H., Haas, Y., Porter, J., Rivier, J. and Vale, W. (1989). The gonadotropin-releasing hormone pituitary receptor interacts with a guanosine triphosphate-binding protein: differential effects of guanyl nucleotides on agonist and antagonist binding. *Endocrinol.* **124**: 798-804.

Petit, L., Lacroix, I., de Coppet, P., Strosberg, A.D. and Jockers, R. (1999). Differential signaling of human Mel1a and Mel1b melatonin receptors through the cyclic guanosine 3'-5'-monophosphate pathway. *Biochem. Pharmacol.* **58**: 633-639.

Recio, J., Gauer, F., Schuster, C., Pevet, P. and Masson-Pevet, M. (1998). Daily and photoperiodic 2-125I-melatonin binding changes in the pars tuberalis of the Syrian hamster (*Mesocricetus auratus*): effect of constant light exposure and pinealectomy. *J. Pineal. Res.* **24**: 162-167.

Reinhart, J., Mertz, L.M. and Catt, K.J. (1992). Molecular cloning and expression of cDNA encoding the murine gonadotropin-releasing hormone receptor. *J. Biol. Chem.* **267**: 21281-21284.

Reppert, S.M., Weaver, D.R., Rivkees, S.A. and Stopa, E.G. (1988). Putative melatonin receptors in a human biological clock. *Science* **242**: 78-81.

Reppert, S.M., Weaver, D.R. and Ebisawa, T. (1994). Cloning and characterization of a mammalian melatonin receptor that mediates reproductive and circadian responses. *Neuron.* **13**: 1177-1185.

Reppert, S.M., Weaver, D.R., Cassone, V.M., Godson, C. and Kolakowski, L.F.J. (1995a). Melatonin receptors are for the birds: molecular analysis of two receptor subtypes differentially expressed in chick brain. *Neuron.* **15**: 1003-1015.

Reppert, S.M., Godson, C., Mahle, C.D., Weaver, D.R., Slaugenhaupt, S.A. and Gusella, J.F. (1995b). Molecular characterization of a second melatonin receptor expressed in human retina and brain: the Mel1b melatonin receptor. *Proc. Natl. Acad. Sci. USA* **92**: 8734-8738.

Reppert, S.M. (1997). Melatonin receptors: molecular biology of a new family of G protein-coupled receptors. *J. Biol. Rhythms* **12**: 528-531.

Rivara, S., Lorenzi, S., Mor, M., Plazzi, P.V., Spadoni, G., Bedini, A. and Tarzia, G. (2005). Analysis of structure-activity relationships for MT2 selective antagonists by melatonin MT1 and MT2 receptor models. *J. Med. Chem.* **48**: 4049-4060.

Roca, A.L., Godson, C., Weaver, D.R. and Reppert, S.M. (1996). Structure, characterization, and expression of the gene encoding the mouse Mel1a melatonin receptor [published erratum appears in *Endocrinol.* 1997; **138**(6):2307]. *Endocrinol.* **137**: 3469-3477.

Ross, A.W., Webster, C.A., Thompson, M., Barrett, P. and Morgan, P.J. (1998). A novel interaction between inhibitory melatonin receptors and protein kinase C-dependent signal transduction in ovine pars tuberalis cells. *Endocrinol.* **139**: 1723-1730.

Roy, D., Angelini, N.L., Fujieda, H., Brown, G.M. and Belsham, D.D. (2001). Cyclical regulation of GnRH gene expression in GT1-7 GnRH-secreting neurons by melatonin. *Endocrinol.* **142**: 4711-4720.

Roy, D. and Belsham, D.D. (2002) Melatonin receptor activation regulates GnRH gene expression and secretion in GT1-7 GnRH neurons. Signal transduction mechanisms. *J. Biol. Chem.* **277**: 251-258.

Seuntjens, E., Hauspie, A., Vankelecom, H. and Denef, C. (2002). Ontogeny of plurihormonal cells in the anterior pituitary of the mouse, as studied by means of hormone mRNA detection in single cells. *J. Neuroendocrinol.* **14**: 611-619.

Shangold, G.A., Murphy, S.N. and Miller, R.J. (1988). Gonadotropin-releasing hormone-induced Ca^{2+} transients in single identified gonadotropes require both intracellular Ca^{2+} mobilization and Ca^{2+} influx. *Proc. Natl. Acad. Sci. USA* **85**: 6566-6570.

Siuciak, J.A., Fang, J.M. and Dubocovich, M.L. (1990). Autoradiographic localization of 2-[125I]iodomelatonin binding sites in the brains of C3H/HeN and C57BL/6J strains of mice. *European J. Pharmacol.* **180**: 387-390.

Skinner, D.C. and Robinson, J.E. (1995). Melatonin-binding sites in the gonadotroph-enriched zona tuberalis of ewes. *J. Reprod. Fertil.* **104**: 243-250.

Slanar, O., Zemkova, H. and Vanecek, J. (1997). Melatonin inhibits GnRH-induced Ca^{2+} mobilization and influx through voltage-regulated channels. *Biol. Signals* **6**: 284-290.

Spadoni, G., Mor, M. and Tarzia, G. (1999). Structure-affinity relationships of indole-based melatonin analogs. *Biol. Signals Recept.* **8**: 15-23.

Stojilkovic, S.S., Iida, T., Merelli, F. and Catt, K.J. (1991). Calcium signaling and secretory responses in endothelin-stimulated anterior pituitary cells. *Mol. Pharmacol.* **39**: 762-770.

Stojilkovic, S.S. and Catt, K.J. (1992). Calcium oscillations in anterior pituitary cells. *Endocr. Rev.* **13**: 256-280.

Stojilkovic, S.S., Kukuljan, M., Iida, T., Rojas, E. and Catt, K.J. (1992). Integration of cytoplasmic calcium and membrane potential oscillations maintains calcium signaling in pituitary gonadotrophs. *Proc. Natl. Acad. Sci. USA* **89**: 4081-4085.

Stojilkovic, S.S., Reinhart, J. and Catt, K.J. (1994). Gonadotropin-releasing hormone receptors: structure and signal transduction pathways. *Endocr. Rev.* **15**: 462-499.

Stojilkovic, S.S. and Catt, K.J. (1995). Novel aspects of GnRH-induced intracellular signaling and secretion in pituitary gonadotrophs. *J. Neuroendocrinol.* **7**: 739-757.

Stojilkovic, S.S. (2005). Ca(2+)-regulated exocytosis and SNARE function. *Trends Endocrinol. Metab.* **16**: 81-83.

Stojilkovic, S.S., Zemkova, H. and Van Goor, F. (2005). Biophysical basis of pituitary cell type-specific Ca(2+) signaling-secretion coupling. *Trends Endocrinol. Metab.* **16**: 152-159.

Stutzin, A., Stojilkovic, S.S., Catt, K.J. and Rojas, E. (1989). Characteristics of two types of calcium channels in rat pituitary gonadotrophs. *Am. J. Physiol.* **257**: C865-C874.

Sugden, D., Pickering, H., Teh, M.T. and Garratt, P.J. (1998). Melatonin receptor pharmacology: toward subtype specificity. *Biology of the Cell* **89**: 531-537.

Svobodova, I., Vanecek, J. and Zemkova, H. (2003). The bidirectional phase-shifting effects of melatonin on the arginine vasopressin secretion rhythm in rat suprachiasmatic nuclei in vitro. *Brain Res. Mol. Brain Res.* **116**: 80-85.

Teh, M.T. and Sugden, D. (1999). The putative melatonin receptor antagonist GR128107 is a partial agonist on *Xenopus laevis* melanophores. *Br. J. Pharmacol.* **126**: 1237-1245.

Tomic, M., Cesnjaj, M., Catt, K.J. and Stojilkovic, S.S. (1994). Developmental and physiological aspects of Ca^{2+} signaling in agonist-stimulated pituitary gonadotrophs. *Endocrinol.* **135**: 1762-1771.

Tse, A. and Hille, B. (1992). GnRH-induced Ca^{2+} oscillations and rhythmic hyperpolarizations of pituitary gonadotropes. *Science* **255**: 462-464.

Tsutsumi, M., Zhou, W., Millar, R.P., Mellon, P.L., Roberts, J.L., Flanagan, C.A., Dong, K., Gillo, B. and Sealfon, S.C. (1992). Cloning and functional expression of a mouse gonadotropin-releasing hormone receptor. *Mol. Endocrinol.* **6**: 1163-1169.

Uchikawa, O., Fukatsu, K., Tokunoh, R., Kawada, M., Matsumoto, K., Imai, Y., Hinuma, S., Kato, K., Nishikawa, H., Hirai, K., Miyamoto, M. and Ohkawa, S. (2002). Synthesis of a novel series of tricyclic indan derivatives as melatonin receptor agonists. *J. Med. Chem.* **45**: 4222-4239.

Valenti, S., Giusti, M., Guido, R. and Giordano, G. (1997). Melatonin receptors are present in adult rat Leydig cells and are coupled through a pertussis toxin-sensitive G-protein. *Eur. J. Endocrinol.* **136**: 633-639.

Van Goor, F., Zivadinovic, D. and Stojilkovic, S.S. (2001). Differential expression of ionic channels in rat anterior pituitary cells. *Mol. Endocrinol.* **15**: 1222-1236.

Vanecek, J., Pavlik, A. and Illnerova, H. (1987). Hypothalamic melatonin receptor sites revealed by autoradiography. *Brain Res.* **435**: 359-362.

Vanecek, J. (1988). The melatonin receptors in rat ontogenesis. *Neuroendocrinol.* **48**: 201-203.

Vanecek, J. and Vollrath, L. (1989). Melatonin inhibits cyclic AMP and cyclic GMP accumulation in the rat pituitary. *Brain Res.* **505**: 157-159.

Vanecek, J. and Vollrath, L. (1990). Melatonin modulates diacylglycerol and arachidonic acid metabolism in the anterior pituitary of immature rats. *Neurosci. Lett.* **110**: 199-203.

Vanecek, J. and Klein, D.C. (1992a). Melatonin inhibits gonadotropin-releasing hormone-induced elevation of intracellular Ca^{2+} in neonatal rat pituitary cells. *Endocrinol.* **130**: 701-707.

Vanecek, J. and Klein, D.C. (1992b). Sodium-dependent effects of melatonin on membrane potential of neonatal rat pituitary cells. *Endocrinol.* **131**: 939-946.

Vanecek, J. and Klein, D.C. (1993). A subpopulation of neonatal gonadotropin-releasing hormone-sensitive pituitary cells is responsive to melatonin. *Endocrinol.* **133**: 360-367.

Vanecek, J. (1998). Cellular mechanisms of melatonin action. *Physiol. Rev.* **78**: 687-721.

Viswanathan, M., Laitinen, J.T. and Saavedra, J.M. (1990). Expression of melatonin receptors in arteries involved in thermoregulation. *Proc. Natl. Acad. Sci. USA* **87**: 6200-6203.

Walaas, S.I., Nairn, A.C. and Greengard, P. (1986). PCPP-260, a Purkinje cell-specific cyclic AMP-regulated membrane phosphoprotein of Mr 260,000. *J. Neurosci.* **6**: 954-961.

Weaver, D.R., Liu, C. and Reppert, S.M. (1996). Nature's knockout: the Mel1b receptor is not necessary for reproductive and circadian responses to melatonin in Siberian hamsters. *Mol. Endocrinol.* **10**: 1478-1487.

Wess, J. (1997). G-protein-coupled receptors: molecular mechanisms involved in receptor activation and selectivity of G-protein recognition. *FASEB J.* **11**: 346-354.

White, B.H., Sekura, R.D. and Rollag, M.D. (1987). Pertussis toxin blocks melatonin-induced pigment aggregation in *Xenopus* dermal melanophores. *J. Comp. Physiol. B, Biochem. Systemic, and Environmental Physiol.* **157**: 153-159.

White, R.B., Eisen, J.A., Kasten, T.L. and Fernald, R.D. (1998). Second gene for gonadotropin-releasing hormone in humans. *Proc. Natl. Acad. Sci. USA* **95**: 305-309.

Williams, L.M. and Morgan, P.J. (1988). Demonstration of melatonin-binding sites on the pars tuberalis of the rat. *Endocrinol.* **119**: R1-R3.

Williams, L.M., Martinoli, M.G., Titchener, L.T. and Pelletier, G. (1991). The ontogeny of central melatonin binding sites in the rat. *Endocrinol.* **128**: 2083-2090.

Williams, L.M., Lincoln, G.A., Mercer, J.G., Barrett, P., Morgan, P.J. and Clarke, I.J. (1997). Melatonin receptors in the brain and pituitary gland of hypothalamo-pituitary disconnected Soay rams. *Neuroendocrinol.* **9**: 639-643.

Zemkova, H. and Vanecek, J. (1997). Inhibitory effect of melatonin on gonadotropin-releasing hormone-induced Ca^{2+} oscillations in pituitary cells of newborn rats. *Neuroendocrinol.* **65**: 276-283.

Zemkova, H. and Vanecek, J. (2000). Differences in gonadotropin-releasing hormone-induced calcium signaling between melatonin-sensitive and melatonin-insensitive neonatal rat gonadotrophs. *Endocrinol.* **141**: 1017-1026.

Zemkova, H. and Vanecek, J. (2001). Dual effect of melatonin on gonadotropin-releasing-hormone-induced Ca(2+) signaling in neonatal rat gonadotropes. *Neuroendocrinol.* **74**: 262-269.

Zemkova, H., Balik, A., Kretschmannova, K., Mazna, P. and Stojilkovic, S.S. (2004). Recovery of Ins(1,4,5)-triphosphate Dependent Calcium Signaling in Neonatal Gonadotrophs. *Cell Calcium* 9636-9641.

Zheng, L., Stojilkovic, S.S., Hunyady, L., Krsmanovic, L.Z. and Catt, K.J. (1994). Sequential activation of phospholipase-C and -D in agonist-stimulated gonadotrophs. *Endocrinol.* **134**: 1446-1454.

Reproductive Biology and Clinical Endocrinology

CHAPTER 1

Roles of Melatonin in Photoperiodic Gonadal Response of Birds

Tomoko Yoshikawa[1,2] Masayuki Iigo[3],
Toshiyuki Okano[1,4] and Yoshitaka Fukada[1,]*

Abstract

Many seasonal activities, such as reproduction, growth, molt, migration and hibernation, are triggered by a change in photoperiod. Seasonal gonadal development of some animal species is one of the photoperiodic responses that have been investigated intensively. In mammals, the duration of the elevated circulating melatonin is critical for determining photoperiodic gonadal responses. On the other hand, the role of melatonin in photoperiodic gonadal response in birds is yet elusive because of inconsistencies among the experimental data reported to date. In non-mammalian vertebrates, photoreceptors in the deep brain region are thought to be responsible for the photoperiodic gonadal response. Cerebrospinal fluid (CSF)-contacting neurons locating in the deep brain express rhodopsin, a canonical rod photopigment in the retina, and these cells are candidates for the photoperiodic photoreceptor. Since the morphology of these cells suggested a possible signal transmission between the cells and the CSF, melatonin levels in the quail and pigeon CSF were measured by using the method of in vivo microdialysis. These studies showed robust circadian oscillations of the melatonin level in the CSF with a profile reflecting the photoperiod, suggesting a possible control by CSF melatonin of gonadotropin-inhibitory hormone (GnIH) and/or of gonadotropin-releasing hormone (GnRH) production, both of which could be the major determinant of avian reproductive activity.

[1]Department of Biophysics and Biochemistry, Graduate School of Science, The University of Tokyo, Tokyo Japan.
[2]Department of Biology, University of Virginia, Charlottesville, VA 22903, USA.
[3]Department of Applied Biochemistry, Faculty of Agriculture, Utsunomiya University, Utsunomiya 321-8505, Japan.
[4] Precursory Research for Embryonic Science and Technology, Japan Science and Technology Agency, 4-1-8 Honcho Kawaguchi, Saitama 332-0012, Japan.
Corresponding author: Department of Biophysics and Biochemistry, Graduate School of Science, The University of Tokyo, Hongo 7-3-1, Bunkyo-Ku, Tokyo 113-0033, Japan.
Tel.: +81 3 5841 4381, Fax: +81 3 5802 8871. E-mail: sfukada@mail.ecc.u-tokyo.ac.jp

INTRODUCTION

Many environmental factors on the earth change periodically. These factors include daily changes of light-dark (LD) and temperature, and seasonal changes in photoperiod and temperature. Many organisms living outside the equatorial region show cyclic reproductive behaviors with a period of a year, so that the offspring grows up in a season that is best for survival. Other activities such as growth, molt, migration, and hibernation are also seasonal for physiological reasons. In many cases, these seasonal events are triggered by a change in photoperiod, which is the most important time cue for living organisms to predict coming seasons because of its precise correlation with the time of year. The physiological mechanism that is driven by the change in photoperiod is called "photoperiodism".

Photoperiodic Responses

The mechanism of photoperiodism is not well established in spite of a considerable amount of research (Goldman, 2001). Two major models, the external coincidence model and the internal coincidence model, have been proposed to describe the possible mechanism (Pittendrigh, 1981). In the external coincidence model, it is presumed that light has two roles. First, light entrains the circadian system intrinsic to the living organism. Second, light given during the photosensitive phase of the organisms induces long-day responses. In the absence of light during the photosensitive phase, the organisms display short-day responses. In the internal coincidence model, on the other hand, light is postulated to entrain two intrinsic circadian oscillators whose phase relationship varies as a function of the photoperiod. This model predicts that the degree of the phase angle difference between the two oscillators determines the type of photoperiodic responses.

In animals, the photoperiodic gonadal response has been investigated intensively. In temperate zones, development and regression of the gonads in the photoperiodic animals correlate with the photoperiod (Fig. 1). The gonads of most birds remain undeveloped during winter and begin to develop with an increase of the photoperiod in spring, leading to the breeding activities during spring and summer. In late summer or fall the gonads start regressing, even though the photoperiod at this time of the year is longer than the photoperiod which triggered the gonadal development in spring. This phenomenon is called "photorefractoriness", and it may prevent further breeding in a season returning to short-day photoperiod.

In mammals, the retina receives light for photoperiodic responses, whereas in non-mammalian vertebrates, the retina is not necessary for the photoperiodic response. Some species of birds still showed photoperiodic gonadal responses after removal of eyes and pineal gland, and local illumination of a limited part of the brain by implanting a small radioluminous pellet, which induced gonadal development in a short-day photoperiod (Oliver and Bayle, 1982). These results suggest that

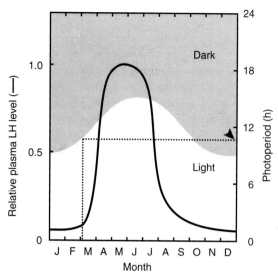

Fig. 1. A relationship between the photoperiod and the plasma level of luteinizing hormone (LH) of birds in the Northern Hemisphere. The arrowhead indicates the critical photoperiod (a minimal length of daytime) triggering the gonadal development in spring.

photoreceptors responsible for the photoperiodic gonadal response may exist in the deep brain region of the birds. Consistently, the expression of rhodopsin gene has been detected in the pigeon lateral septum (Wada et al., 1998). It is not known how the light information is converted into the hypothalamic endocrine signal that leads to the gonadal development.

Control of Melatonin Secretion

Melatonin secretion from the pineal gland shows a circadian rhythm in a variety of vertebrates (Underwood et al., 2001). This melatonin rhythm reflects changes in the activity of arylalkylamine N-acetyltransferase, a key enzyme that regulates melatonin synthesis (Klein et al., 1997). In mammals, the light signal received by the retina is transmitted to the suprachiasmatic nucleus (SCN) where the central circadian oscillator is located, and the multisynaptic neural transmission from the SCN regulates the melatonin synthesis in the pineal gland. In many non-mammalian vertebrates, on the other hand, the pineal gland is photosensitive, and the synthesis of melatonin is controlled by the circadian oscillator and photoreceptive molecule(s) that are intrinsic to the gland (Zatz, 1996). In most vertebrates, melatonin synthesis and its secretion are high during the night and low during the day, and the duration of the high level secretion is directly affected by the duration of the night (dark period). This suggests a possibility that melatonin plays a role as the mediator of photoperiod, and numerous studies have been performed in order to show the role of melatonin in photoperiodic gonadal responses (Mayer et al., 1997 Malpaux et al., 2001).

Roles of Melatonin in Photoperiodic Responses

Three strategies have been adopted to determine if melatonin plays a role as a neuroendocrine messenger for the photoperiodism.

  (i) Manipulation of the endogenous melatonin rhythm by exposing the animal to a long-day photoperiod (causing a short peak duration of the melatonin secretion), a short-day photoperiod (causing a long peak duration) and constant light (suppressing the melatonin secretion at a constitutively low level).

 (ii) Surgical removal of melatonin-producing tissues: pinealectomy and enucleation.

(iii) Exogenous melatonin administration: timed-injection of melatonin (extending the duration of the elevated melatonin level), timed-infusion of melatonin (extending the duration or creating an artificial rhythm of melatonin) and implantation of a melatonin-releasing capsule (causing a constitutively high level of melatonin).

Many studies have been performed with combinations of these methods, and it was concluded that the duration of elevated circulating melatonin is critical for determination of photoperiodic gonadal responses in mammals (Malpaux et al., 2001). It remains yet to be determined if melatonin plays the same role in birds.

Melatonin and Photoperiodic Gonadal Response in Birds

Roles of melatonin in photoperiodic gonadal responses have been studied extensively in a variety of domestic and wild-captured birds, but the results are variable among studies even on the same species (Mayer et al., 1997). The Japanese quail has been most commonly used due to its highly reproducible and sensitive response to a long-day stimulus (Follett and Sharp, 1969). There was no significant difference between intact and pinealectomized quail in the patterns of gonadal regression and recrudescence that were induced by changes of the photoperiods (Siopes and Wilson, 1974). Simpson et al. (1983) failed to show any effect of pinealectomy on the quantitative relationship between luteinizing hormone (LH) secretion and the photoperiod, although the pinealectomy reduced the nighttime plasma concentration of melatonin to a level of 46% of the intact one (Underwood et al., 1984). These results suggest that melatonin released from non-pineal tissues including the retina, should affect the gonadal response in the pinealectomized quail. In fact, a low-dose of melatonin implant (1 μg) had an anti-gonadal effect in the quail (Homma et al., 1967). Upon pinealectomy combined with enucleation of the quail, the nighttime plasma melatonin remarkably decreased, but was kept at 13% level of intact animals (Underwood et al., 1984), and the melatonin content in the diencephalon was kept at one-third level of the intact quail (Ubuka et al., 2005). The remaining melatonin is considered to originate from non-pineal and non-retinal tissue(s), which is yet to be determined.

Juss et al. (1993) extended the duration of the high nocturnal melatonin level in the quail plasma by injecting melatonin in order to determine if the duration correlates with the gonadal response. The injection at neither dawn nor dusk caused a short-day effect (i.e., suppression of the gonadal development) in the quail entrained to LD 12:12 cycles, which stimulates the gonadal development. Rather, it is interesting to note that the melatonin injection had a small but significant stimulatory effect on the gonadal growth. Consistently, a continuous administration of melatonin from implanted Silastic capsules also had a stimulatory effect on the gonads in one half of the birds tested (Juss et al., 1993).

A possible contribution of melatonin to the photoperiodic response was examined in nighttime interruption experiments (Meyer, 1998), in which a light pulse was given during the nighttime to suppress the endogenous melatonin level. In the absence of exogenous administration of melatonin, a light pulse given at the photosensitive phase, 14 hours after lights-on, induced a gonadal development of the quail maintained in a short-day photoperiod. A similar experiment was performed with supplementation of melatonin prior to the onset of the light pulse. Such a supplementation of melatonin was expected to prevent the photostimulatory effect of the pulse, if melatonin is a mediator translating the length of the night (the dark period) as in mammals. However, the supplementation gave no effect (Meyer, 1998). The apparent inconsistency between the results of Meyer (1998) and Juss et al. (1993) may be ascribed to the doses of melatonin for the injection. In these studies, the melatonin injection could have caused elevation of the plasma melatonin concentration to a level that exceeded a range of physiological melatonin concentration. It should be noted in particular that Meyer (1998) used a dose of melatonin two to three orders higher than that used by Juss et al. (1993). The extremely elevated melatonin level could have had unexpected effects, which did not occur when the melatonin level was kept in the physiological range (refer following paragraphs).

The other approach has been made by Ohta et al. (1989) who used antiserum against melatonin to lower the circulating melatonin level in the quails that were maintained in a non-stimulatory photoperiod (LD 8:16). Daily injections of the antiserum at lights-off, which may delay the elevation of circulating melatonin levels, caused significant gonadal growth, while the daily injections four hours before lights-on to advance the timing of the decline of the nighttime melatonin level had no effect on the gonadal development. These results suggest that the timing of the rise in the circulating melatonin level as well as the duration, is of critical importance for the photoperiodic gonadal development. Here it should be emphasized that melatonin-binding capacity of melatonin receptors in the avian brain shows a circadian change (Yuan and Pang, 1990; Brooks and Cassone, 1992). Therefore, the entire duration of the elevated level of melatonin should be detected precisely by the receptors only when the elevation occurs at the appropriate time of the day. Photoperiodic history and the gonadal status of the experimental animals at the beginning of experiments are also important, because the combination of the photoperiod and gonadal circulating steroids influences the expression level of melatonin receptors in some brain areas of the quail (Canonaco et al., 1994). Melatonin receptors in the brain of some species of rodents are regulated directly by melatonin (Gauer et al., 1994; Schuster et al., 2001). The density of melatonin receptors in the rat pars tuberalis, a putative central site for photoperiodicity in mammals, was inversely related to the circulating melatonin concentration, while the receptor density in the SCN was independent of melatonin (Gauer et al., 1994). In Syrian and Siberian hamsters, melatonin receptor densities in both pars tuberalis and SCN were regulated by melatonin, although other factors, such as the circadian clock and LD cycles, also contribute to the regulation (Schuster et al., 2001). Importantly, Yuan and Pang (1992) reported that the melatonin-binding capacity in the chicken brain is regulated by endogenous melatonin and influenced by exogenously administered melatonin. Thus, it is possible that the melatonin-dependent regulation of melatonin receptors is involved in the photoperiodic gonadal response in birds.

Taking the previous data into account, a possible mechanism for the photoperiodic gonadal development in birds can be hypothesized (Fig. 2). In this model, the gonadal development is switched on if the duration of melatonin receptor activation is shorter than a certain length (T; maximum nighttime length that does not inhibit gonadal development, in Fig. 2). The binding capacity of the melatonin receptors (Mel-R) is assumed to decrease in response to a high melatonin level during the dark period or upon administration of melatonin. Under the normal condition (Fig. 2A), the duration of nocturnal melatonin secretion (X_A) is the same as the dark period (Z), and, therefore, the duration of elevated melatonin level that is detected by melatonin receptors (Y_A) is also the same as the dark period. In this condition, the gonadal development is stimulated when the dark period becomes shorter than T. When a lower dose of melatonin is exogenously administered, the nocturnal melatonin concentration is kept at a physiological level for a longer period (X_B) than T (Fig. 2B). In this case, the duration of the reduced level of the binding capacity of melatonin receptors (Y_B) might be extended upon melatonin administration, but the entire duration of high melatonin level is detected by the melatonin receptors to block the effect of the long-day photoperiod. However, when an extremely high dose of melatonin is administered and circulating melatonin level is elevated to a non-physiological level, the binding capacity of melatonin receptors would be decreased to a level extremely lower than the physiological one (Fig. 2C). Even though the circulating melatonin level is very high, it should be undetectable until circulating melatonin returns to the physiological level so as to allow the receptor level to recover from the strong suppression. As a result, the duration for which the melatonin receptors can detect high melatonin level (Y_C) may become shorter than T. In this case, the melatonin administration does not inhibit the effect of the long-day photoperiod, or

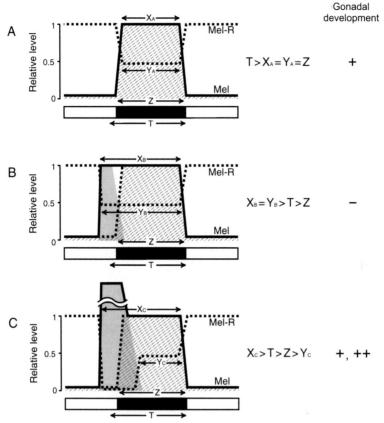

Gonadal
development

$T > X_A = Y_A = Z$ +

$X_B = Y_B > T > Z$ −

$X_C > T > Z > Y_C$ +, ++

Fig. 2. Possible relationships between circulating melatonin level (Mel) and the binding capacity of melatonin receptors (Mel-R) in various daily conditions. (A) Normal long-day photoperiod condition. (B) A condition when melatonin was exogenously administered (shaded area) in the afternoon to extend the peak duration at a physiological dose. (C) A condition when melatonin was administered in the afternoon to raise the circulating melatonin to an extremely high level. Solid line, circulating melatonin; hatched area, endogenous melatonin (estimated); broken line, the binding capacity of melatonin receptors; X, a duration of time when the circulating melatonin level is high; Y, a duration of time when the melatonin level is recognized as high by melatonin receptors; Z, a duration of dark period; T, a maximum dark period that does not inhibit gonadal development.

rather induces even larger gonadal development. For future evaluation of this model, it is important to study not only the localization of the melatonin target sites for the photoperiodic gonadal response, but also the regulatory mechanism of melatonin receptor expression. A study using the ovine pars tuberalis cells suggested that cyclic adenosine monophosphate (cAMP)-dependent signaling pathway partially regulated the expression of melatonin receptors (Barrett et al., 1996), but other factors involved in the regulatory mechanism remain uncharacterized (Witt-Enderby et al., 2003).

Target Site of Melatonin in Photoperiodic Gonadal Response

In vitro autoradiography (Siuciak et al., 1991; Canonaco et al., 1994) and *in situ* hybridization analyses (Reppert et al., 1995) revealed a wide range of distribution of melatonin receptors in the avian brain. Until recently, target sites of melatonin that induce photoperiodic gonadal response remained elusive. Ubuka et al. (2005) reported co-expression of messenger ribonucleic acid (mRNA) for a melatonin receptor subtype, Mel_{1c}, with gonadotropin-inhibitory hormone (GnIH) in the paraventricular nucleus (PVN) of the quail diencephalon. Furthermore, the GnIH expression in the PVN decreased in pinealectomized-enucleated birds, and the decrease was reversed by melatonin injection (Ubuka et al., 2005), suggesting a possible signal cascade from melatonin to GnIH secretion via Mel_{1c} receptor. Consistently, GnIH concentration in the diencephalon was low in a long-day photoperiod and high in a short-day photoperiod, correlating negatively to the melatonin level. These results strongly suggest that melatonin carries photoperiodic information into the PVN where it is translated to the secretion of GnIH. Secreted GnIH, in turn, affects anterior pituitary hormones, which induce seasonal variations in reproductive activity.

Melatonin binding sites are also found in reproductive organs of the quail (Pang et al., 1993), and it was reported that melatonin directly acts on the ovarian granulosa cells of the hen to lower the responsiveness to LH (Murayama et al., 1997). Therefore, it is possible that the reproductive organs respond to circulating melatonin directly to modify the photoperiodic gonadal response.

Deep Brain Photoreceptor Cell and Cerebrospinal Fluid

The cerebrospinal fluid (CSF) of animals circulates with relatively high velocity in a particular direction, and in the brain ventricular system of the goat, for example, it would be renewed every 30 minutes (Kanematsu et al., 1989). It has been shown that the CSF contains a variety of hormones, e.g., gonadotropin-releasing hormone (GnRH), vasopressin, growth hormone and melatonin (Nicholson, 1999; Lehman and Silver, 2000). Melatonin was found in the CSF of several mammalian species, and its concentration was higher than the circulating plasma levels (rhesus monkey, Reppert et al., 1979; cat, Reppert et al., 1982; goat, Kanematsu et al., 1989; sheep, Shaw et al., 1989). Notably, the melatonin concentrations within the CSF of the sheep were one order higher in the lateral ventricle than in the cisterna magna (Shaw et al., 1989). In the ewe, the melatonin concentration in the third ventricle was seven times higher than in the lateral ventricle (Skinner and Malpaux, 1999). These regional differences in melatonin concentration suggest that the CSF is an important vehicle of melatonin for transporting from its origin(s) such as the pineal gland to the putative target sites.

Photoreceptors in the deep brain regions are thought to play an important role for photoperiodic gonadal response (Oliver and Bayle, 1982). Foster et al. (1985) showed involvement of rhodopsin-like photopigment in the response.

Immunohistochemical analyses using antibodies against visual and non-visual opsins have been performed to localize deep brain photoreceptor cells, and the immunopositive cells were detected in the brains of various vertebrate species (Wada et al., 1998; Yoshikawa et al., 1998; Foster et al., 1994; Yoshikawa and Oishi, 1998). Although the immunopositive cells were localized in different nuclei of the brain depending on species, most of the cells had common morphological characteristics of CSF-contacting neurons (Vigh and Vigh-Teichmann, 1998). These cells were found in regions surrounding the ventricles and each cell extended a process into the ventricle where the knob-like terminal of the process made a direct contact with the CSF (Fig. 3C). This unique morphology suggests their functional importance as the deep brain photoreceptor cells, possibly mediating a signal transmission between the knob-like terminals and the CSF.

The putative photoreceptive CSF-contacting neurons in the deep brain have morphological features similar to those of the retinal and pineal

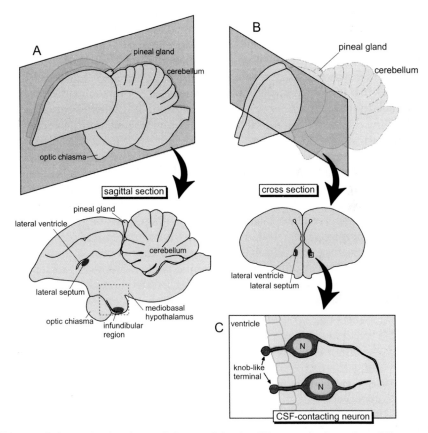

Fig. 3. Schematic drawings of the quail brain. (A) a sagittal section, (B) a cross section, and (C) CSF-contacting neurons.

photoreceptors (Vigh and Vigh-Teichmann, 1998). The outer segments of the retinal and pineal photoreceptors are rich in photopigments, and the knob-like terminal of the CSF-contacting neuron (Fig. 3C) probably corresponds to the outer segment (Vigh and Vigh-Teichmann, 1998). Considering the ability of both the retina and pineal gland to secrete melatonin in a light-sensitive manner (Underwood et al., 2001), deep brain photoreceptor cells may also secrete melatonin into the CSF depending on LD cycles. As the deep brain photoreception is important for the photoperiodic gonadal response in birds as described above, it could be hypothesized that melatonin secreted from the deep brain photoreceptor cells might transmit the photoperiodic information to the target sites via the CSF.

To examine this possibility, *in vivo* microdialysis technique has a great advantage, because it enables detection of local changes in concentration of dialyzable compounds such as melatonin in the CSF and collection of the samples continuously from each living animal (Ebihara et al., 1997). Previously, the expression of rhodopsin mRNA in the pigeon lateral septum was detected, and rhodopsin-like immunoreactivities were observed in the CSF-contacting neurons in the same region (Fig. 3; Wada et al., 1998). For these reasons, a change of the melatonin concentration was measured in the CSF near the lateral septum of the pigeon (in parallel with the quail). The dialysis probe was placed in the lateral ventricle of the pigeon reared in LD cycles (LD 18:6 long-day, and LD 6:18 short-day). A robust circadian rhythm of the melatonin level was observed in continuously perfused samples, and the rhythmicities were maintained in the following constant darkness (DD) after the long-day (Fig. 4A) and the short-day conditions (Fig. 4B). Since similar localization of rhodopsin-like immunoreactivities was reported in the quail lateral septum (Fig. 3; Silver et al., 1988), the same technique was applied to the quail lateral ventricle. The quail also showed circadian rhythms of melatonin in LD cycles of the long-day (Fig. 4C) and short-day conditions (Fig. 4D), but the variation became vague in the subsequent DD after the short-day photoperiod (Fig.4D), when compared to that observed in DD after the long-day photoperiod (Fig. 4C). The peak duration, a period for which the melatonin level exceeded the half level of the peak value, was shorter in the long-day photoperiod (Fig. 4A, C) than that in the short-day photoperiod (Fig. 4B, D). Neither the peak level nor the total melatonin secreted within a day was significantly different between the long-day and short-day conditions in both species. The fluctuation of the melatonin concentration in the CSF was also measured in the enucleated and enucleated-pinealectomized

Fig. 4. Daily fluctuations of the melatonin levels in the microdialyzed CSF collected from the lateral ventricles of the pigeon and the quail. Pigeons (*Columba livia*) captured in Tokyo (Japan) or four-week-old male quail (*Coturnix coturnix japonica*) obtained from a local breeder were kept in the long-day (LD 18:6) or short-day (LD 6:18) photoperiod for at least two weeks (for pigeons) or four weeks (for quail) before

Fig. 4 Contd. ...

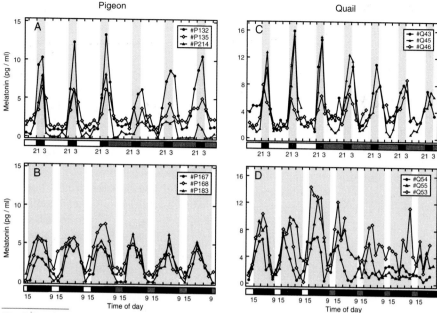

Fig. 4 Contd. ...

the experiments were started. The microdialysis probe containing acrylonitrile-sodium methallyl sulfonate copolymer (AN69, Hospal) was placed in the left lateral ventricle so as to face toward the knob-like terminals of rhodopsin-immunoreactive CSF-contacting neurons. Ringer solution (147 mM NaCl, 2.3 mM $CaCl_2$, 4 mM KCl, pH 6.5) was continuously perfused by an infusion pump through the probe at a rate of 1.0 µl/min. The microdialyzed samples were collected every 3 h, and melatonin concentration of the samples were determined by RIA using anti-melatonin serum (G280, Bühlmann Laboratories AG) and 2-[^{125}I]iodomelatonin (2,200 Ci/mmol, New England Nuclear) as a tracer. Standard curves for known amount of melatonin (0.125-250 pg/ml) were drawn by using serial two-fold dilutions of authentic melatonin dissolved in PBS containing 1% bovine serum albumin (BSA-PBS). The listed components were added to the tube in the following order; 100 µl of sample, 100 µl of BSA-PBS, 100 µl of the anti-melatonin serum (1:50,000 diluted in PBS containing 50 mM EDTA and 1% normal goat serum) and 100 µl of 2-[^{125}I]iodomelatonin (approximately 10,000 cpm in BSA-PBS). After incubation for 20 h at 4°C, rabbit antiserum against goat IgG (GI-010, Shibayagi, 1:40 dilution) in EDTA-PBS (200 µl) was added to the mixture, and incubated for 20 h at 4°C. Then, 500 µl of ice-cold saturated ammonium sulfate solution (pH 7.4) was added to the mixture in order to precipitate the antibodies with bound melatonin. After incubation for 45 min on ice, the tubes were centrifuged (2,000 g, 30 min, 4°C) and the supernatant was removed by aspiration. Radioactivity of the resultant pellet was then determined with a gamma counter, and the bound melatonin was quantified. The solid and open bars at the bottom represent the light (day) and dark (night) periods, respectively. The birds were released into constant darkness after the LD cycles, and the gray bars at the bottom represent the periods that correspond to the daytime in the preceding LD cycles. Three typical profiles from three individual birds were shown for each condition. Panels A and B represent data from the pigeon, and panels C and D from the quail. The LD cycles are LD 18:6 in A and C, and LD 6:18 in B and D.

quail to examine any contribution of the eyes and the pineal gland to the CSF melatonin level. Enucleated (EX) quail showed clear CSF melatonin rhythms (Fig. 5A, B) which were not significantly different from control birds, suggesting that neither ocular melatonin nor the light signals received in the eyes play a major role for regulation of the CSF melatonin level in the lateral ventricle. On the other hand, pinealectomy in addition to the enucleation (EX + PX) remarkably reduced the melatonin level (Fig. 5C, D), indicating a significant contribution of the pineal gland to the CSF melatonin rhythm. Importantly, a small amount of melatonin was still detected in the CSF of the pinealectomized-enucleated quail (Fig. 5C, D). This result may support the hypothesis that deep brain photoreceptor cells secrete melatonin into the CSF.

Another possible role may be considered for the knob-like terminal of the deep brain photoreceptor cells, i.e., the terminal could receive a signal from the CSF, instead of sending a signal to the CSF. This idea is supported by Saldanha et al. (1994) and Kiyoshi et al. (1998), who showed immunohistochemical data, indicating a direct contact between VIP and GnRH neurons in the lateral septum where putative deep brain photoreceptor

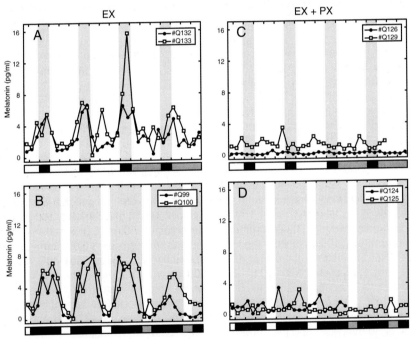

Fig. 5. Daily fluctuations of the melatonin levels in the microdialyzed CSF from the lateral ventricles of the enucleated and nucleated-pinealectomized quail. Enucleation and pinealectomy were performed just before the implantation surgery of a microdialysis probe. Two typical profiles from two individual birds were shown for each condition. Panels A and B represent data from the enucleated (EX) quail, and panels C and D from the enucleated-pinealectomized (EX + PX) quail. The LD cycles are LD 18:6 in A and C, and LD 6:18 in B and D.

cells were localized. Furthermore, a subset of the VIP neurons in the area showed rhodopsin-like immunoreactivities (Silver et al., 1988). Taken together, it is possible that the deep brain photoreceptor cells in the lateral septum may receive not only light, but also chemical signals such as melatonin from the CSF, and transmit the signal to GnRH neurons where light information could be converted to the endocrine signal. There are no experimental data identifying the chemical signal that the deep brain photoreceptor cells receive from the CSF, and therefore, such a signal transmission from the CSF to the photoreceptor cells is still open to question.

Storage of Photoperiodic Information in Circadian System

There are several lines of evidence that photoperiodic information is stored in circadian systems. The pituitary gland of the quail retains for a few days a pattern of gonadotropin secretion, characteristic of being exposed to a stimulatory photoperiod even after being transferred to a non-stimulatory photoperiod (Follett et al., 1991). The house sparrow entrained to a long-day photoperiod showed plasma melatonin rhythm with a higher amplitude and a shorter peak duration when compared to those of birds entrained to a short-day photoperiod, and these differences in pattern of melatonin rhythms were detectable after two days in DD (Brandstatter et al., 2000).

To examine whether the photoperiodic information is stored in the circadian oscillator which controls the CSF melatonin level, the melatonin rhythms in the CSF of the pigeon's lateral ventricle were compared between the birds entrained to long-day and short-day photoperiods (Fig. 6). Daily fluctuation in the melatonin level was recorded for four consecutive days, i.e., the last day in LD cycles (LD-last), the transition day from LD to DD (LD-DD), the first day in DD (DD-1) and the second day in DD (DD-2). Fig. 6 shows the four data sets superimposed on the single 24-hour abscissa. First, in the both photoperiods, the peak values in melatonin concentration were significantly higher in the first two days (LD-last and LD-DD) than in the last two days (DD-1 and DD-2), indicating the strong effect of the light on the peak value of melatonin. Second, when the birds were entrained to the short-day photoperiod, no difference was found in the peak duration of the melatonin rhythm among the four-day profiles (Fig. 6A). When the birds were reared in the long-day photoperiod, the peak duration was short as the melatonin level decreased significantly just after the lights-on (asterisk in Fig. 6B, $p < 0.001$, vs. the previous time point in students' t-test) probably due to the acute suppression of melatonin synthesis by light. On the other hand, such a decrease was not observed in later days in constant darkness, hence the peak durations became longer in those days. However, the peak duration in DD-1 (13.2 ± 0.5 h) and DD-2 (12.3 ± 0.3 h) after the short-day photoperiod were still longer than those after the long-day photoperiod (8.6 ± 1.1 h in DD-1 and 10.2 ± 0.2 h in DD-2). This observation suggests that the photoperiodic history is reflected in the CSF melatonin release at least for a few days when being placed in DD condition.

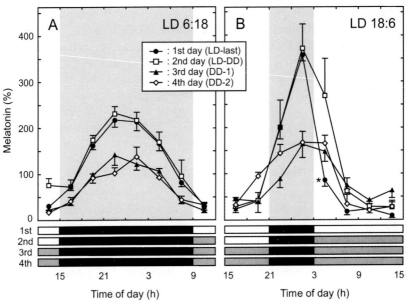

Fig. 6. Superimposed patterns of melatonin levels during a transition from LD-entrained to DD condition in the pigeon. Daily melatonin rhythms recorded for four consecutive days were superimposed on each other. The four days are: the last day in LD (LD-last; closed circle), the transition day from LD to DD (LD-DD; open square), the first day in DD (DD-1; closed triangle) and the second day in DD (DD-2; open diamond). The melatonin levels were plotted as relative values (%) after they were normalized against the average value of the fluctuated melatonin concentrations measured during the six days (the last two days in LD and the subsequent four days in DD) for each bird. Mean values +/-Standard Error of Mean (SEM) of eight (A) or four (B) birds were plotted. Lighting conditions are shown on the bottom.

Photoperiodic Time Measurement and Mediobasal Hypothalamus

The mediobasal hypothalamus (MBH; Fig. 3A) is an important center controlling photoperiodic gonadal response, because a partial lesion of the MBH blocks the photoperiodic response of the quail (Sharp and Follett, 1969; Ohta and Homma, 1987), even though GnRH neurons have been left intact (Juss, 1993). Yoshimura et al. (2003) found that the expression of a gene for type 2-iodothyronine deiodinase (*Dio2*) is induced in the quail MBH by a light pulse given at the photo-inducible phase. This result agreed well with the previous studies that showed the important role of thyroid hormone in the photoperiodism (Dawson et al., 2001), because Dio2 catalyzes the intracellular deiodination of thyroxine (T4) prohormone to convert it to the active 3,5,3'-triiodothyronine (T3). Thyroid hormones are known to play critical roles in the development, plasticity and function of the central nervous system (Bernal, 2002). Consistently, GnRH nerve terminals in the quail median eminence showed seasonal morphological changes, which suggest that photoperiodic GnRH release is controlled at the GnRH terminals by neuro-glial plasticity (Yamamura et al., 2004). Taken together, it is suggested that Dio2 and thyroid hormone in the MBH control GnRH release, and that this mechanism may underlie the photoperiodic gonadal response in the quail.

CONCLUSION

The role of melatonin in photoperiodic gonadal response in birds has been an interesting issue but is still open to debate. The key function left unrevealed in the photoperiodic gonadal response is how light information is converted to hypothalamic endocrine signals leading to changes of gonadotropin secretion. As described in the present article, it is suggested that the photoperiodic gonadal response of birds may contain two regulatory systems:

 (i) GnIH system in the PVN, and (ii) GnRH system in the MBH.
 (ii) The deep brain photoreceptor cells may receive light, and secrete melatonin into the CSF in a manner reflecting the photoperiods. The locally released melatonin would reach melatonin-receptor–expressing cells in the PVN via the CSF and regulate the expression of GnIH that controls gonadotropin secretion.
 (iii) Light received by the deep brain photoreceptor cells may change *Dio2* expression through which GnRH secretion in the MBH could be controlled. Further studies would be required for better understanding of how these systems interact and balance with each other to regulate the photoperiodic gonadal response *in vivo*.

ACKNOWLEDGMENTS

The authors would thank to Drs. A. Adachi and S. Ebihara (Nagoya Univ.) for their technical help and suggestions. Ms. L.S. Harris provided valuable comments in the early stage of the manuscript. This work was supported in part by a Grant-in-Aid for Scientific Research from the Japanese Ministry of Education, Culture, Sports, Science and Technology (14104003). Tomoko Yoshikawa was supported by Fellowship from the Japan Society for the Promotion of Science for Young Scientists.

REFERENCES

Barrett, P., MacLean, A., Davidson, G. and Morgan, P.J. (1996). Regulation of the Mel1a melatonin receptor mRNA and protein levels in the ovine pars tuberalis: evidence for a cyclic adenosine 3',5'-monophosphate-independent Mel1a receptor coupling and an autoregulatory mechanism of expression. *Mol. Endocrinol.* **10**: 892-902.

Brooks, D.S. and Cassone, V.M. (1992). Daily and circadian regulation of 2-[^{125}I]iodomelatonin binding in the chick brain. *Endocrinol.* **131**: 1297-1304.

Bernal, J. (2002). Action of thyroid hormone in brain. *J. Endocrinol. Invest.* **25**: 268-288.

Brandstatter, R., Kumar, V., Abraham, U. and Gwinner, E. (2000). Photoperiodic information acquired and stored in vivo is retained in vitro by a circadian oscillator, the avian pineal gland. *Proc. Natl. Acad. Sci. USA* **97**: 12324-12328.

Canonaco, M., Tavolaro, R., Facciolo, R.M., Artero, C. and Franzoni, M.F. (1994). Combined gonadal and photic influences on 2-[^{125}I] iodomelatonin binding level changes in some brain areas of the quail. *J. Exp. Zool.* **269**: 383-388.

Dawson, A., King, V.M., Bentley, G.E. and Ball, G.F. (2001). Photoperiodic control of seasonality in birds. *J. Biol. Rhythms* **16**: 365-380.

Ebihara, S., Adachi, A., Hasegawa, M., Nogi, T., Yoshimura, T. and Hirunagi, K. (1997). In vivo microdialysis studies of pineal and ocular melatonin rhythms in birds. *Biol. Signals* **6**: 233-240.

Follett, B.K. and Sharp, P.J. (1969). Circadian rhythmicity in photoperiodically induced gonadotrophin release and gonadal growth in the quail. *Nature* **223**: 968-971.

Follett, B.K., Robinson, J.E., Simpson, S.M. and Harlow, C.R. (1991). In: *Biological Clocks in Seasonal Reproductive Cycles* (Follett, B.K. and Follett, D.E., eds.), Scientechnica, 185-201.

Foster, R.G., Follett, B.K. and Lythgoe, J.N. (1985). Rhodopsin-like sensitivity of extra-retinal photoreceptors mediating the photoperiodic response in quail. *Nature* **313**: 50-52.

Foster, R.G., Grace, M.S., Provencio, I., DeGrip, W.J. and Garcia-Fernandez, J.M. (1994). Identification of vertebrate deep brain photoreceptors. *Neurosci. Biobehav. Rev.* **18**: 541-546.

Gauer, F., Masson-Pevet, M., Stehle, J. and Pevet, P. (1994). Daily variations in melatonin receptor density of rat pars tuberalis and suprachiasmatic nuclei are distinctly regulated. *Brain Res.* **641**: 92-98.

Goldman, B.D. (2001). Mammalian photoperiodic system: formal properties and neuroendocrine mechanisms of photoperiodic time measurement. *J. Biol. Rhythms* **16**: 283-301.

Homma, K., MacFarlanf, L.Z. and Wilson, W.O. (1967). Response of the reproductive organs of the Japanese quail to pinealectomy and melatonin injections. *Poult. Sci.* **46**: 314-319.

Juss, T.S. (1993). Neuroendocrine and neural changes associated with the photoperiodic control of reproduction. In: *Avian Endocrinology* (Sharp, P.J., ed.). Society for Endocrinology, *Bristol.* 47-60.

Juss, T.S., Meddle, S.L., Servant, R.S. and King, V.M. (1993). Melatonin and photoperiodic time measurement in Japanese quail (*Coturnix coturnix japonica*). *Proc. R. Soc. Lond. B.* **254**: 21-28.

Kanematsu, N., Mori, Y., Hayashi, S. and Hoshino, K. (1989). Presence of a distinct 24-hour melatonin rhythm in the ventricular cerebrospinal fluid of the goat. *J. Pineal Res.* **7**: 143-152.

Kiyoshi, K., Kondoh, M., Hirunagi, K. and Korf, H. (1998). Confocal laser scanning and electron-microscopic analyses of the relationship between VIP-like and GnRH-like-immunoreactive neurons in the lateral septal-preoptic area of the pigeon. *Cell. Tissue Res.* **293**: 39-46.

Klein, D.C., Coon, S.L., Roseboom, P.H., Weller, J.L., Bernard, M., Gastel, J.A., Zatz, M., Iuvone, P.M., Rodriguez, I.R., Begay, V., Falcon, J., Cahill, G.M., Cassone, V.M. and Baler, R. (1997). The melatonin rhythm-generating enzyme: molecular regulation of serotonin N-acetyltransferase in the pineal gland. *Rec. Prog. Horm. Res.* **52**: 307-357.

Lehman, M. and Silver, R. (2000). CSF signaling in physiology and behavior. *Prog. Brain Res.* **125**: 415-433.

Meyer, W. (1998). Melatonin supplementation does not prevent photostimulatory effects of night interruption lighting in Japanese quail. *J. Pineal Res.* **24**: 102-107.

Mayer, I., Bornestaf, C. and Borg, B. (1997). Melatonin does not prevent long photoperiod stimulation of secondary sexual characters in the male three-spined stickleback *Gasterosteus aculeatus*. *Comp. Biochem. Physiol.* **118A**: 515-531.

Malpaux, B., Migaud, M., Tricoire, H. and Chemineau, P. (2001). Biology of mammalian photoperiodism and the critical role of the pineal gland and melatonin. *J. Biol. Rhythms* **16**: 336-347.

Murayama, T., Kawashima, M., Takahashi, T., Yasuoka, T., Kuwayama, T. and Tanaka, K. (1997). Direct action of melatonin on hen ovarian granulosa cells to lower responsiveness to luteinizing hormone. *Proc. Soc. Exp. Biol. Med.* **215**: 386-392.

Nicholson, C. (1999). Signals that go with the flow. *Trends Neurosci.* **22**: 143-145.

Ohta, M. and Homma, K. (1987). Detection of neural connection to the infundibular complex by partial or complete hypothalamic deafferentation in male quail. *Gen. Comp. Endocrinol.* **68**: 286-292.

Ohta, M., Kadota, C. and Konishi, H. (1989). A role of melatonin in the initial stage of photoperiodism in the Japanese quail. *Biol. Reprod.* **40**: 935-941.

Oliver, J. and Bayle, J.D. (1982). Brain photoreceptors for the photo-induced testicular response in birds. *Experientia* **38**: 1021-1029.

Pang, S.F., Cheng, K.M., Pang, C.S., Wang, Z.P., Yuan, H. and Brown, G.M. (1993). Differential effects of short photoperiod on 2-[^{125}I]iodomelatonin binding in the testis and brain of quail. *Biol. Signals* **2**: 146-154.

Pittendrigh, C.S. (1981). Circadian organization and the photoperiodic phenomena. In: *Biological Clocks in Seasonal Reproductive Cycles* (Follett, B.K. and Follett, D.E., eds.). Wright, Bristol. 1-35.

Reppert, S.M., Perlow, M.J., Tamarkin, L. and Klein, D.C. (1979). A diurnal melatonin rhythm in primate cerebrospinal fluid. *Endocrinol.* **104**: 295-301.

Reppert, S.M., Coleman, R.J., Heath, H.W. and Keutmann, H.T. (1982). Circadian properties of vasopressin and melatonin rhythms in cat cerebrospinal fluid. *Am. J. Physiol.* **243**: 489-498.

Reppert, S.M., Weaver, D.R., Cassone, V.M., Godson, C. and Kolakowski, L.F. Jr. (1995) Melatonin receptors are for the birds: molecular analysis of two receptor subtypes differentially expressed in chick brain. *Neuron.* **15**: 1003-1015.

Saldanha, C.J., Leak, R.K. and Silver, R. (1994). Detection and transduction of daylength in birds. *Pcychoneuroendocrinology* **19**: 641-656.

Schuster, C., Gauer, F., Malan, A., Recio, J., Pevet, P. and Masson-Pevet, M. (2001). The circadian clock, light/dark cycle and melatonin are differentially involved in the expression of daily and photoperiodic variations in mt(1) melatonin receptors in the Siberian and Syrian hamsters. *Neuroendocrinol.* **74**: 55-68.

Sharp, P. and Follett, B.K. (1969). The effect of hypothalamic lesions on gonadotrophin release in Japanese quail (*Coturnix coturnix japonica*). *Neuroendocrinol.* **5**: 205-218.

Shaw, P.F., Kennaway, D.J. and Seamark, R.F. (1989). Evidence of high concentrations of melatonin in lateral ventricular cerebrospinal fluid of sheep. *J. Pineal Res.* **6**: 201-208.

Silver, R., Witkovsky, P., Horvath, P., Alones, V., Barnstable, C.J. and Lehman, M.N. (1988). Coexpression of opsin- and VIP-like-immunoreactivity in CSF-contacting neurons of the avian brain. *Cell. Tissue Res.* **253**: 189-198.

Simpson, S.M., Urbanski, H.F. and Robinson, J.E. (1983). The pineal gland and the photoperiodic control of luteinizing hormone secretion in intact and castrated Japanese quail. *J. Endocrinol.* **99**: 281-287.

Siopes, T.D. and Wilson, W.O. (1974). Extraocular modification of photoreception in intact and pinealectomized coturnix. *Poult. Sci.* **53**: 2035-2041.

Siuciak, J.A., Krause, D.N. and Dubocovich, M.L. (1991). Quantitative pharmacological analysis of 2-125I-iodomelatonin binding sites in discrete areas of the chicken brain. *J. Neurosci.* **11**: 2855-2864.

Skinner, D.C. and Malpaux, B. (1999). High melatonin concentrations in third ventricular cerebrospinal fluid are not due to Galen vein blood recirculating through the choroid plexus. *Endocrinol.* **140**: 4399-4405.

Ubuka, T., Bentley, G.E., Ukena, K., Wingfield, J.C. and Tsutsui, K. (2005). Melatonin induces the expression of gonadotropin-inhibitory hormone in the avian brain. *Proc. Natl. Acad. Sci. USA* **102**: 3052-3057.

Underwood, H., Binkley, S., Siopes, T. and Mosher, K. (1984). Melatonin rhythms in the eyes, pineal bodies, and blood of Japanese quail (*Coturnix coturnix japonica*). *Gen. Comp. Endocrinol.* **56**: 70-81.

Underwood, H., Steele, C.T. and Zivkovic, B. (2001). Circadian organization and the role of the pineal in birds. *Microsci. Res. Tech.* **53**: 48-62.

Vigh, B. and Vigh-Teichmann, I. (1998). Actual problems of the cerebrospinal fluid-contacting neurons. *Microsci. Res. Tech.* **41**: 57-83.

Wada, Y., Okano, T., Adachi, A., Ebihara, S. and Fukada, Y. (1998). Identification of rhodopsin in the pigeon deep brain. *FEBS Lett.* **424**: 53-56.

Witt-Enderby, P.A., Bennett, J., Jarzynka, M.J., Firestine, S. and Melan, M.A. (2003). Melatonin receptors and their regulation: biochemical and structural mechanisms. *Life Sci.* **72**: 2183-2198.

Yamamura, T., Hirunagi, K., Ebihara, S. and Yoshimura, T. (2004). Seasonal morphological changes in the neuro-glial interaction between gonadotropin-releasing hormone nerve terminals and glial endfeet in Japanese quail. *Endocrinol.* **145**: 4264-4267.

Yoshikawa, T. and Oishi, T. (1998). Extraretinal photoreception and circadian systems in nonmammalian vertebrates. *Comp. Biochem. Physiol.* **119B**: 65-72.

Yoshikawa, T., Okano, T., Oishi, T. and Fukada, Y. (1998). A deep brain photoreceptive molecule in the toad hypothalamus. *FEBS Lett.* **424**: 69-72.

Yoshimura, T., Yasuo, S., Watanabe, M., Iigo, M., Yamamura, T., Hirunagi, K. and Ebihara, S. (2003). Light-induced hormone conversion of T4 to T3 regulates photoperiodic response of gonads in birds. *Nature* **426**: 178-181.

Yuan, H. and Pang, S.F. (1990). [125I]melatonin binding sites in membrane preparations of quail brain: characteristics and diurnal variations. *Acta Endocrinol. (Copenh).* **122**: 633-639.

Yuan, H. and Pang, S.F. (1992). [125I]iodomelatonin binding sites in the chicken brain: diurnal variation and effect of melatonin injection or pinealectomy. *Biol. Signals* **1**: 208-218.

Zatz, M. (1996). Melatonin rhythms: trekking toward the heart of darkness in the chick pineal. *Semin. Cell Dev. Biol.* **7**: 811-120.

The Role of Prenatal Androgen Excess on the Development of Polycystic Ovary Syndrome

Agathocles Tsatsoulis* and Nectaria Xita

Abstract

Polycystic ovary syndrome (PCOS) is a common endocrine disorder in women of reproductive age characterized by anovulation, hyperandrogenism and metabolic aberrations manifested by central adiposity and insulin resistance.

Although PCOS usually becomes clinically manifest during adolescence, the natural history of PCOS may originate in very early development, even in intrauterine life. According to the developmental origin hypothesis of PCOS, foetal exposure to androgen excess simultaneously "programmes" multiple organ-systems that will later manifest the phenotype of PCOS. This hypothesis was based on experimental animal studies and clinical observations. Animal studies have shown that experimentally induced prenatal androgen excess results in LH hypersecretion, impaired insulin secretion and action accompanied by hyperandrogenism, anovulation and polycystic appearing ovaries in adulthood. Clinical observations also support a foetal origin of PCOS. Women exposed to foetal 21-hydroxylase deficiency (despite the normalization of the adrenal androgen excess after birth) or congenital foetal virilizing tumour (despite the removal of the tumour at birth) develop features characteristic of PCOS. Although the molecular genetic basis of the origin of androgens during foetal development of PCOS is unclear, recent studies have indicated that a number of genes, involved in the production/action of androgens or genes that influence sex hormone-binding globulin (SHBG) production and aromatase enzyme activity may be associated with PCOS and increased androgen levels. These genetic variants may also contribute to excess androgens during intrauterine life and thus provide the genetic link to the developmental origin of PCOS. Other genetic and environmental factors or the interaction between the two may further influence the expression of the PCOS phenotype in adult life in genetically predisposed individuals.

Department of Endocrinology, University of Ioannina, Ioannina, 45110, Greece.
*Corresponding author: Tel.: +3026510-99625, Fax: +3026510-46617. E-mail: atsatsou@uoi.gr

INTRODUCTION

Polycystic ovary syndrome (PCOS) is a common endocrine and metabolic disorder affecting 6-10% of women in their reproductive life (Knochenhauer et al., 1998; Diamanti-Kandarakis et al., 1999; Asuncion et al., 2000; Azziz et al., 2004). The endocrine abnormalities of PCOS include excess androgen production of ovarian and/or adrenal origin (hyperandrogenism) and arrested follicular development leading to chronic oligo- or anovulation (Franks, 1995). The metabolic aberration is characterized by insulin resistance and hyperinsulinaemia, which further exacerbate the hyperandrogenism and ovulatory dysfunction, and in association with other features of the Metabolic Syndrome, impose an increased risk for type 2 diabetes and cardiovascular disease (Dunaif et al., 1989; Wild, 2002). A common feature of PCOS, resulting from aberrant folliculogenesis, is the accumulation of small subcortical follicular cysts with increased ovarian stromal volume yielding a characteristic morphology on ultrasound (Franks, 1995). Often, women with PCOS have increased luteinizing hormone (LH) secretion and decreased serum sex hormone-binding globulin (SHBG) levels, which further contribute to hyperandrogenaemia and increased tissue availability of free androgens (Franks, 1995).

The spectrum of the above endocrine and metabolic abnormalities may vary among affected women, creating a heterogeneous clinical and biochemical phenotype with implications for the diagnosis of the syndrome. Recently, an International Consensus Group proposed that PCOS can be diagnosed after exclusion of other conditions that cause hyperandrogenism and if at least two of the following three criteria are present: (i) oligoovulation or anovulation (manifesting as oligomenorrhoea or amenorrhoea), (ii) elevated serum androgen levels or clinical manifestations of androgen excess (hirsutism, acne or androgenic alopecia), and (iii) polycystic ovaries, as defined by ultrasound examination (The Rotterdam ESHRE/ASRM-Sponsored PCOS Consensus Workshop Group, 2004).

The aetiology of the syndrome is largely unknown but there is increasing evidence of a strong genetic basis (Franks et al., 2002; Xita et al., 2002). PCOS usually manifests clinically during adolescence, along with maturation of the hypothalamic-pituitary-gonadal axis (Franks, 2002). However, recent evidence suggests that the natural history of PCOS may originate in very early development, even in intrauterine life (Abbott et al., 2002).

In this chapter, the evidence from clinical, experimental and genetic research supporting the view for the developmental origin hypothesis of PCOS is analyzed. This is preceded by a short review of the pathophysiology of the syndrome.

Pathophysiology of PCOS

Several pathophysiological pathways are implicated for the spectrum of abnormalities of PCOS. These include neuroendocrine defects leading to increased LH secretion, dysregulation of androgen biosynthesis resulting in enhanced ovarian and/or adrenal androgen production and defects in insulin secretion, and action leading to hyperinsulinaemia and insulin resistance (Tsilchorozidou et al., 2004).

Even though PCOS is considered to be the result of a vicious cycle where all the above mechanisms are involved, the most consistent biochemical feature is excess androgen production with the ovary being the main source of hyperandrogenaemia (Gilling-Smith et al., 1997). It has been shown that ovarian theca cells in women with PCOS are more efficient at converting androgenic precursors to testosterone than are normal theca cells. This is due to increased activity of multiple steroidogenic enzymes in PCOS theca cells, resulting in raised androgen production (Nelson et al., 1999, 2001). Intraovarian androgen excess appears to stimulate excessive growth of small follicles while impairing the selection of the dominant preovulatory follicle, leading to accumulation of many cystic follicles and also causing thecal and stromal hyperplasia (Rosenfield, 1999).

More recent molecular analysis has indicated that a number of genes are overexpressed in PCOS theca cells compared to normal theca cells. These include aldehyde dehydrogenase 6, retinol dehydrogenase 2, and the transcription factor GATA6 (Wood et al., 2004a). In addition, an alteration in Wnt signalling proteins in PCOS theca cells has been documented (Wood et al., 2004a) as well as aberrant expression of growth differentiation factor-9 in oocytes of women with PCOS (Filho et al., 2002). These findings support the notion that PCOS theca cells have an altered molecular phenotype, suggestive of a genetic alteration or an epigenetic effect (Wood et al., 2004a).

The relative proportion of LH and FSH (follicle stimulating hormone) released from the pituitary gonadotrophs is determined, in part, by the frequency of hypothalamic GnRH (gonadotrophin releasing hormone). Increased pulse frequency of GnRH favours transcription of the β-subunit of LH over the β-subunit of FSH, leading to increased LH/FSH ratio. Since women with PCOS have an increased LH pulse frequency, it has been inferred that GnRH pulse frequency must also be accelerated in this syndrome. Whether this is due to an intrinsic abnormality of the GnRH pulse generator or is caused by alteration of the steroid feedback mechanism, is not clear (Ehrmann, 2005).

Besides the "first hit" that is related to enhanced ovarian and/or adrenal androgen biosynthesis, a "second hit" that further exacerbates the hyperandrogenism comes from hyperinsulinaemia that is frequently present in women with PCOS. Insulin acts synergistically with LH to enhance androgen production by theca cells and also lowers hepatic SHBG synthesis and this may indirectly increase the proportion of biologically available free testosterone (Ehrmann, 2005). Recently, a unifying hypothesis for the developmental origin of PCOS has been proposed on the basis of clinical observations and evidence from animal models (Abbott et al., 2002).

Developmental Origin Hypothesis of PCOS

Evidence suggests that alterations in the nutritional and endocrine milieu during foetal development can result in permanent structural and functional changes, predisposing an individual to metabolic and cardiovascular disease in adult life (Barker, 1998). This "foetal origin hypothesis" of adult diseases is based on epidemiological and experimental evidence.

Epidemiological studies have shown that perturbed foetal growth, as reflected by low birth weight, is associated with an increased risk of coronary heart disease and its risk factors, including obesity, type 2 diabetes and hypertension (Barker, 1998). These epidemiological associations have been replicated by human observational studies and animal experiments (Ong and Dunger, 2002). The above associations are thought to be the consequence of "programming", whereas a stimulus or insult at a crucial, sensitive period of early life has permanent effects on structure, physiology and metabolism (Lucas, 2005).

Recent evidence suggests that foetal growth may also be a modulator of adrenarche. Thus, prospective studies in girls have shown that low birth weight is associated with premature adrenarche and subsequent ovarian hyperandrogenism, hyperinsulinism and dyslipidaemia during puberty (Ibanez et al., 1996, 1998). It has been suggested that the relationship of premature adrenarche and the subsequent risk of developing PCOS with low birth weight might invoke a common early origin (Ibanez et al., 1998)

Besides prenatal growth, increased prenatal androgen exposure is thought to play an important role in foetal programming of adult disease. The developmental origin of PCOS was recently proposed on the basis of experimental and clinical observations (Abbott et al., 2002). According to this hypothesis, PCOS may originate during intrauterine development. Foetal exposure to androgen excess simultaneously "programmes" multiple organ-systems that will later manifest the heterogeneous phenotype of PCOS. In the following section the experimental and clinical data in support of the foetal origin hypothesis of PCOS is analyzed. Evidence from genetic association studies for a possible genetic link to the developmental origin of PCOS is also provided.

Experimental Studies

Experimental evidence strongly suggests the hypothesis that the intrauterine environment may affect the phenotypic expression of PCOS. Animal studies have shown that experimentally induced androgen excess during foetal life may lead to reproductive and metabolic abnormalities in later life that resemble those found in women with PCOS.

Experiments in rats, 40 years ago, showed that the pattern of gonadotrophin-hormone release is programmed at the hypothalamus by the concentration of androgens during early development (Barraclough and Gorski, 1961). In addition, female rats exposed to androgen excess may develop anovulatory sterility and polycystic ovaries (Vom Saal and Bronson, 1980).

Studies in sheep showed that *in utero* exposure to androgens reduces the sensitivity of GnRH neural network to progesterone negative feedback, causes increased LH secretion, and affects ovarian follicular development in the female offspring (Padmanabhan et al., 1998; Robinson et al., 1999). This is postulated to be caused by androgen-induced reprogramming of neural development such that GnRH neurons are desensitized to steroid feedback. In this connection, a recent study showed that prenatal androgenization alters GABAergic drive to GnRH neurons in adults, implicating that the GABAergic system may be reprogrammed by androgens *in utero* (Sullivan and Moenter, 2004). This may explain why GnRH pulse frequency remains high in PCOS, inducing the preferential release of LH release and impairing the release of FSH, and thus follicular maturation, which is characteristic of the syndrome. In addition, the intrafollicular availability of activin is altered in prenatally androgenized lambs (West et al., 2001). The decrease in the paracrine actions of activin may also contribute to the selection defect in folliculogenesis.

Furthermore, prenatal androgen exposure of sheep not only leads to multifollicular ovarian development and early reproductive failure, but also results in intrauterine growth retardation and postnatal catch up growth (Manikkam et al., 2004). This observation may provide the link for the association of low birth weight with premature adrenarche and subsequent pubertal hyperandrogenism (Ibanez et al., 1998).

A more appropriate model for understanding the reproductive outcome of female prenatal androgen excess is the non-human primate. Studies have shown that female rhesus monkeys exposed *in utero* to levels of testosterone equivalent to those found in foetal males, develop clinical and biochemical characteristics in adult life resembling those observed in women with PCOS, i.e., they have delayed menarche, irregular anovulatory cycles and develop enlarged ovaries with multiple medium-sized antral follicles (Abbott et al., 1998). The monkeys also exhibit increased ovarian and adrenal androgen secretion with reduced steroid negative feedback on LH release leading to LH hypersecretion (Dumesic et al., 1997; Abbott et al., 1998; Eisner et al., 2002).

In addition to its impact on the reproductive axis, prenatal androgenization of female rhesus monkeys may alter adipose tissue distribution as well as insulin secretion and action. Thus, female rhesus monkeys exposed to prenatal androgen excess manifest a preferential accumulation of abdominal and visceral fat during adulthood, which is independent of obesity (Eisner et al., 2003). It is postulated that this may reflect a masculinized pattern of visceral fat accumulation, because body fat distribution is a sexually dimorphic trait.

Prenatal androgen excess may also disturb the insulin-glucose homeostasis in female rhesus monkeys with androgen excess in early gestation impairing pancreatic β-cell function, whereas in late gestation altering insulin sensitivity (Eisner et al., 2000) The same authors have recently reported that adult male monkeys exposed to excess androgens *in utero,* also develop insulin resistance and impaired insulin secretion to glucose stimulation in a manner similar to their female counterparts (Bruns et al., 2004).

The development of insulin resistance in prenatally androgenized female and male rhesus monkeys may be related to visceral adiposity (adiposity-dependent insulin resistance). On the other hand, the insulin secretory defect may result from reprogramming of the potassium-dependent ATP (adenosine triphosphate) channels on the pancreatic β-cells by the androgen excess (Bruns et al., 2004). The impaired glucose stimulated β-cell response, coupled with insulin resistance, increase the likelihood of developing type 2 diabetes in later adulthood. In a similar manner, these two abnormalities may lead to an increased risk of type 2 diabetes in women with PCOS and their male relatives (Legro et al., 1999). Since male and female rhesus monkeys exposed to the same altered intrauterine environment induced by exogenous androgen develop similar metabolic abnormalities, it is postulated that the same abnormalities seen in PCOS families may also be caused by altered intrauterine environment. The latter could be due to genetic factors, environmental exposures or the interaction of both (Bruns et al., 2004).

Clinical Observations

The hypothesis for the developmental origin of PCOS was conceived on the basis of astute clinical observations and this triggered off the experimental studies on the prenatally androgenized animal models discussed above. The clinical observations were made on women with classical 21-hydroxylase deficiency, which in adulthood develops features closely resembling those found in PCOS. Despite the normalization of androgen excess after birth, these women manifest anovulatory cycles, ovarian hypernadrogenism, LH hypersecretion, polycystic appearing ovaries and insulin resistance (Hague et al., 1990; Barnes et al., 1994). Similarly, the PCOS phenotype is prevalent in women with congenital foetal androgen-secreting tumours, despite their removal after birth (Barnes et al., 1994).

The possible role of PCOS itself as a cause for prenatal androgen excess has been evaluated in a recent study (Sir-Peterman et al., 2002). The authors reported that pregnant PCOS women have significantly higher concentrations of androgens than normal pregnant women. The origin of the androgen excess during pregnancy in PCOS women is uncertain, but it could be due to increased androgen produced by the maternal theca cells or by the placenta stimulated by human chorionic gonadotrophin (hCG). Normally, maternal androgens or androgens produced in the placenta are rapidly converted to oestrogens by the activity of the placental aromatase. However, when the enzyme activity is inhibited, these androgens could be increased. Insulin has been shown to inhibit aromatase activity in human cytotrophoblasts and stimulate 3β-HSD activity (Nestler, 1987, 1989, 1990). Therefore, in PCOS women in whom insulin levels are increased, this could partly explain the reason for the high androgen levels observed in these patients during pregnancy and the potential exposure of their female offsprings to androgens (Sir-Peterman et al., 2002).

In addition to increased androgen levels during pregnancy, the same authors reported that PCOS women might also deliver newborns of small size for their gestational age at a higher prevalence than control mothers (Sir-Peterman et al., 2002). Taking into account that foetal growth retardation may be related to prenatal androgen excess, the prenatal exposure to androgens in the offspring of PCOS mothers may provide the stimulus for both the low birth weight and their frequent development of the PCOS phenotype later in life.

Genetic Studies

Both the clinical observations and experimental research presented above, suggest a common prenatal aetiology for the multiple manifestations of PCOS that may be programmed *in utero* or early postnatal life by androgen excess.

The question that arises concerns the potential origin of the excess androgens during intrauterine life. Theoretically, this could be from maternal (ovarian or adrenal) sources, placental production or endogenous foetal (ovarian or adrenal) origin. As discussed above, it is unlikely that an excess of maternal androgens is implicated because the human foetus is normally protected from the effects of maternal androgens by a combination of increased placental aromatase activity and high levels of sex hormone binding protein (SHBG) present during normal pregnancy. However, this balancing effect may be overcome if placental aromatase activity is inhibited or the production of SHBG is reduced by genetic or other factors (for example, hyperinsulinaemia in the case of pregnant PCOS women).

On the other hand, although it is thought that the human foetal ovary is quiescent in terms of sex steroid production, it is possible that genetic factors may influence the steroidogenic activity *in-utero* in response to hCG or during infancy by the burst of hypothalamic-pituitary secretion. Indeed, the high prevalence of PCO among first degree relatives of women with this syndrome suggests a common genetic origin (Franks et al., 1997; Azziz and Kashar-Miller, 2000).

In recent years, molecular and genetic association studies have provided evidence in support of this notion. The observation that cultured human theca cells from PCOS women exhibit increased steroidogenic activity, prompted the consideration of genes encoding steroidogenic enzymes as possible candidate genes for the aetiology of PCOS. A pentanucleotide repeat polymorphism in the promoter region of the CYP11A (encoding P450 cholesterol side-chain cleavage enzyme) has been associated with hyperadrogenism in women with PCOS (Gharani et al., 1997; Diamanti-Kandarakis et al., 2000). However, subsequent studies failed to show association or linkage of CYP11A gene to PCOS phenotype (Gaasenbeek et al., 2004).

In addition, human theca cells have been shown to express androgen receptors (ARs), which may act as transcriptional factors in molecular pathways involved in cellular differentiation and function (Horie et al., 1992) A recent study showed that women expressing a polymorphic variant in the AR gene encoding for a more active receptor, display higher serum androgen levels (Westberg et al., 2001). This may reflect an enhanced androgenic influence on the androgen producing theca cells in the ovary. In the same study, a polymorphic variant of the oestrogen receptor β (ERβ) gene was also associated with high androgen and low SHBG levels (Westberg et al., 2001). Since both AR and ERβ genes are possibly involved in the prenatal sexual differentiation (Patterson et al., 1994; Couse and Korach, 1999), the possibility that a relatively pronounced androgenic effect during development may influence the differentiation of sexually dimorphic organs should be considered. Furthermore, a recent case-control study revealed that PCOS women were more frequently carriers of G allele of the ApaI variant of the IGF-2 gene (San Millan et al., 2004). G alleles of the ApaI variant in the IGF-2 may increase IGF-2 expression and IGF-2 may stimulate adrenal and ovarian androgen secretion. Overall, the findings of the above genetic association studies indicate that the steroidogenic activity of ovarian and/or adrenal androgen producing cells may be genetically determined.

An additional way that may contribute to prenatal androgenization of the female foetus, regardless of the source of androgens (maternal or foetal), would be by reduced binding of androgens by SHBG or reduced aromatization, leading to increased availability of free androgens or a high androgen to oestrogen ratio respectively, as mentioned earlier.

With regard to SHBG, serum levels are often low in women with PCOS and in individuals at risk for diabetes and heart disease (Lapidus et al., 1986; Lindstedt et al., 1991). The reason for the low SHBG levels in many women with PCOS is unclear, but it was thought to be the result of hyperinsulinaemia and hyperandrogenaemia (Nestler et al., 1991; Toscano et al., 1992). However, it has also been suggested that circulating SHBG levels may in part be genetically determined (Hogeveen et al., 2002).

Recently a (TAAAA)n pentanucleotide repeat polymorphism at the promoter of the human SHBG gene was described and reported to influence its transcriptional activity *in vitro* (Hogeveen et al., 2001). It has been suggested that this functional polymorphism could contribute to individual differences in plasma SHBG levels and thereby influence the access of sex steroids to their target tissues.

In the light of this evidence, the hypothesis that the SHBG (TAAAA)n repeat polymorphism is related to SHBG levels in hyperandrogenic women with PCOS was tested (Xita et al., 2003). The association of this particular polymorphism with PCOS and its relation to SHBG and free androgen levels among Greek women with PCOS (n=185) and ethnically matched controls (n=324) were examined. Women with PCOS were more frequently carriers of longer (TAAAA)n repeat alleles compared to controls. Among women with

PCOS, it was also found that carriers of long repeat genotypes had lower SHBG levels and higher free androgen indices compared to women with short genotypes. This difference, however, was not observed in the control group of women. A separate study by Cousin et al. (2004) also examined whether the (TAAAA)n repeat polymorphism influences SHBG levels among white French women with hirsutism. The authors observed that women with long repeat genotypes have lower SHBG levels than women with short genotypes. In a more recent study among young men, Wood et al. (2004b) also reported that carriers of longer repeat allele have lower SHBG levels.

These studies suggest that there may be a genetic contribution to decreased SHBG levels frequently seen in women with PCOS. Those individuals with genetically determined low SHBG levels may be exposed to higher free androgen levels throughout life but, more importantly, during foetal life when programming of the developing foetus is taking place.

Apart from the SHBG gene, another candidate gene that has been investigated for possible association with PCOS and increased androgen levels is the aromatase (CYP19) gene. Aromatase catalyzes the conversion of androgens to oestrogens under the influence of FSH and is present in a number of tissues including the ovary and adrenals, placenta, skin, adipose tissue and nervous system (Simpson et al., 1994). Reduced aromatase activity may lead to the development of PCOS, since women with rare loss-of-function mutations causing aromatase deficiency may manifest features of PCOS (Harada et al., 1992; Belgorosky et al., 2003). Furthermore, antral follicles from women with PCOS exhibit low aromatase activity (Takayama et al., 1996).

Although previous linkage and association studies failed to find an association between aromatase and PCOS, probably due to low statistical power, a recent association study showed a strong link between genetic variations in CYP19 and androgen excess in females (Petry et al., 2005). In particular, the study demonstrated that a common genetic variation (an intronic SNP close to exon 3 of CYP19, the SNP50) was associated with premature adrenarche, pubertal ovarian hyperadrogenism and variation in PCOS symptom score in both girls and young women. The additional association with circulating testosterone levels in both female groups could explain these associations (Petry et al., 2005).

In summary, the above association studies have demonstrated a relationship between genetic variations in both the SHBG and aromatase genes with androgen excess in women with PCOS and girls with precocious adrenarche, suggesting that hyperandrogenism in these conditions may be partly genetically determined.

CONCLUSION

Initial clinical observations noted that women with a history of 21-hydroxylase deficiency under treatment, or women with a history of congenital foetal virilizing tumours that were removed after birth, could develop the biochemical and clinical features of PCOS in adult life. These astute observations prompted the conduct of controlled experimental studies using prenatally-androgenized female rhesus monkeys or sheep as models. The experimental research convincingly demonstrated that exposure of the female foetus to androgen excess during intrauterine life programmes the various organ-systems in such a way that in adult life, all the characteristic features of the PCOS phenotype are manifested.

It appears that both the endocrine and the metabolic features of the PCOS phenotype are common between the prenatally androgenized non-human primate model and women with PCOS, and could have a common origin in prenatal life programmed by the androgen excess. The endocrine abnormalities include two distinct components common for the primate model and human PCOS. The first is LH hypersecretion, suggestive of an altered hormonal negative feedback on the hypothalamic-pituitary axis that may be induced prenatally by the androgen excess. The second and more important key component is the hyperandrogenic phenotype of the theca cells. This may also reflect reprogramming of the theca cells during differentiation by the altered sex steroid milieu. This view is further supported by observations on the aromatase "knockout" (ArKO) mice. Female ArKO mice have undetectable levels of oestrogens in the presence of high androgen, LH and FSH levels. These ArKO mice are infertile with folliculogenesis arrested in the antral stage. Their ovaries exhibit an increased diffused interstitium with the presence of cells morphologically resembling Leydig cells (Britt and Findlay, 2002). Thus, the altered androgen to oestrogen milieu may influence the differentiation of the ovarian theca cells towards a male-type phenotype.

The metabolic features also have two components that are common to prenatally-androgenized female primates and women with PCOS. The first is visceral fat accumulation associated with insulin resistance and hyperinsulinaemia, and the second is impaired pancreatic β-cell insulin secretion to glucose stimulation. Since body fat distribution in humans is sexually dimorphic, the enhanced accumulation of abdominal adiposity in PCOS may, partly, reflect a masculinized pattern of fat distribution. Interestingly, the metabolic abnormalities of insulin resistance and impaired β-cell function are also observed in prenatally androgenized male rhesus monkeys (Bruns et al., 2004). As both female and male primates, who are exposed to the same intrauterine environment are characterized by androgen excess and develop similar metabolic abnormalities, it is possible that an altered intrauterine environment induced by androgen excess may also cause the metabolic deficits seen in families with PCOS.

On the basis of experimental research, it is suggested that both key abnormalities of PCOS (endocrine-reproductive and metabolic abnormalities) may result from foetal reprogramming of key endocrine and metabolic tissues by androgen excess. Androgens present during differentiation may act as gene transcriptional factors or induce other critical transcription factors in these tissues enhancing gene expression and permanently programming their future phenotype.

The molecular genetic basis of the origin of androgens during foetal development of PCOS is not clear. However, recent association studies have indicated that a number of genes may be associated with PCOS and increased androgen levels in affected women. These are genes involved in the production or action of androgens or genes that influence SHBG production and aromatase enzyme activity. Common genetic variants of these genes have been related to androgen levels in women with PCOS. These genetic variants may also contribute to excess androgens during intrauterine life and thus provide the genetic link to the developmental origin of PCOS. Other genetic factors and environmental exposure or the interaction between the two may further influence the expression of the PCOS phenotype in adult life in genetically predisposed individuals (Fig. 1).

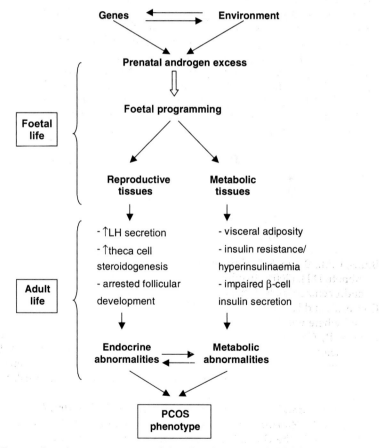

Fig. 1.   Prenatal androgen excess and the development of PCOS phenotype.

In conclusion, genetically determined prenatal androgenization may programme key reproductive and metabolic tissues of the developing female fetus during critical periods of their differentiation leading to the PCOS phenotype in adult life.

REFERENCES

Abbott, D.H., Dumesic, D.A., Eisner, J.R., Colman, R.J. and Kemnitz, J.W. (1998). Insights into the development of PCOS from studies of prenatally androgenized female rhesus monkeys. *Trends Endocrinol. Metab.* **9**: 62-67.

Abbott, D.H., Dumesic, D.A. and Franks, S. (2002). Developmental origin of polycystic ovary syndrome-a hypothesis. *J. Endocrinol.* **174**: 1-5.

Asuncion, M., Calvo, R.M., San Millan, J.L., Sancho, J., Avila, S. and Escobar-Morreale, H.F. (2000). A prospective study of the prevalence of the polycystic ovary syndrome in unselected Caucasian women from Spain. *J. Clin. Endocrinol. Metab.* **85**: 2434-2438.

Azziz, R. and Kashar-Miller, M.D. (2000). Family history as a risk factor for the polycystic ovary syndrome. *J. Pediatr. Endocrinol. Metab.* **13(5)**: 1303-1306.

Azziz, R., Woods, K.S., Reyna, R., Key, T.J., Knochenhauer, E.S. and Yildiz, B.O. (2004). The prevalence and features of the polycystic ovary syndrome in an unselected population. *J. Clin. Endocrinol. Metab.* **89**: 2745-2749.

Barker, D.J.P. (1998). In utero programming of chronic disease. *Clin. Sci.* **95**: 115-128.

Barnes, R.B., Rosenfield, R.L., Ehrmann, D.A., Cara, J.F., Cuttler, L., Levitsky, L.L. and Rosenthal, I.M. (1994). Ovarian hyperandrogenism as a result of congenital adrenal virilizing disorders: evidence for perinatal masculinization of neuroendocrine function in women. *J. Clin. Endocrinol. Metab.* **79**: 1328-1333.

Barraclough, C.A. and Gorski, R.A. (1961). Evidence that the hypothalamus is responsible for androgen-induced sterility in the female rat. *Endocrinol.* **68**: 68-79.

Belgorosky, A., Pepe, C., Marino, R., Guercio, G., Saraco, N., Vaiani, E. and Rivarola, M.A. (2003). Hypothalamic–pituitary–ovarian axis during infancy, early and late prepuberty in an aromatase-deficient girl who is a compound heterozygote for two new point mutations of the CYP19 gene. *J. Clin. Endocrinol. Metab.* **88**: 5127-5131.

Britt, K.L. and Findlay, J.K. (2002). Estrogen actions in the ovary revisited. *J. Endocrinol.* **175**: 269-276.

Bruns, C.M., Baum, S.T., Colman, R.J., Eisner, J.R., Kemnitz, J.W., Weindruch, R. and Abbott, D.H. (2004). Insulin resistance and impaired insulin secretion in prenatally androgenized male rhesus monkeys. *J. Clin. Endocrinol. Metab.* **89**: 6218-6223.

Couse, J.F. and Korach, K.S. (1999). Estrogen receptor null mice: what have we learned and where will they lead us? *Endocr. Rev.* **20**: 358-417.

Cousin, P., Calemard-Michel, L., Lejeune, H., Raverot, G., Yessaad, N., Emptoz-Bonneton, A., Morel, Y. and Pugeat, M. (2004). Influence of SHBG gene pentanucleotide TAAAA repeat and D327N polymorphism on serum sex hormone-binding globulin concentration in hirsute women. *J. Clin. Endocrinol. Metab.* **89**: 917-924.

Diamanti-Kandarakis, E., Kouli, C.R., Bergiele, A.T., Filandra, F.A., Tsianateli, T.C., Spina, G.G., Zapanti, E.D. and Bartzis, M.I. (1999). A survey of the polycystic ovary syndrome in the Greek island of Lesbos: hormonal and metabolic profile. *J. Clin. Endocrinol. Metab.* **84**: 4006-4011.

Diamanti-Kandarakis, E., Batzis, M.I., Bergiele, A.T., Tsianateli, T.C. and Kouli, C.R. (2000). Microsatellite polymorphism (tttta) at –528 base pairs of gene CYP11a influences hyperandrogenemia in patients with polycystic ovary syndrome. *Fertil. Steril.* **73**: 735-741.

Dumesic, D.A., Abbott, D.H., Eisner, J.R. and Goy, R.W. (1997). Prenatal exposure of female rhesus monkeys to testosterone propionate increases serum luteinizing hormone levels in adulthood. *Fertil. Steril.* **67**: 155-163.

Dunaif, A., Segal, K.R., Futterfeit, W. and Dobrjansky, A. (1989). Profound peripheral insulin resistance, independent of obesity in polycystic ovary syndrome. *Diabetes.* **38**: 1165-1174.

Ehrmann, D.A. (2005). Polycystic ovary syndrome. *N. Engl. J. Med.* **352**: 1223-1236.

Eisner, J.R., Dumesic, D.A., Kemnitz, J.W. and Abbott, D.H. (2000). Timing of prenatal androgen excess determines differential impairment in insulin secretion and action in adult female rhesus monkeys. *J. Clin. Endocrinol. Metab.* **85**: 1206-1210.

Eisner, J.R., Barnett, M.A., Dumesic, D.A. and Abbott, D.H. (2002). Ovarian hyperandrogenism in adult female rhesus monkeys exposed to prenatal androgen excess. *Fertil. Steril.* **77**: 167-172.

Eisner, J.R., Dumesic, D.A., Kemnitz, J.W., Colman, R.J. and Abbott, D.H. (2003). Increased adiposity in female rhesus monkeys exposed to androgen excess during early gestation. *Obes. Res.* **11**: 279-286.

Filho, T., Baracat, E.C., Lee, T.H., Suh, C.S., Matsui, M., Chang, R.J., Shimasaki, S. and Erickson, G.F. (2002). Aberrant expression of growth differentiation factor-9 in oocytes of women with polycystic ovary syndrome. *J. Clin. Endocrinol. Metab.* **87**: 1337-1344.

Franks, S. (1995). Polycystic ovary syndrome. *N. Engl. J. Med.* **333**: 853-861.

Franks, S., Gharani, N., Waterworth, D., Batty, S., White, D., Williamson, R. and McCarthy, M. (1997). The genetic basis of polycystic ovary syndrome. *Hum. Reprod.* **12**: 2641-2648.

Franks, S. (2002). Adult polycystic ovary syndrome begins in childhood. *Best Pract. Res. Clin. Endocrinol. Metab.* **16**: 263-272.

Gaasenbeek, M., Powell, B.L., Sovio, U., Haddad, L., Gharani, N., Bennett, A., Groves, C.J., Rush, K., Goh, M.T., Conwdy, G.S., Ruokonen, A., Martikainen, H., Pouta, A., Taponen, S., Harticainen, A.L., Halford, S., Javvelin, M.R., Franks, S. and McCarthy, M.I. (2004). Large-scale analysis of the relationship between CYP11A promoter variation, polycystic ovarian syndrome and serum testosterone. *J. Clin. Endocrinol. Metab.* **89**: 2408-2413.

Gharani, N., Waterworth, D.M., Batty, S., White, D., Gilling-Smith, C., Conway, G.S., McCarthy, M., Franks, S. and Williamson, R. (1997). Association of the steroid synthesis gene CYP11a with polycystic ovary syndrome and hyperandrogenism. *Hum. Mol. Genet.* **6**: 397-402.

Gilling-Smith, C., Story, H., Rogers, V. and Franks, S. (1997). Evidence for a primary abnormality of thecal cell steroidogenesis in the polycystic ovary syndrome. *Clin. Endocrinol.* **47**: 93-99.

Hague, W.M., Adams, J., Rodda, C., Brook, C.G., de Bruyn, R., Grant, D.B. and Jacobs, H.S. (1990). The prevalence of polycystic ovaries in patients with congenital hyperplasia and their close relatives. *Clin. Endocrinol.* **33**: 501-510.

Harada, N., Ogawa, H., Shozu, M. and Yamada, K. (1992). Genetic studies to characterize the origin of the mutation in placental aromatase deficiency. *Am. J. Hum. Genet.* **51**: 666-672.

Hogeveen, K.N., Talikka, M. and Hammond, G.L. (2001). Human sex hormone-binding globulin promoter activity is influenced by a $(TAAAA)_n$ repeat element within an Alu sequence. *J. Biol. Chem.* **276**: 36383-36390.

Hogeveen, K.N., Cousin, P., Pugeat, M., Dewailly, D., Soudan, B. and Hammond, G.L. (2002). Human sex hormone-binding globulin variants associated with hyperandrogenism and ovarian dysfunction. *J. Clin. Invest.* **109**: 973-981.

Horie, K., Takakura, K., Fujiwara, H., Suginami, H., Liao, S. and Mori, T. (1992). Immunohistochemical localization of androgen receptor in the human ovary throughout the menstrual cycle in relation to oestrogen and progesterone receptor expression. *Hum. Reprod.* **7**: 184-190.

Ibanez, L., Hall, J.E., Potau, N., Carrascosa, A., Prat, N. and Taylor, A.E. (1996). Ovarian 17-hydroxyprogesterone hyperresponsiveness to gonadotropin-releasing hormone (GnRH) agonist challenge in women with polycystic ovary syndrome is not mediated by luteinizing hormone hypersecretion – evidence from GnRH agonist and human chorionic gonadotropin stimulation testing. *J. Clin. Endocrinol. Metab.* **81**: 4103-4107.

Ibanez, L., Potau, N., Francois, I. and DeZegher, F. (1998). Precocious pubarche, hyperinsulinism, and ovarian hyperandrogenism in girls: relation to reduced fetal growth. *J. Clin. Endocrinol. Metab.* **83**: 3558-3562.

Knochenhauer, E.S., Key, T.J., Kahsar-Miller, M., Waggoner, W., Boots, L.R. and Azziz, R. (1998). Prevalence of the polycystic ovary syndrome in unselected black and white women of the south eastern United States: a prospective study. *J. Clin. Endocrinol. Metab.* **83**: 3078-3082.

Lapidus, L., Lindstedt, G., Lundberg, P.A., Bengtsson, C. and Gredmark, T. (1986). Concentrations of sex-hormone-binding-globulin and corticosteroid binding globulin in serum in relation to cardiovascular risk factors and to 12-year incidence of cardiovascular disease and overall mortality in postmenopausal women. *Clin. Chem.* **32**: 146-152.

Legro, R.S., Kunselman, A.R., Dodson, W.C. and Dunaif, A. (1999). Prevalence and predictors of risk for type 2 diabetes mellitus and impaired glucose tolerance in polycystic ovary syndrome: a prospective, controlled study in 254 affected women. *J. Clin. Endocrinol. Metab.* **84**: 165-169.

Lindstedt, G., Lundberg, P.A., Lapidus, L., Lundgren, H., Bengtsson, C. and Bjorntorp, P. (1991). Low sex-hormone-binding globulin concentration as independent risk factor for development of NIDDM. 12-year follow up of population study of women in Gothenburg, Sweden. *Diabetes* **40**: 123-128.

Lucas, A. (2005). Long-term programming effects of early nutrition-implications for the preterm infant. *J. Perinatol.* **25**: S2-S6.

Manikkam, M., Crespi, E.J., Doop, D.D., Herkimer, C., Lee, J.S., Yu, S., Brown, M.B., Foster, D.L. and Padmanabhan, V. (2004). Fetal programming: prenatal testosterone excess leads to fetal growth retardation and postnatal catch-up growth in sheep. *Endocrinol.* **145**: 790-798.

Nelson, V.L., Legro, R.S., Strauss, J.F. III and McAllister, J.M. (1999). Augmented androgen production is a stable steroidogenic phenotype of propagated theca cells from polycystic ovaries. *Mol. Endocrinol.* **13**: 946-957.

Nelson, V.L., Qin Kn, K.N., Rosenfield, R.L., Wood, J.R., Penning, T.M., Legro, R.S., Strauss, J.F. 3rd and McAllister, J.M. (2001). The biochemical basis for increased testosterone production in theca cells propagated from patients with polycystic ovary syndrome. *J. Clin. Endocrinol. Metab.* **86**: 5925-5933.

Nestler, J.E. (1987). Modulation of aromatase and P450 cholesterol side-chain cleavage enzyme activities of human placental cytotrophoblasts by insulin and insulin-like growth factor I. *Endocrinol.* **121**: 1845-1852.

Nestler, J.E. (1989). Insulin and insulin-like growth factor-I stimulate the 3 beta-hydroxysteroid dehydrogenase activity of human placental cytotrophoblasts. *Endocrinol.* **125**: 2127-2133.

Nestler, J.E. (1990). Insulin-like growth factor II is a potent inhibitor of the aromatase activity of human placental cytotrophoblasts. *Endocrinol.* **127**: 2064-2070.

Nestler, J.E., Powers, L.P., Matt, D.W., Steingold, K.A., Plymate, S.R., Rittmaster, R.S., Clore, J.N. and Blackard, W.G. (1991). A direct effect of hyperinsulinaemia on serum sex-hormone-binding globulin levels in obese women with the polycystic ovary syndrome. *J. Clin. Endocrinol. Metab.* **72**: 83-89.

Ong, K.K. and Dunger, D.B. (2002). Perinatal growth failure: the road to obesity insulin resistance and cardiovascular disease in adults. *Best Pract. Res. Clin. Endocrinol. Metabol.* **16**: 191-207.

Padmanabhan, V., Evans, N., Taylor, J.A. and Robinson, J.E. (1998). Prenatal exposure to androgens leads to the development of cystic ovaries in the sheep. *Biol. Reprod.* **56**(1): 194.

Patterson, M.N., McPhaul, M.J. and Hughes, I.A. (1994). Androgen insensitivity syndrome. In: Bailliere's Clinical Endocrinology and Metabolism: Hormones, Enzymes and Receptors. (Sheppard, M.C. and Stewart, P.M., eds.) Bailliere Tindall, London, 379-404.

Petry, C.J., Ong, K.K., Michelmore, K.F., Artigas, S., Wingate, D.L., Balen Ah De Zegher, F., Ibanez, L. and Dunger, D.B. (2005). Association of aromatase (CYP 19) gene variation with features of hyperandrogenism in two populations of young women. *Hum. Reprod.* (in press).

Robinson, J.E., Forsdike, R.A. and Taylor, J.A. (1999). In utero exposure of female lambs to testosterone reduces the sensitivity of the gonadotropin-releasing hormone neuronal network to inhibition by progesterone. *Endocrinol.* **140**: 5797-5805.

Rosenfield, R.L. (1999). Ovarian and adrenal function in polycystic ovary syndrome. *Endocrinol. Metab. Clin. North. Am.* **28**: 265-293.

San Millan, J.L., Corton, M., Villuendas, G., Sancho, J., Peral B. and Escobar-Morreale, H.F. (2004). Association of the polycystic ovary syndrome with genomic variants related to insulin resistance, type 2 diabetes mellitus, and obesity. *J. Clin. Endocrinol. Metab.* **89**: 2640-2646.

Simpson, E.R., Mahendroo, M.S., Means, G.D., Kilgore, M.W., Hinshelwood, M.M., Graham-Lorence, S., Amarneh, S.B., Ito, Y., Fisher, C.R., Michael, M.D., Mendelson, C.R. and Bulun, S.E. (1994). Aromatase cytochrome P450, the enzyme responsible for estrogen biosynthesis. *Endocr. Rev.* **15**: 342-355.

Sir-Peterman, T., Maliqueo, M., Angel, B., Lara, H.E., Perez-Bravo, F. and Recabarren, S.E. (2002). Maternal serum androgens in pregnant women with polycystic ovarian syndrome: possible implications in prenatal androgenization. *Hum. Reprod.* **17**: 2573-2579.

Sullivan, S.D. and Moenter, S.M. (2004). Prenatal androgens alter GABAergic drive to gonadotropin-releasing hormone neurons: Implications for a common fertility disorder. *PNAS* **101**: 7129-7134.

Takayama, K., Fukaya, T., Sasano, H., Funayama, Y., Suzuki, T., Takaya, R., Wada, Y. and Yajima, A. (1996). Immunohistochemical study of steroidogenesis and proliferation in polycystic ovarian syndrome. *Hum. Reprod.* **11**: 1387-1392.

The Rotterdam ESHRE/ASRM-Sponsored PCOS Consensus Workshop Group. (2004). Revised 2003 consensus on diagnostic criteria and long-term health risks related to polycystic ovary syndrome (PCOS). *Hum. Reprod.* **19**: 41-47.

Toscano, V., Balducci, R., Bianchi, P., Guglielmi, R., Mangiantini, A. and Sciarra, F. (1992). Steroidal and non-steroidal factors in plasma sex hormone-binding globulin regulation. *J. Steroid. Biochem. Mol. Biol.* **43**: 431-437.

Tsilchorozidou, T., Overton, C. and Conway, G.S. (2004). The pathophysiology of polycystic ovary syndrome. *Clin. Endocrinol.* **60**: 1-17.

Vom Saal, F.S. and Bronson, F.H. (1980). Sexual characteristics of adult female mice are correlated with their blood testosterone levels during prenatal development. *Science* **208**: 597-599.

West, C., Foster, D.L., Evans, N.P., Robinson, J. and Padmanabhan, V. (2001). Intra-follicular activin availability is altered in prenatally-androgenized lambs. *Mol. Cell. Endocrinol.* **185**: 51-59.

Westberg, L., Baghaei, F., Rosmond, R., Hellstrand, M., Landen, M., Jansson, M., Holm, G., Bjorntorp, P. and Eriksson, E. (2001). Polymorphisms of the androgen receptor gene and the estrogen receptor beta gene are associated with androgen levels in women. *J. Clin. Endocrinol. Metab.* **86**: 2562-2568.

Wild, R.A. (2002). Long-term health consequences of PCOS. *Hum. Reprod. Update* **8**: 231-241.

Wood, J.R., Ho, C., Nelson, V.L., McAllister, J.M. and Strauss, J.F. (2004a). The molecular signature of polycystic ovary syndrome (PCOS) theca cells defined by gene expression profiling. *J. Reprod. Immunol.* **63**: 51-60.

Wood, K., Kopp, P., Colangelo, L.A., Liu, K. and Gapstur, S.M. (2004b). A cross-sectional analysis of the association between serum sex hormone-binding globulin (SHBG) levels and the *SHBG* gene pentanucleotide TAAAA repeat: the CARDIA Male Hormone Study. Program of the 86th Annual Meeting of The Endocrine Society, New Orleans, LA, 539-540.

Xita, N., Georgiou, I. and Tsatsoulis, A. (2002). The genetic basis of polycystic ovary syndrome. *Eur. J. Endocrinol.* **147**: 717-725.

Xita, N., Tsatsoulis, A., Chatzikyriakidou, A. and Georgiou, I. (2003). Association of the (TAAAA)n repeat polymorphism in the sex hormone-binding globulin (SHBG) gene with polycystic ovary syndrome and relation to SHBG levels. *J. Clin. Endocrinol. Metab.* **88**: 5976-5980.

CHAPTER 3

Effects of Antidiabetic Drugs on Reproductive System, Life Span and Tumor Development

Vladimir N. Anisimov

Abstract

Studies in mammals have led to the suggestion that hyperglycemia and hyperinsulinemia are important factors both in aging and in the development of cancer. Insulin/insulin-like growth factor 1 (IGF-1) signaling molecules that have been linked to longevity include DAF-2 and InR (insulin receptor) and their homologues in mammals, and inactivation of the corresponding genes followed by the increase in life span in nematodes, fruit flies and mice. It is possible that the life-prolonging effect of caloric restriction is due to decreasing IGF-1 levels. A search of pharmacological modulators of insulin/IGF-1 signaling pathway mimetic effects of life span extending mutations or calorie restriction could be a perspective direction in regulation of longevity. Some observations suggest that antidiabetic drugs could be promising candidates for both the life span extension and the prevention of cancer.

INTRODUCTION

Caloric restriction (CR) is the only known intervention in mammals that has been shown consistently to increase life span, reduce incidence and retard the onset of age-related diseases, including cancer and diabetes. CR has also been shown to increase the resistance to stress and toxicity, and maintain the function and vitality in laboratory mammals of younger ages (Weindruch and Walford, 1988; Masoro, 2000). Studies in CR rhesus monkeys have produced physiological responses strikingly similar to those observed in rodents (Roth et al., 1999; Mattson et al., 2003). Emerging data from these studies suggest that long-term CR will reduce morbidity and mortality in primates, and thus

Department of Carcinogenesis and Oncogerontology, N.N. Petrov Research Institute of Oncology, Pesochny-2, St.Petersburg 197758, Russia. Tel.: +7 812 596 8607, Fax: +7 812 596 8947. E-mail: aging@mail.ru (V.N. Anisimov)

may exert beneficial "anti-aging" effects in humans (Roth et al., 1999; Lee et al., 2001).

The crucial event of the action of CR is the reduction in the levels of insulin and insulin-like growth factor-1 (IGF-1) and also an increased insulin sensitivity in rodents (Weindruch and Walford, 1988) and monkeys (Lane et al., 2000). In *C. elegans* and *D. melanogaster*, the mutation modification of genes operating in the signal transduction from the insulin receptor to transcription factor *daf-16* (*age-1*, daf-2, CHICO, InR, etc.) is strongly associated with longevity (Kenyon, 2001; Clancy et al., 2001; Tatar et al., 2001). Whole-genome analysis of gene expression during aging of nematode worm *C. elegans* provided a new evidence on the role of insulin homologue genes and *SIR2* homologues in longevity by interacting with the *daf-2/age-1* insulin-like signaling pathway and regulating downstream targets (Lund et al., 2002).

DAF-2 and InR are structural homologues of tyrosine kinase receptors in vertebrates that include the insulin receptor and the IGF-1 receptor. It was shown that in vertebrates insulin receptor regulates energy metabolism, whereas the IGF-1 receptor promotes a growth. During the last few years, a series of elegant experiments on mice and rats in which a number of key elements of the insulin/IGF-1 signaling pathway were genetically modified, provide an evidence of the involvement of the system in the control of mammalian aging and longevity (Coschigano et al., 2000; Bartke and Turyn, 2001; Dominici et al., 2000, 2002; Flurkey et al., 2001; Shimokawa et al., 2002; Hsieh et al., 2002a, b; Bartke et al., 2003; Bluher et al., 2003; Holzenberger et al., 2003; Tatar et al., 2003).

Recently it was shown that the incidence of mutations in insulin response element (IRE) of APO C-III T-455 C directly correlates with longevity in humans. This is the evidence showing that mutation located downstream to *daf-16* in insulin signal transduction system is associated with longevity (Anisimov et al., 2001). It is worth mentioning that centenarians display a lower degree of resistance to insulin and lower degree of oxidative stress, as compared with elderly people below 90 years (Paolisso et al., 1996; Barbieri et al., 2003). The authors suggest that centenarians may have been selected for appropriate insulin regulation as well as for the appropriate regulation of tyrosine hydroxylase (TH) gene, whose product is the rate limiting in the synthesis of catecholamines, stress-response mediators. It was shown that catecholamine might increase free radical production through induction of the metabolic rate and autooxidation in diabetic animals (Singal et al., 1983). A recent study on aging parameters of young (up to 39) and old (over 70) individuals having similar IGF-1 serum levels provides evidence of the important role of this peptide for life potential (Ruiz-Torres and Soares de Melo Kirzner, 2002). Roth et al. (2002) analyzed data from the Baltimore Longitudinal Study of Aging and reported that survival was greater in men who maintained lower insulin level. In women, genetic variation causing reduced insulin/IGF-1 signaling pathway activation is beneficial for old age survival (Van Heemst et al., 2005).

Hyperglycemia is an important aging factor involved in generations of advanced glycosylation end products (AGEs) (Facchini et al., 2000; Ulrich and Cerami, 2001; Elahi et al., 2002). Untreated diabetics with elevated glucose levels suffer many manifestations of accelerated aging, such as impaired wound healing, cataracts, vascular and microvascular damage (Dilman, 1994). The accumulation of the AGE, pentosidine, is accelerated in diabetics and has been suggested to be a reliable biomarker of aging (Ulrich and Cerami, 2001). The action of insulin provides the major modulator of glucose storage and utilization. It is important to stress that hyperinsulinemia is also an important factor in the development of cancer (Dilman, 1978, 1994; Colangelo et al., 2002; Gupta et al., 2002).

The concept of CR mimetics is now being intensively explored (Hadley et al., 2001; Mattson et al., 2001; Weindruch et al., 2001). CR mimetics involves interventions that produce physiological and anti-aging effects similar to CR. Reviewing the available data on the benefits and adverse effects of calorie restriction and genetic modifications, Longo and Finch (2003) suggested three categories of drugs which may have potential to prevent or postpone age-related diseases and extend life span: drugs that (1) stimulate dwarf mutations and therefore decrease pituitary production of growth hormone (GH); (2) prevent IGF-1 release from the liver, or (3) decrease IGF-1 signaling by the action on either extracellular or intracellular targets.

ANTIDIABETIC DRUGS AND LONGEVITY

Several years ago, it was suggested to use biguanide antidiabetics as a potential anti-aging treatment (Dilman, 1971, 1978; Dilman and Anisimov, 1980). The antidiabetic drugs, phenformin, buformin, and metformin, were observed to reduce hyperglycemia and produce the following effects: improved glucose utilization; reduced free fatty acid utilization, gluconeogenesis, serum lipids, insulin and IGF-1, and reduced body weight both in humans and experimental animals (Dilman, 1994). At present, phenformin is not used in clinical practice due to its adverse side effects observed in patients with non-compensated diabetes. It is worth mentioning that during more than 10 years, long experience of administration of phenformin for patients without advanced diabetes, Dilman (1994) observed no case of lactic acidosis or any other side effects. It is believed that the analysis of results of long-term administration of (1) this drug, (2) antidiabetic biguanides (buformin and metformin), and (3) antidiabetic drugs, to non-diabetic animals is very important for better understanding the links between insulin and longevity.

Buformin was supplemented to nutrient medium in various concentrations (from 1.0 to 0.00001 mg/ml) during the larval stage and over the life span of *C. elegans*. The drug given at the concentration of 0.1 mg/ml increased the mean life span of the nematoda by 23.4% (p < 0.05) and the maximum life span by 26.1%, as compared to the controls (Bakaev, 2002).

The available data on the effects of antidiabetic biguanides on life span in mice and rats is summarized in the Table 1. Female C3H/Sn mice, kept from the age of 3.5 months at standard *ad libitum* diet, were given phenformin five times a week orally at a single dose of 2 mg/mouse until a natural death (Dilman and Anisimov, 1980). The treatment with phenformin prolonged the mean life span of mice by 21% (p < 0.05), the mean life span of last 10% survivors by 28%, and the maximum life span by 5.5 months (by 26%), in comparison with the control. At the time of death of the last mice in the control group, 42% of phenformin-treated mice were alive.

Table 1. Effects of antidiabetic biguanides on mortality rate in mice and rats

Strain	Treatment	No. of Animals	Life Span (days)			Reference
			Mean	Last 10% of Survivors	Maximum	
Mice						
C3H/Sn	Control	30	450 ± 23.4	631 ± 11.4	643	Dilman and
	Phenformin	24	545 ± 39.2 (+21.1%)	810 ± 0 ** (+28.4%)	810 (+26%)	Anisimov (1980)
NMRI						
	Control	50	346 ± 11.9	480 ± 9.2	511	Popovich et al.
	Diabenol	50	369 ± 12.9	504 ± 6.4* (+5.9%)	518	(2005)
HER-2/	Control	30	264 ± 3.5	297 ± 7.3	311	Anisimov et al.
neu	Metformin	24	285 ± 5.2 (+8%)	336 ± 2.7 (+17.9%)	340 (+9.3%)	(2005)
Rats						
LIO	Control	41	652 ± 27.3	885 ± 11.3	919	Anisimov
	Phenformin	44	652 ± 28.7	974 ± 16.2** (+10.1%)	1,009 (+9.8%)	(1982)
	Control	74	687 ± 19.2	925 ± 22.5	1,054	Anisimov
	Buformin	42	737 ± 26.4 (+7.3%)	1036± 38.9* (+12%)	1,112 (+5.5%)	(1980)

The difference with control is significant:* p < 0.05; ** p < 0.01 (Student's *t* test).

Phenformin was given five times a week to female outbred LIO rats starting from the age of 3.5 months until a natural death in a single dose of 5 mg/rat/day orally (Anisimov, 1982, 1987). Administration of phenformin failed to influence the mean life span in rats. At the same time, the mean life span of the last 10% survivors was increased by 10% (p <0.005), and the maximum life span was increased by 3 months (+ 10%) in comparison with the controls (Table 1). The treatment with phenformin slightly decreased the body weight of rats in comparison with the control (p > 0.05). The disturbances in the estrus function was observed in 36% of 15-16-month-old rats of the control group, and only in 7% of rats in phenformin-treated group (p < 0.05).

Buformin was given five times a week to female LIO rats starting from the age of 3.5 months until a natural death in a single dose of 5 mg/rat/day orally (Anisimov, 1980). The treatment slightly increased the mean life span of rats (by 7%; $p > 0.05$). The mean life span of the last 10% survivors increased by 12% ($p < 0.05$) and the maximum life span increased by 2 months (+5.5%) as compared with controls. The body weight of rats treated with buformin was slightly (5.2 to 9.4%), but statistically significantly ($p < 0.05$), decreased in comparison with the control from the age of 12 months to 20 months ($p < 0.05$). At the age of 16-18 months, 38% of control rats revealed the disturbances in the estrus cycle persistent estrus (repetitive pseudopregnancies or anestrus), whereas in females treated with buformin these disturbances were observed only in 9% of rats ($p < 0.05$).

Recently it was found that metformin, like buformin and phenformin, significantly increases the life span of rats (G.S. Roth, personal communication).

There are only a few data on the effect of other than biguanides antidiabetic drugs on life span of animals. It is shown that treatment with Diabenol® [9-β-diethylaminoethyl-2,3-dihydroimidazo-(1,2-α) benzimidazol dihydrochloride] (Spasov et al., 1997) failed to influence body weight gain dynamics, food and water consumption, body temperature, slowed down age-related disturbances in estrous function, and increased life span of all the 10% most long-living NMRI mice. The treatment with Diabenol inhibited spontaneous tumor incidence and increased the mammary tumor latency in these mice. Diabenol treatment slowed down the age-related changes in estrous function in HER-2/neu mice, failed to influence survival of these mice, and slightly inhibited the incidence and decreased the size of mammary adenocarcinoma metastases into the lung (Popovich et al., 2005).

It was shown that hypoglycemic activity of Diabenol was 1.5 times higher than that of maninil (glibenclamide) and equal to the effect of glyclazide in rats, rabbits and dogs (Spasov et al., 1997; Anisimov et al., 2002). Hypoglycemic effect of Diabenol included both pancreatotropic and extrapancreatic pathways. Diabenol restores physiological profile of insulin secretion, decreases tissue resistance to insulin, and prolongs hypoglycemic effect of insulin. It increases glucose utilization in glucose loading test in the old obese rats. It was suggested that Diabenol influences insulin receptors in peripheral tissues. Diabenol increases uptake of glucose by isolated rat diaphragm *in vitro* both without supplementation of insulin into the medium or with supplemented insulin. Diabenol also decreases platelet and erythrocyte aggregation and blood viscosity, inhibits mutagenic effect of 2-acetylaminofluorene and has antioxidant activity (Spasov et al., 1997; Mezheritskiet et al., 1998; Anisimova et al., 2002). Thus, these results suggest that like biguanides, Diabenol has a potential to increase the life span and inhibit carcinogenesis.

ANTIDIABETIC DRUGS AND REPRODUCTIVE SYSTEM

Several other effects of treatment with antidiabetic biguanides related to reproduction and aging, are known from earlier studies. For example, it decreased hypothalamic threshold of the sensitivity to feedback inhibition by estrogens (Anisimov and Dilman, 1975; Dilman and Anisimov, 1979), which is one of the most important mechanisms regulating age-related decline and switch-off of the reproductive function (Dilman and Anisimov, 1979; Rossmanith, 2001; Hung et al., 2003; Yaghmaie et al., 2003). It is worth noting that another antidiabetic biguanide, metformin, may improve menstrual regularity, leading to spontaneous ovulation, and enhances the induction of ovulation with clomiphene citrate in women with polycystic ovary syndrome (Awartani and Cheung, 2002; Nestler et al., 2002). The treatment with phenformin also decreased hypothalamic threshold sensitivity to feedback regulation by glucocorticoids and by metabolic stimuli (glucose and insulin) (Dilman et al., 1979, 1979a; Dilman, 1994). It was recently shown that elements involved in the insulin/IGF-1 signaling pathway are regulated at the expression and/or functional level in the central nervous system. This regulation may play a role in the brain's insulin resistance (Fernandes et al., 2001), in the control of ovarian follicular development and ovulation (Richards et al., 2002), and the brain's control of life span (Chiba et al., 2002; Mattson, 2002; Mattson et al., 2002). Antidiabetic biguanides also alleviated age-related metabolic immunodepression (Dilman, 1994). These mechanisms can be involved in geroprotective effect of biguanides. Treatment with chromium picolinate, which elevated the insulin sensitivity in several tissues, including hypothalamus, significantly increased the mean life span and decreased the development of age-related pathology in rats (McCarty, 1994). It is hypothesized that antidiabetic biguanides, and possibly chromium picolinate, regulate thyrosine hydroxilase and insulin/IGF-1 signaling pathway genes, both associated with longevity (De Benedictis et al., 1998, 2001; Kenyon, 2001). It was shown that the polymorphism at TH-INS locus affects non-insulin dependent type 2 diabetes (Huxtable et al., 2000), and is associated with hypothalamic obesity (Weaver et al., 1992), polycystosis ovary syndrome (Waterworth et al., 1997), hypertriglyceridemia and atherosclerosis (Tybiaerg-Hansen et al., 1990).

EFFECTS OF ANTIDIABETIC DRUGS ON CARCINOGENESIS

Long-term treatment with phenformin significantly inhibited (by 4.0-fold, $p < 0.01$) the incidence of spontaneous mammary adenocarcinomas in female C3H/Sn mice (Dilman and Anisimov, 1980). The tumor yield curve rise was also significantly slowed down as a result of the treatment. The treatment with phenformin was followed by 1.6-fold decrease in total spontaneous tumor incidence in rats (Anisimov, 1982), whereas the total tumor incidence was decreased by 49.5% in buformin-treated rats (Anisimov, 1980).

The anticarcinogenic effects of antidiabetic biguanides have been demonstrated in several models of induced carcinogenesis (Table 2). Daily oral administration of phenformin or buformin suppressed 7,12-dimethylbenz(a)anthracene (DMBA)-induced mammary tumor development in rats (Dilman et al., 1978; Anisimov et al., 1980, 1980a). Phenformin-treated rats revealed a tendency towards a decrease in serum insulin level. The treatment with phenformin normalized the tolerance to glucose, serum insulin and IGF-1 level in rats exposed to intravenous injections of N-nitrosomethylurea (NMU) and inhibited mammary carcinogenesis in these animals (Anisimov et al., 1980). Treatment of rats with 1,2-dimethylhydrazine (DMH) (once a week during four weeks) caused the decrease in the level of biogenic amines, particularly of dopamine in the hypothalamus, the decrease of glucose tolerance, and the increase of the blood level of insulin and triglycerides. The exposure to DMH also caused the inhibition of lymphocyte blastogenic response to phytohemagglutinin and lipopolysaccharide, the

Table 2. Effects of antidiabetic biguanides on carcinogenesis in mice and rats

Species	Drug	Carcinogen	Main Target(s)	Effect of Treatment	Reference
Mice					
	Phenformin	Spontaneous	Mammary gland	Inhibition	Dilman and Anisimov (1980)
	Metformin	Spontaneous (HER-2/neu)	Mammary gland	Inhibition	Anisimov et al. (2005)
	Diabenol	Spontaneous	Mammary gland, lymphoma	Inhibition	Popovich et al. (2005)
	Phenformin	MCA	Subcutaneous tissue	Inhibition	Vinnitski and Iakimenko (1981)
Rats					
	Buformin	Spontaneous	Total incidence	Inhibition	Anisimov (1980)
	Phenformin	Spontaneous	Total incidence	Inhibition	Anisimov (1982)
	Phenformin	DMBA	Mammary gland	Inhibition	Dilman et al. (1978)
	Phenformin	NMU	Mammary gland	Inhibition	Anisimov et al. (1980)
	Buformin	DMBA	Mammary gland	Inhibition	Anisimov et al. (1980a)
	Buformin	NMU, transplacentally	Nervous system	Inhibition	Alexandrov et al. (1980)
	Phenformin	NEU, transplacentally	Nervous system, kidney	Inhibition	Bespalov and Alexandrov (1985)
	Phenformin	DMH	Colon	Inhibition	Anisimov et al (1980b)
	Phenformin	DMH	Colon	Inhibition	Dilman et al. (1977)
	Diabenol	DMH	Colon	Inhibition	Popovich et al. (2005)
	Phenformin	X-rays	Total incidence	Inhibition	Anisimov et al. (1982)
Hamster	Metformin	NBOPA	Pancreas	Inhibition	Schneider et al. (2001)

Abbreviations: DMBA – (7,12 dimethylbenz(a)anthracene) DMH – 1,2-dimethylhydrazine; MCA – 20-methylcholanthrene; NBOPA – N-nitrosobis-(2-oxopropyl)amine; NEU – N-nitrosoethylurea; NMU – N-nitrosomethylurea; X-rays – total-body X-ray irradiation.

decrease in the level of antibody produced against sheep erythrocytes and the decrease in phagocytic activity of macrophages (Dilman et al., 1977). Administration of phenformin started from the first injection of the carcinogen which restored all the above mentioned immunological indices and inhibited DMH-induced colon carcinogenesis (Dilman et al., 1977; Anisimov et al., 1980b). It is important to note that colon 38-adenocarcinoma growth was significantly inhibited in liver-specific IGF-1-deficient mice, whereas injections with recombinant human IGF-1 displayed adequately promoted tumor growth and metastasing (Wu et al., 2002).

A decrease of glucose utilization in the oral glucose tolerance test was found in the 3-month-old female progeny of rats exposed to NMU on the 21st day of pregnancy (Alexandrov et al., 1980). The serum insulin level did not differ from the control, but the cholesterol level was higher in the offspring of NMU-treated rats as compared with the control. Postnatal treatment with buformin, which started from the age of two months, significantly inhibited the development of malignant neurogenic tumors in rats transplancentally exposed to NMU (Alexandrov et al., 1980). Similar results were observed in rats exposed transplacentally to N-nitrosoethylurea (NEU) and postnatally to phenformin (Bespalov and Alexandrov, 1985). Authors observed the decrease of development of nervous system and renal tumors induced transplacentally with NEU. The treatment with phenformin also inhibited the carcinogenesis induced by a single total-body X-ray irradiation in rats (Anisimov et al., 1982).

Vinnitski and Iakimenko (1981) have shown that treatment with phenformin increased the immunological reactivity and inhibited carcinogenesis induced by subcutaneous (s.c.) administration of 20-methyl-cholanthrene in BALB/c mice. In high fat-fed hamsters, the treatment with N-nitrosobis-(2-oxopropyl)amine was followed by the development of pancreatic malignancies in 50% of the cases, whereas no tumors were found in the hamsters treated with the carcinogen and metformin (Schneider et al., 2001).

Thus, anticarcinogenic effect of antidiabetic drugs has been demonstrated in relation to spontaneous carcinogenesis in mice and rats, in different models of chemical carcinogenesis in mice, rats and hamsters, and in radiation carcinogenesis model in rats. Phenformin administered orally to mice potentiated the antitumor effect of cytostatic drug cyclophosphamide on transplantable squamous cell cervical carcinoma, hepatoma-22a and Lewis lung carcinoma. Administration of phenformin to rats with transplanted Walker 256 carcinoma enhanced the antitumor effect of hydrazine sulfate (Dilman and Anisimov, 1979a). It was observed that phenformin inhibits proliferation and induced enhanced and transient expression of the cell cycle inhibitor p21 and apoptosis in human tumor cells lines (Caraci et al., 2003).

Metformin decreased cell proliferation in breast cancer cells *in vitro* (Zakikhani et al., 2006). Mechanism includes AMP kinase activation. The growth inhibition also associated with decreased mammalian target of rapamycin (mTOR) and S6 kinase activation and a general decrease in mRNA translation. Metformin activates AMP-activated protein kinase (AMPK) via tumor suppressor gene LKB1. Mutation of this gene leads to different kind of cancers (Shen et al., 2002). It was found that the LKB1 overexpression can result in G1 cell cycle arrest (Giardiello et al., 2002). The reduction of the serum level of cholesterol and beta-lipoproteins and the increase in the expression of granzyme B and perforins in mammary tumors was involved in inhibitory effect of metformin on mammary carcinogenesis in transgenic HER-2/neu female mice (Anisimov et al., 2005).

The comparative study of 10 years' results of metabolic rehabilitation (included restriction in fat and carbohydrate diet and treatment with antidiabetic biguanides) of cancer patients had shown significant increase in the survival rate of patients with breast and colorectal cancer, increase in the length of cancer-free period, and decrease in the incidence of metastasis as compared with control patients (Berstein et al., 1992, 2004; Berstein, 2005). It was recently reported that reduced risk of cancer in patients with type II diabetes treated with metformin compared with those taking sulfonureas (Evans et al., 2005; Bowker et al., 2006).

In rats exposed to 1,2-dimethylhydrazine, treatment with Diabenol significantly inhibited multiplicity of all colon tumors, decreased by 2.2 times the incidence of carcinomas in ascending colon and by 3.1 times their multiplicity. Treatment with Diabenol was followed by higher incidence of exophytic and well-differentiated colon tumors, as compared with the control rats exposed to the carcinogen alone (76.3% and 50%, and 47.4% and 14.7%, respectively) (Popovich et al., 2005).

It was suggested that antidiabetic biguanides inhibit metabolic immunodepression, developed in animals exposed to carcinogenic agents, which is similar to the immunodepression inherent to normal aging and specific age-related pathology (Dilman, 1978, 1994). If immunodepression is one of the important factors in carcinogenesis, then the elimination of metabolic immunodepression, which arises in the course of normal aging or under the influence of chemical carcinogens or ionizing radiation, can provide an anticarcinogenic prophylactic effect (Dilman et al., 1977; Dilman, 1978, 1994; Anisimov, 1987).

Although it is known that free radicals are produced during metabolic reactions, it is largely unknown which factor(s), of physiological or pathophysiological significance, modulate their production *in vivo*. It has been suggested that hyperinsulinemia may increase free radicals and, therefore, promote aging, independent of glycemia (Dilman, 1971, 1994; Facchini et al., 2000, 2001). Plasma levels of lipid hydroperoxides are higher, and antioxidant factors are lower in individuals who are resistant to insulin-stimulated glucose disposal but otherwise glucose tolerant, nonobese, and normotensive (Facchini et al., 2000a). This finding indicates that enhanced oxidative stress is present before diabetes ensues and, therefore, cannot simply be explained by overt hyperglycemia. There is substantial evidence supporting the hypothesis that selective resistance to insulin-stimulated (muscle) glucose disposal and the consequential compensatory hyperinsulinemia trigger a variety of metabolic effects, probably resulting in accelerated oxidative stress and aging (Dilman, 1994; Facchini et al., 2000).

The antidiabetics biguanides inhibit fatty acid oxidation, gluconeogenesis of the liver, increase the availability of insulin receptors, decrease monoamine oxidase activity (Muntoni, 1974, 1999), increase sensitivity of hypothalamo-pituitary complex to negative feedback inhibition, reduce excretion of glucocorticoid metabolites and dehydroepiandrosterone-sulfate (Dilman, 1994). These drugs have been proposed for the prevention of the age-related increase of cancer and atherosclerosis, and for retardation of the aging process (Dilman, 1971, 1994). It has been shown that administration of antidiabetic biguanides to patients with hyperlipidemia lowers the level of blood cholesterol, triglycerides, and β-lipoproteins. Bioguanides also inhibits the development of atherosclerosis, and reduces hyperinsulinemia in men with coronary artery disease. It increases hypothalamo-pituitary sensitivity to inhibition by dexamethasone and estrogens, causes restoration of estrous cycle in persistent-estrous old rats, improves cellular immunity in atherosclerotic and cancer patients, lowers blood IGF-1 levels in cancer and atherosclerotic patients with Type IIb hyperlipoproteinemia, (Dilman et al., 1979, 1982, 1988; Dilman, 1994). Recently it was shown that metformin decreases platelet superoxide anion production in diabetic patients (Gargiulo et al., 2002). There are observations that Diabenol has similar effects (Popovich et al., 2005).

CONCLUSION

The striking similarities have been described between insulin/IGF-1 signaling pathways in yeast, worms, flies, and mice (Kenyon, 2001). Many characteristics of mice that are long-lived due to genetic modifications, resemble the effects of caloric restriction in wild-type (normal) animals (Anisimov, 2003; Bartke et al., 2003). A comparison of characteristics of animals exposed to these endogenous and exogenous influences shows a number of similarities but also some dissimilarities. Effects of antidiabetic biguanides seem to be more adequate in the prevention of age-related deteriorations in glucose metabolism and in insulin signaling pathway, as well as in important parameters for longevity, such as fertility and resistance to oxidative stress and tumorigenesis, than those induced by caloric restriction and genetic manipulations.

ACKNOWLEDGMENTS

This work was supported in part by Grant # 05-04-48110 from the Russian Foundation for Basic Research, and by Grant # NSh-5054.2006.4 from the President of Russian Federation.

REFERENCES

Alexandrov, V.A., Anisimov, V.N., Belous, N.M., Vasilyeva, I.A. and Mazon, V.B. (1980). The inhibition of the transplacental blastomogenic effect of nitrosomethylurea by postnatal administration of buformin to rats. *Carcinogenesis* **1**: 975-978.

Anisimov, V.N. and Dilman, V.M. (1975). The increase of hypothalamic sensitivity to the inhibition by estrogens induced by the treatment with L-DOPA, diphenylhydantoin or phenformin in old rats. *Bull. Exp. Biol. Med.* **80**(11): 96-98.

Anisimov, V.N., Belous, N.M., Vasilyeva, I.A. and Dilman, V.M. (1980). Inhibitory effect of phenformin on the development of mammary tumors induced by N-nitroso-methylurea in rats. *Exp. Onkol.* **2**(3): 40-43.

Anisimov, V.N. (1980). Effect of buformin and diphenylhydantoin on life span, estrus function and spontaneous tumor incidence in female rats. *Vopr. Onkol.* **26**(6): 42-48.

Anisimov, V.N., Ostroumova, M.N. and Dilman, V.M. (1980a). Inhibition of blastomogenic effect of 7,12-dimethylbenz(a)anthracene in female rats by buformin, dipheninhydantoin, polypeptide pineal extract and L-DOPA. *Bull. Exp. Biol. Med.* **89**: 723-725.

Anisimov, V.N., Pozharisski, K.M. and Dilman, V.M. (1980b). Effect of phenformin on the blastomogenic action of 1,2-dimethylhydrazine in rats. *Vopr. Onkol.* **26**(8): 54-58.

Anisimov, V.N., Belous, N.M. and Prokudina, E.A. (1982). Inhibition by phenformin of the radiation carcinogenesis in female rats. *Exp. Onkol.* **4**(6): 26-29.

Anisimov, V.N. (1982). Effect of phenformin on life span, estrus function and spontaneous tumor incidence in rats. *Farmakol Toksikol.* **45**(4):127.

Anisimov, V.N. (1987). *Carcinogenesis and Aging.* Vol. 2. CRC Press, Boca Raton.

Anisimov, S.V., Volkova, M.V., Lenskaya, L.V., Khavinson, V.Kh., Solovieva, D.V. and Schwartz, E.I. (2001). Age-associated accumulation of the Apolipoprotein C-III gene T-455C polymorphism C allele in a Russian population. *J. Gerontol. Biol. Sci.* **56A**: B27-B32.

Anisimov, V.N. (2003). Insulin/IGF-1 signaling pathway driving aging and cancer as a target for pharmacological intervention. *Exp. Gerontol.* **38**: 1041-1049.

Anisimov, V.N., Berstein, L.M., Egormin, P.A., Piskunova, T.S., Popovich, I.G., Zabezhinski, M.A., Kovalenko, I.G., Poroshina, T.E., Semenchenko, A.V., Provinciali, M., Re, F. and Franceschi, C. (2005). Effect of metformin on life span and on the development of spontaneous mammary tumors in HER-2/neu transgenic mice. *Exp. Gerontol.* **40**: 685-693.

Awartani, K.A. and Cheung, A.P. (2002). Metformin and polycystic ovary syndrome: a literature review. *J. Obstet. Gynecol. Can.* **24**: 393-401.

Bakaev, V.V. (2002). Effect of 1-butylbiguanide hydrochloride on the longevity in the nematoda *Caenorhabditis elegans*. *Biogerontol.* **3**(1): 23-24.

Barbieri, M., Rizzo, M.R., Manzella, D., Grella, R., Ragno, E., Carbonella, M., Abbatecola, A.M. and Paolisso, G. (2003). Glucose regulation and oxidative stress in healthy centenarians. *Exp. Gerontol.* **38**: 137-143.

Bartke, A. and Turyn, D. (2001). Mechanisms of prolonged longevity: mutants, knock-outs, and caloric restriction. *J. Anti-Aging Med.* **4**: 197-203.

Bartke, A., Chandrashekar, V., Dominici, F., Tutyn, D., Kinney, B., Steger, R. and Kopchick, J.J. (2003). Insulin-like growth factor 1 (IGF-1) and aging: controversies and new insights. *Biogerontol.* **4**: 1-8.

Berstein, L.M., Evtushenko, T.P., Tsyrlina, E.V., Bobrov, Yu.F., Ostroumova, M.N., Kovalenko, I.G., Semiglazov, V.F., Simonov, N.N. and Dilman, V.M. (1992).

Comparative study of 5- and 10-year-long results of the metabolic rehabilitation of cancer patients. In: *Neuroendocrine System, Metabolism, Immunity and Cancer (Clinical Aspects)* (Hanson, K.P. and Dilman, V.M., eds.). N.N. Petrov Research Institute of Oncology Publ., St. Petersburg, 102-112.

Berstein, L.M., Kvatchevskaya, Ju.O., Poroshina, T.E., Kovalenko, I.G., Tsyrlina, E.V., Zimarina, T.S., Ourmantcheeva, A.F., Ashrafian, L. and Thijssen, J.H.H. (2004). Insulin resistance, its consequences for clinical course of the disease and possibilities of correction in endometrial cancer. *J. Cancer Res. Clin. Oncol.* **130**: 687-693.

Berstein, L.M. (2005). Clinical usage of hypolipidemic and antidiabetic drugs in the prevention and treatment of cancer. *Cancer Lett.* **224**: 203-212.

Bespalov, V.G. and Alexandrov, V.A. (1985). Influence of anticarcinogenic agents on the transplacental carcinogenic effect of N-nitroso-N-ethylurea. *Bull. Exp. Biol. Med.* **100**: 73-76.

Bluher, M., Kahn, B.B. and Kahn, C.R. (2003). Extended longevity in mice lacking the insulin receptor in adipose tissue. *Science* **299**: 572-574.

Bowker, S.L., Majumdar, S.R., Veugelers, P. and Johnson, J.A. (2006). Increased cancer-related mortality for patients with type 2 diabetes who use sulfonylureas or insulin. *Diabetes Care* **9**: 254-258.

Caraci, F., Chisari, M., Frasca, G., Chiecho, S., Salomone, S., Pinto, A., Sortino, M.A. and Bianchi, A. (2003). Effects of phenformin on the proliferation of human tumor cell lines. *Life Sci.* **74**: 643-650.

Chiba, T., Yamaza, H., Higami, Y. and Shimokawa, I. (2002). Anti-aging effects of caloric restriction: Involvement of neuroendocrine adaptation by peripheral signaling. *Microsc. Res. Tech.* **59**: 317-324.

Clancy, D.J., Gems, D., Harshman, L.G., Oldham, S., Stocker, H., Hafen, E., Leevers, S.J. and Partridge, L. (2001). Extension of life-span by loss of CHICO, a *Drosophila* insulin receptor substrate protein. *Science* **292**: 104-106.

Colangelo, L.A., Gapstur, S.M., Gann, P.H., Dyer, A.R. and Liu, K. (2002). Colorectal cancer mortality and factors related to the insulin resistance syndrome. *Cancer Epidemiol. Biomarkers Prev.* **11**: 385-391.

Coschigano, K.T., Clemmons, D., Bellush, L.L. and Kopchick, J.J. (2000). Assessment of growth parameters and life span of GHR/BP gene-disrupted mice. *Endocrinol.* **141**: 2608-2613.

De Benedictis, G., Carotenuto, L., Carrieri, G., De Luca, M., Falcone, E., Rose, G., Calvalcanti, S., Corsonello, F., Feraco, E., Baggio, G., Bertolini, S., Mari, D., Mattace, R., Yashin, A.I., Bonafe, M. and Frenceschi, C. (1998). Gene/longevity association studies at four autosomal loci (REN, THO, PARP, SOD2). *Eur. J. Human Gen.* **6**: 534-541.

De Benedictis, G., Tan, Q., Jeune, B., Christensen, K., Ukraintseva, S.V., Bonafe, M., Franceschi, C., Vaupel, J.W. and Yashin, A.I. (2001). Recent advances in human gene-longevity association studies. *Mech. Ageing Dev.* **122**: 909-920.

Dilman, V.M. (1971). Age-associated elevation of hypothalamic threshold to feedback control and its role in development, aging and disease. *Lancet* **1**: 1211-1219.

Dilman, V.M., Sofronov, B.N., Anisimov, V.N., Nazarov, P.G. and L'vovich, E.G. (1977). Phenformin elimination of the immunodepression caused by 1,2-dimethyl-hydrazine in rats. *Vopr. Onkol.* **23**(8): 50-54.

Dilman, V.M. (1978). Ageing, metabolic immunodepression and carcinogenesis. *Mech. Ageing Dev.* **8**: 153-173.

Dilman, V.M., Berstein, L.M., Zabezhinski, M.A., Alexandrov, V.A. and Pliss, G.B. (1978). Inhibition of DMBA-induced carcinogenesis by phenformin in the mammary gland of rats. *Arch. Geschwulstforsch.* **48**: 1-8.

Dilman, V.M. and Anisimov, V.N. (1979). Hypothalamic mechanisms of ageing and of specific age pathology–I. Sensitivity threshold of hypothalamo-pituitary complex to homeostatic stimuli in the reproductive system. *Exp. Gerontol.* **14**: 161-174.

Dilman, V.M., Bobrov, J.F., Ostroumova, M.N., Lvovich, E.G., Vishnevsky, A.S., Anisimov, V.N. and Vasiljeva, I.A. (1979). Hypothalamic mechanisms of ageing and of specific age pathology–III. Sensitivity threshold of hypothalamo-pituitary complex to homeostatic stimuli in energy system. *Exp. Gerontol.* **14**: 217-224.

Dilman, V.M., Ostroumova, M.N. and Tsyrlina, E.V. (1979a). Hypothalamic mechanisms of ageing and of specific age pathology–II. Sensitivity threshold of hypothalamo-pituitary complex to homeostatic stimuli in adaptive homeostasis. *Exp. Gerontol.* **14**: 175-181.

Dilman, V.M. and Anisimov, V.N. (1979a). Potentiation of antitumor effect of cyclophosphamide and hydrazine sulfate by treatment with the antidiabetic agent, 1-phenylethylbiguanide (phenformin). *Cancer Lett.* **7**: 357-361.

Dilman, V.M. and Anisimov, V.N. (1980). Effect of treatment with phenformin, dyphenylhydantion or L-DOPA on life span and tumor incidence in C3H/Sn mice. *Gerontology* **26**: 241-245.

Dilman, V.M., Berstein, L.M., Ostroumova, M.N., Fedorov, S.N., Poroshina, T.E., Tsyrlina, E.V., Buslaeva, V.P., Semiglazov, V.F., Seleznev, I.K., Bobrov, Yu.F., Vasilieva, I.A., Kondratiev, V.B., Nemirovsky, V.S. and Nikiforov, Y.F. (1982). Metabolic immunodepression and metabolic immunotherapy: an attempt of improvement in immunologic response in breast cancer patients by correction of metabolic disturbances. *Oncology* **39**: 13-19.

Dilman, V.M., Berstein, L.M., Yevtushenko, T.P., Tsyrlina, Y.V., Ostroumova, M.N., Bobrov, Yu.F., Revskoy, S.Yu., Kovalenko, I.G. and Simonov, N.N. (1988). Preliminary evidence on metabolic rehabilitation in cancer patients. *Arch. Geschwulstforsch.* **58**: 175-183.

Dilman, V.M. (1994). *Development, Aging and Disease. A New Rationale for an Intervention.* Harwood Academic Publ., Chur.

Dominici, F.P., Arosegui Diaz, G., Bartke, A., Kopchik, J.J. and Turyn, D. (2000). Compensatory alterations of insulin signal transduction in liver of growth hormone receptor knockout mice. *J. Endocrinol.* **166**: 579-590.

Dominici, F.P., Hauck, S., Argention, D.P., Bartke, A. and Turyn, D. (2002). Increased insulin sensitivity and upregulation of insulin receptor, insulin receptor substrate (ISR)-1 and IRS-2 in liver of Ames dwarf mice. *J. Endocrinol.* **173**: 81-94.

Elahi, D., Muller, D.C., Egan, J.M., Andres, R., Veldhuist, J. and Meneilly, G.S. (2002). Glucose tolerance, glucose utilization and insulin secretion in aging. *Novartis Found. Symp.* **242**: 222-242.

Evans, J.M., Donnelly, L.A., Emslie-Smith, A.M., Alessi, A.R. and Morris, A.D. (2005). Metformin and reduced risk of cancer in diabetic patients. *BMJ* **330**: 1304-1305.

Facchini, F.S., Hua, N.W., Reaven, G.M. and Stoohs, R.A. (2000). Hyperinsulinemia: the missing link among oxidative stress and age-related diseases? *Free Radicals Biol. Med.* **29**: 1302-1306.

Facchini, F.S., Humphreys, M.H., Abbasi, F., DoNascimento, C.A. and Reaven, G.M. (2000a). Relation between insulin resistance and plasma concentrations of lipid hydroperoxides, carotenoids, and tocopherols. *Am. J. Clin. Nutr.* **72**: 776-779.

Facchini, F.S., Hua, N., Abbasi, F. and Reaven, G.M. (2001). Insulin resistance as a predictor of age-related diseases. *J. Clin. Endocrinol. Metab.* **86**: 3574-3578.

Fernandes, M.L., Saad, M.J. and Velloso, L.A. (2001). Effect of age on elements of insulin-signaling pathway in central nervous system of rats. *Endocrine* **16**: 227-234.

Flurkey, K., Papaconstantinou, J., Miller, R.A. and Harrison, D.E. (2001) Life-span extension and delayed immune and collagen aging in mutant mice with defects in growth hormone production. *Proc. Natl. Acad. Sci. USA* **98**: 6736-6741.

Gargiulo, P., Caccese, D., Pignatelli, P., Brufani, C., De Vito, F., Marino, R., Lauro, R., Violi, F., Di Mario, U. and Sanguigni, V. (2002). Metformin decreases platelet superoxide anion production in diabetic patients. *Diabetes Metab. Res. Rev.* **18**: 156-159.

Giardiello, F.M., Brensinger, J.D., Tersmette, A.C., Goodman, S.N., Petersen, G.M., Booker, S.V., Cruz-Correa, M. and Offerhaus, J.A. (2000). Very high risk of cancer in familial Peutz-Jeghers syndrome. *Gastroenterology* **119**: 1447-1453.

Gupta, K., Krishnaswamy, G., Karnad, A. and Peiris, A.N. (2002). Insulin: a novel factor in carcinogenesis. *Am. J. Med. Sci.* **323**: 140-145.

Hadley, E.C., Dutta, C., Finkelstein, J., Harris, T.B., Lane, M.A., Roth, G.S., Sherman, S.S. and Starke-Reed, P.E. (2001). Human implications of caloric restriction's effect on laboratory animals: An overview of opportunities for research. *J. Gerontol. Ser. A.* **56A** (Special Issue I): 5-6.

Holzenberger, M., Dupond, J., Ducos, B., Leneuve, P., Gefoen, A., Even, P.C., Cervera, P. and Le Bouc, Y. (2003). IGF-1 receptor regulates lifespan and resistance to oxidative stress in mice. *Nature* **421**: 182-187.

Hsieh, C.C., DeFord, J.H., Flurkey, K., Harrison, D.E. and Papaconstantinou, J. (2002). Implications for the insulin signaling pathway in Snell dwarf mouse longevity: a similarity with the C. elegans longevity paradigm. *Mech. Ageing Dev.* **123**: 1229-1244.

Hsieh, C.C., DeFord, J.H., Flurkey, K., Harrison, D.E. and Papaconstantinou, J. (2002a). Effects of the Pit1 mutation on the insulin signaling pathway: implications on the longevity of the long-lived Snell dwarf mouse. *Mech. Ageing Dev.* **123**: 1245-1255.

Hung, A.J., Stanbury, M.G., Shanabrough, M., Horvath, T.L., Garcia-Segura, L.M. and Naftolin, F. (2003). Estrogen, synaptic plasticity and hypothalamic reproductive aging. *Exp. Gerontol.* **38**: 53-59.

Huxtable, S.J., Saker, P.J., Haddad, L., Walker, M., Frayling, T.M., Levy, J.C., Hitman, G.A., O'Rahilly, S., Hattersley, A.T. and McCarthy, M.I. (2000). Analysis of parent-offspring trios provides evidence for linkage and association between the insulin gene and type 2 diabetes mediated exclusively through paternally transmitted class III variable number tandem repeat alleles. *Diabetes* **49**: 126-130.

Kenyon, C. (2001). A conserved regulatory system for aging. *Cell* **105**: 165-168.

Lane, M.A., Tilmont, E.M., De Angelis, H., Handy, A., Ingram, D.K., Kemnitz, J.W. and Roth, G.S. (2000). Short-term calorie restriction improves disease-related markers in older male rhesus monkeys (*Macaca mulatta*). *Mech. Ageing Dev.* **112**: 185-196.

Lee, C.K., Allison, D.B., Brand, J., Weindruch, R. and Prolla, T.A. (2002). Transcriptional profiles associated with aging and middle age-onset caloric restriction in mouse hearts. *Proc. Natl. Acad. Sci. USA* **99**: 14988-14993.

Longo, V.D. and Finch, C.E. (2003). Evolutionary medicine: from dwarf model systems to healthy centenarians? *Science* **299**: 1342-1346.

Lund, J., Tedesco, P., Duke, K., Wang, J., Kim, S.K. and Johnson, T.E. (2002). Transcriptional profile of aging in C. elegans. *Curr. Biol.* **12**: 1566-1573.

Masoro, E.J. (2006). Caloric restriction and ageing: controversial issues. *J. Gerontol. Biol. Sci.* **61A**: 14-19.

Mattson, M.P., Duan, W., Lee, J., Guo, Z., Roth, G.S., Ingram, D.K. and Lane, M.A. (2001). Progress in the development of caloric restriction mimetic dietary supplements. *J. Anti-Aging Med.* **4**: 225-232.

Mattson, M.P. (2002). Brain evolution and life span regulation: conservation of signaling transduction pathways that regulate energy metabolism. *Mech. Ageing Dev.* **123**: 947-953.

Mattson, M.P., Duan, W. and Maswood, N. (2002). How does the brain control lifespan? *Ageing Res. Rev.* **1**: 155-165.

Mattson, J.A., Lane, M.A., Roth, G.S. and Ingram, D.K. (2003). Calorie restriction in rhesus monkeys. *Exp. Gerontol.* **38**: 35-46.

McCarty, M.F. (1994). Longevity effect of chromium picolinate – 'rejuvenation' of hypothalamic function? *Medical Hypotheses* **3**: 253-265.

Muntoni, S. (1974). Inhibition of fatty acid oxidation by biguanides: implication for metabolic physiopathology. *Adv. Lipid Res.* **12**: 311-377.

Muntoni, S. (1999). Metformin and fatty acids. *Diabetes Care* **22**: 179-180.

Nestler, J.E., Stovall, D., Akhther, N., Iorno, M.J. and Jakubowicz, D.J. (2002). Strategies for the use of insulin-sensitizing drugs to treat infertility in women with polycystic ovary syndrome. *Fertil. Steril.* **77**: 209-215.

Paolisso, G., Gambardella, A., Ammendola, S., D'Amore, A. and Varrichio, M. (1996). Glucose tolerance and insulin action in healthy centenarians. *Am. J. Physiol.* **270**: E890-E896.

Popovich, I.G., Zabezhinski, M.A., Egormin, P.A., Tyndyk, M.L., Anikin, I.V., Spasov, A.A., Semenchenko, A.V. and Anisimov, V.N. (2005). Insulin in aging and cancer: new antidiabetic drug Diabenol as geroprotector and anticarcinogen. *Int. J. Biochem. Cell Biol.* **37**: 1117-1129.

Richards, J.S., Russell, D.L., Ochsner, S., Hsieh, M., Doyle, K.H., Falender, A.E., Lo, Y.K. and Sharma, S.C. (2002). Novel signaling pathways that control ovarian follicular development, ovulation, and luteinization. *Recent Prog. Horm. Res.* **57**: 195-220.

Rossmanith, W.G. (2001). Neuroendocrinology of aging in the reproductive system: gonadotropin secretion as an example. In: *Follicular Growth, Ovulation and Fertilization: Molecular and Clinical Basis* (Kumar, A. and Mukhopadhayay, A.K., eds.). Narosa Publ. House, New Delhi, India, 15-25.

Roth, G.S., Ingram, D.K. and Lane, M.A. (1999). Calorie restriction in primates: will it work and how will we know? *J. Am. Geriatr. Soc.* **46**: 869-903.

Roth, G.S., Lane, M.A., Ingram, D.K., Mattison, J.A., Elahi, D., Tobin, J.D., Muller, D. and Metter, E.J. (2002). Biomarkers of caloric restriction may predict longevity in humans. *Science* **297**: 811.

Ruiz-Torres, A. and Soares de Melo Kirzner, M. (2002). Ageing and longevity are related to growth hormone/insulin-like growth factor-1 secretion. *Gerontol.* **48**: 401-407.

Schneider, M.B., Matsuzaki, H., Harorah, J., Ulrich, A., Standlop, J., Ding, X.Z., Adrian, T.E. and Pour, P.M. (2001). Prevention of pancreatic cancer induction in hamsters by metformin. *Gastroenterology* **120**: 1263-1270.

Shen, Z., Wen, X.-F., Lan, F., Shen, Z.Z. and Shao, Z.-M. (2002). The tumor suppressor gene LKB1 is associated with prognosis in human breast carcinoma. *Clin. Cancer Res.* **8**: 2085-2090.

Shimokawa, I., Higami, Y., Utsuyama, M., Tuchiya, T., Komatsu, T., Chiba, T. and Yamaza, H. (2002). Life span extension by reducing in growth hormone-insulin-growth factor-1 axis in a transgenic rat model. *Am. J. Pathol.* **160**: 2259-2265.

Singal, P.K., Beamish, R.E. and Dhalla, N.S. (1983). Potential oxidative pathways of catecholamine in the formation of lipid peroxides and genesis of heart disease. *Adv. Exp. Biol. Med.* **161**: 391-401.

Tatar, M., Kopelman, A., Epstein, D., Tu, M.P., Yin, C.M. and Garofalo, R.S. (2001). A mutant *Drosophila* insulin receptor homolog that extends life-span and impairs neuroendocrine function. *Science* **292**: 107-110.

Tatar, M., Bartke, A. and Antebi, A. (2003). The endocrine regulation of aging by insulin-like signals. *Science* **299**: 1346-1351.

Tybaierg-Hansen, A., Gerdes, I.U., Overgaard, K., Ingerslev, J., Faergeman, O. and Nerup, J. (1990). Polymorphysm in 5′ flanking region of human insulin gene. Relationships with atherosclerosis, lipid levels and age in three samples from Denmark. *Arteriosclerosis* **10**: 372-378.

Ulrich, P. and Cerami, A. (2001). Protein glycation, diabetes, and aging. *Recent Prog. Horm. Res.* **56**: 1-21.

Van Heemst, D., Beekman, M., Mooijaart, S.P., Hejmans, B.T., Brandt, B.W., Zwaan, B.J., Slagboom, P.E. and Westerndorp, R.G.J. (2005). Reduced insulin/in IGF-1 signaling and human longevity. *Aging Cell* **4**: 79-85.

Vinnitski, V.B. and Iakimenko, V.A. (1981). Effect of phenformin, L-DOPA and para-chlorophenylalanine on the immunological reactivity and chemical carcinogenesis in BALB/c mice. *Vopr. Onkol.* **27** (6): 45-50.

Waterworth, D.M., Bennett, S.T., Gharani, N., McCarthy, M.I., Hague, S., Batty, S., Conway, G.S., White, D., Todd, J.A., Franks, S. and Williamson, R. (1997). Linkage and association of insulin gene VNTR regulatory polymorphism with polycystic ovary syndrome. *Lancet* **349**: 986-990.

Weaver, J.U., Kopelman, P.G. and Hitman, G.A. (1992). Central obesity and hyperinsulinemia in women are associated with polymorphism in the 5′ flanking region of the human insulin gene. *Eur. J. Clin. Invest.* **22**: 265-270.

Weindruch, R. and Walford, R. (1988). *The Retardation of Aging and Disease by Dietary Restriction*. Springfield, Ill., C.C. Thomas.

Weindruch, R., Keenan, K.P., Carney, J.M., Fernandes, G., Feuers, R.J., Halter, J.B., Ramsey, J.J., Richardson, A., Roth, G.S. and Spindler, S.R. (2001). Caloric restriction mimetics: metabolic intervention. *J. Gerontol. Biol. Sci.* **56A** (Special Issue 1): 20-33.

Wu, Y., Yakar, S., Zhao, L., Hennighausen, L. and LeRoith, D. (2002). Circulating insulin-like growth factor-1 levels regulate colon cancer growth and metastasis. *Cancer Res.* **62**: 1030-1035.

Yaghmaie, F., Garan, S.A., Massaro, M. and Timiras, P.S. (2003). A comparison of estrogen receptor-alpha immunoreactivity in the arcuate hypothalamus of young and middle-aged C57BL6 female mice. *Exp. Gerontol.* **38**: 220.

Zakikhani, M., Dowling, R., Fantus, I.G., Sonenberg, N. and Pollak, M. (2006). Metformin is an AMP kinase-dependent growth inhibitor for breast cancer cells. *Cancer Res.* **66**: 10269-10273.

CHAPTER 4

Obesity and Male Infertility

Ahmad Hammoud[1] and Douglas T. Carrel[2],*

Abstract

The male factor is responsible for 25–30% of cases of infertility; in addition, it contributes to another 30% in combination with a female factor. The causes of male infertility are various. Possible etiologies include cryptorchidism, testicular torsion or trauma, varicocele, seminal tract infections, antisperm antibodies, hypogonadotropic hypogonadism, gonadal dysgenesis and obstruction of the reproductive channels (Oehninger, 2000). With the increasing prevalence of sedentary life, diet changes and weight gain, obesity is emerging as an important cause of adverse health effects including male infertility (Magnusdottir et al., 2005). The mechanisms of association between obesity and male infertility are detailed below. The subsequent sections in this chapter evaluate and list the potential therapeutic interventions for obese infertile men.

INCIDENCE OF OBESITY AND MALE INFERTILITY

Overweight is defined as body mass index between 25 and 29.9 kg/m², and obesity as body mass index $\geq$ 30 kg/m² (United States Center for Disease Control United (CDC). The incidence of obesity is increasing to pandemic proportions. In the United States, the incidence of obesity increased from 12 to 17.9% in the period from 1991 to 1998 (Mokdad et al., 1999). During the same period, the incidence of obese males increased from 11.7 to 17.9% (Mokdad et al., 1999). A more recent study estimated the incidence of obese individuals to be 30.6% (Hedley et al., 2004). These numbers are in parallel to a reported increase in male infertility. There is an increasing belief that obesity is a chronic disease that has severe consequences on the physical and psychological well-being as well as on fecundity. It is estimated that in the

[1] Division of Reproductive Endocrinology & Infertility, Department of Obstetrics & Gynecology, University of Utah School of Medicine, 30 N. 1900 E. Suite 2B200, Salt Lake City, Utah 84132. E-mail: ahmad.hammoud@hsc.utah.edu

[2] Associate Professor of Surgery (Urology), Physiology, and OB-GYN, Director of IVF and Andrology Laboratories, University of Utah School of Medicine, 675 Arapeen Dr., Suite 205, Salt Lake City, Utah 84108.

* *Corresponding author*: Tel.: +801 581 3740, Fax: +801 581 6127.
E-mail: douglas.carrell@hsc.utah.edu

United States the sperm count may decrease by as much as 1.5% each year (Swan et al., 2000). This decline is accentuated the most in the western world where obesity is becoming more prevalent. This change seems to be recent, which may indicate a link with life style changes and the increased occurrence of obesity (Jensen et al., 2004).

RELATIONSHIP BETWEEN OBESITY AND MALE INFERTILITY

It has been shown that sperm concentration is correlated to subfertiltiy and infertility. Among men, the probability of causing a pregnancy increases with increasing sperm concentration up to 55 million/mL (Slama et al., 2002). In couples trying to conceive for the first time, the probability of succeeding increases with sperm count up to 40 million/mL (Bonde et al., 1998). In a recent report that included couples who visited a reproduction clinic, men with male fertility were shown to have three times higher prevalence of obesity than the controls. Moreover, sperm count and sperm concentration were negatively correlated to the body mass index in patients with normal sperm analysis at enrollment (Magnusdottir et al., 2005). In another report that included a total of 1,558 young Danish men, overweight men had fewer normal forms, lower sperm concentration and lower total sperm count than men with body mass index between 20 and 25 kg/m^2 (Jensen et al., 2004).

The relation between obesity and male infertility is complex. Obesity can affect male fertility through multiple mechanisms that act individually or in synergism. Obesity can cause male infertility by altered spermatogenesis or by "mechanical factor". The altered spermatogenesis is mainly due to hypogonadism and the deleterious effect of increased levels of estrogens. The mechanical cause is manifested by the erectile dysfunction resulting from hypogonadism or cardiovascular risk factors related to obesity.

OBESITY AND HYPOGONADISM

It is known that obese males have a chronic state of hypogonadism. In obese men, both total and free testosterones were shown to be decreased. The decrease in androgen levels is proportional to the degree of obesity (Giagulli et al., 1994). Adrenal androgens are diminished as well (Tchernof et al., 1995). Low androgen levels result from increased levels of circulating estrogens or obesity associated comorbidities like diabetes mellitus, insulin resistance and sleep apnea.

a) Hyperestrogenic Hypogonadotropic Hypogonadism

This entity is almost exclusive to the obese male. As it implies, males with this condition have hypogonadism (low testosterone), low gonadotropins and elevated estrogen levels. The hypoandrogenism is attributed to the increase in circulating estrogens. In fact, estrone and estradiol are both increased in obese males compared to controls (Schneider et al., 1979). The role of estrogen in male reproductive health was highlighted with the growing public concern that exposures to environmental chemicals with estrogenic activity may impact human reproductive health (Oliva et al., 2001). In humans, Pavlovich et al., identified a subset of infertile men who were found to have decreased serum testosterone-to-estradiol ratios. In this study, men with severe male infertility had significantly lower testosterone and higher estradiol than fertile control subjects, resulting in a decreased testosterone-to-estradiol ratio (Pavlovich et al., 2001). The aromatization of C19 androgens, i.e., testosterone and androstenedione, is a key step in estrogen biosynthesis and is catalyzed by the aromatase enzyme, which is a product of the *CYP*19 gene. It is believed that the increase in estrogens in obese males is due to conversion of adrenal and testicular androgens by the aromatase present in the fatty tissue (De Boer et al., 2005).

Estrogen production by the adipose tissue is dependent on the availability of androgenic precursors in the circulation (Simpson et al., 1999). Estrogens affect spermatogenesis directly within the testis or by alterations in pituitary secretion of gonadotropins. There is evidence that estrogen has a biologic activity in the hypothalamus and the pituitary. Estrogen receptors (ERα and ERβ) are present in several hypothalamic nuclei and in pituitary gonadotropes, indicating that estrogen regulates the hypothalamus-pituitary axis. Increased circulating estrogen exerts a negative feedback mechanism on the production of gonadotropins in obese males. Estrogen acts on the hypothalamus to affect GnRH (gonadotropin releasing hormone) pulses, and at the pituitary level to regulate gonadotropin [FSH (follicle stimulating hormone) and LH (luteinizing hormone)] secretion (Akingbemi et al., 2005). In severe obesity, the pituitary function appears suppressed with normal or decreased levels of LH in the presence of decreased levels of testosterone (Giagulli et al., 1994). This hormonal profile in obese males is usually described as hyperestrogenic hypogonadotropic hypogonadism.

It also appears that excess estrogen has a direct deleterious effect on spermatogenesis. This conclusion was supported by observations in rats treated with high doses of estrogen (DES (diethylstilbesterol) 10 and 1 mcg). Daily sperm production (DSP)/testis, absolute and relative weights of the testis, epididymis, and seminal vesicle and sperm numbers in both regions of the epididymis declined significantly in a dose-dependent manner after DES treatment. Among rats in the high-dose DES group, none could impregnate a female (Goyal et al., 2003).

b) Insulin Resistance and Hypogonadism

Obesity is associated with low levels of testosterone through a different mechanism. Total body fat, intra-abdominal fat and subcutaneous fat were all associated with low levels of total and free testosterone (Tsai et al., 2004). It appears that central obesity in particular is more associated with the decrease in circulating androgen levels. Insulin resistance has been reported to be associated with low testosterone levels. Age adjusted fasting insulin and

C-peptide were shown to be inversely correlated to total and free testosterone (Tsai et al., 2004). This association is confounded by the independent relation between the SHBG (sex hormone binding globulin) and insulin resistance (Stellato et al., 2000). After adjusting for SHBG levels, low testosterone was shown to be correlated to insulin resistance (Tsai et al., 2004). Whether the association between glucose intolerance and low androgen levels is an independent relation, remains a matter of debate. Both can be consequences of the metabolic imbalance associated with obesity.

c) Sleep Apnea and Hypogonadism

Obesity is commonly associated with sleep apnea. Patients with sleep apnea often have a fragmented sleep course due to repetitive episodes of upper airway obstructions and hypoxia, followed by arousal (Young et al., 2004). It has been demonstrated that patients with fragmented sleep have a blunted nocturnal rise of testosterone (Luboshitzky et al., 2001). In particular, patients with obstructive sleep apnea have lower mean testosterone and LH values compared to both young and middle aged controls. Morning testosterone levels were also lower in patients with obstructive sleep apnea (Luboshitzky et al., 2005). Finally, Patients with obstructive sleep apnea that loose weight, increase their testosterone levels (Semple et al., 1984). The decrease in testosterone levels associated with sleep apnea is primarily during the night and morning time. These alterations in testosterone levels are likely to contribute to the hypogonadism experienced by obese males.

OBESITY AND ERECTILE DYSFUNCTION

Erectile dysfunction is defined as the persistent or recurrent inability to achieve or maintain an erection sufficient for satisfactory sexual intercourse. It is known that erectile dysfunction is associated with infertility. In a recent study where participants answered the Sexual Health Inventory questionnaire, 27% of infertile men reported abnormal score (erectile dysfunction) compared to 11% of the control fertile group (O'Brien et al., 2005). In a survey for health professionals, obesity was associated with a 1.3 relative risk for erectile dysfunction (Bacon et al., 2003). In men reporting symptoms of erectile dysfunction, overweight or obesity was found in 79% of the subjects (Feldman et al., 2000).

The relation between obesity and erectile dysfunction can be partly explained by the elevated levels of several proinflammatory cytokines in obese individuals. These markers of inflammation are positively associated with endothelial dysfunction that is linked directly to male erectile dysfunction through the nitric oxide pathway (Sullivan et al., 1999). It is not clear if the association is due to an independent effect or due to cardiovascular risk factors that are commonly associated with obesity (Chung et al., 1999). In fact, well-recognized risk factors for cardiovascular diseases, such as smoking, diabetes, hypertension, and dyslipidemia are strong epidemiological links to erectile dysfunction (Feldman et al., 2000). Hypogonadism also contributes to the sexual dysfunction found in obese males (Seftel, 2005). Whether obesity is coupled with erectile dysfunction independently or through cardiovascular risk factors or hypogonadism, it is evident that obese men have a higher incidence of erectile dysfunction that affects their sexual life and fertility.

GENETIC DISORDERS ASSOCIATED WITH OBESITY AND INFERTILITY

Multiple chromosomal and genetic defects result in syndromes that include both obesity and infertility in males. Some of these entities are common, others are less frequent but their knowledge is essential since obesity or infertility can be the presenting symptom. Moreover, detecting these genetic aberrations allows couples to be informed about the potential to transmit genetic abnormalities to their offspring.

a. Prader Willi Syndrome

This syndrome is the most common syndromal cause of human obesity. It is characterized by diminished fetal activity, hypotonia, mental retardation, short stature, hypogonadotropic hypogonadism and morbid obesity (Farooqi, 2005). This is an imprinting disease caused by the absence of the paternal segment 15q11.2-q12, either through deletion (75%) or through loss of the entire paternal chromosome with the presence of two maternal homologues (22%) of patients. Prometaphase banding examination can visualize deletions of chromosome 15 that accounts for 70–80% of cases (Goldstone, 2004). Cryptorchidism is found in 80–100% of patients with this syndrome. In addition, hypoplastic external genitalia and delayed or incomplete pubertal growth (Suzuki et al., 2002). Testicular biopsy shows undifferentiated Leydig cells and arrested spermatogenesis at the late spermatid level (Diemer and Desjardins, 1999).

b. Laurence-Moon and Bardet-Biedel Syndromes

These are rare autosomal recessive disorders. They consist of common multiple defects including mental retardation, retinal pigmentary dystrophy, hypogonadism and morbid obesity (Mohsin et al., 2003). Polydactylia is found mainly in Laurence-Moon syndrome. The differentiation between these two syndromes is not unanimous, for many, they constitute different expressions of one entity "the Laurence-Moon-Bardet-Biedel Syndrome" (Yamada et al., 2000). Known genetic abnormalities include defects in the four loci BBS1 to BBS4. Infertility is frequent, but some patients have normal testicular function and spermatogenesis (Diemer and Desjardins, 1999).

c. Klinfelter Syndrome

Klinfelter syndrome is the most common chromosomal disease causing infertility in men. This syndrome results forms a numerical aberration in sex chromosome. The typical karyotype is 47,XXY. Phenotypically, patients can be obese with a feminine pattern of fat distribution and gynecomastia. They have tall stature, small testis and hypogonadism with severe alteration is spermatogenesis, practically azoospermia (Diemer and Desjardins, 1999). In adult patients, about 70% complain of decreased libido and impotence. The numerical chromosome aberrations in this syndrome are due to non-disjunction predominantly during meiotic divisions occurring in germ-cell development or less frequently in early embryonic mitotic cell divisions. It is not known if the defect in the testis arises from an intrinsic problem in germ cells, or is due to lack of support from the Sertoli cells (Lanfranco, 2004). The diagnosis is confirmed by a karyotype analysis of blood lymphocytes (Kamischke et al., 2003).

EVALUATION OF THE OBESE INFERTILE PATIENT

The medical history of these patients should focus on ruling out other etiologies for male infertility. Relevant history includes prior fertility, childhood illnesses such as viral orchitis or cryptorchidism, genital trauma or prior pelvic or inguinal surgery, infections such as epididymitis or urethritis, gonadotoxin exposures such as prior radiation therapy/chemotherapy, recent fever or heat exposure, current medications, family history of birth defects, mental retardation, reproductive failure or cystic fibrosis, prior medical problem in particular severe liver disease, history of smoking, use of alcohol, illegal drug use, anabolic steroids, or over the counter sperm enhancement formula (Jarow et al., 2002). The clinician should also look for symptoms of hypogonadism including sexual dysfunction, such as reduced libido, erectile dysfunction, diminished penile sensation, difficulty in attaining orgasm as well as reduced ejaculation with orgasm, reduced energy, depressed mood or diminished sense of well-being, increased irritability, difficulty in concentrating, and other cognitive problems (Seftel, 2005).

Physical examination should note the weight and height for BMI (body mass index) calculation. The exam should also determine the testis size and consistency, consistency of the epididymides, presence of a varicocele, and signs of hypogonadism that include anemia, muscle wasting, absence or regression of secondary sex characteristics including body habitus, hair distribution and gynecomastia (Jarow et al., 2002).

A hormonal profile consisting of serum estradiol, total testosterone, free testosterone index, levels of FSH, LH, prolactin and SHBG should be obtained. Hypogonadotropic obese males usually have low serum testosterone levels (<300 ng/dl), high estrogen, normal or low gonadotropins and decreased SHBG. The semen analysis is the cornerstone of the laboratory evaluation of these infertile men. At least two properly performed semen analyses are necessary to confirm the diagnosis. If possible, the two semen analyses should be separated by a period of at least one month. The specimen could be collected at home or in the laboratory. It could also be collected by masturbation or by intercourse, using special semen collection condoms that do not contain substances such as latex rubber that are detrimental to sperm. The semen specimen should be maintained at room or body temperature and examined within one hour of collection. The semen analysis parameters analyzed included volume of ejaculate (ml.), sperm concentration (million per ml.), motility (percent) and morphology (percent) (WHO criteria). The motility index (10^{-6} motile sperm per ejaculate) or total motile sperm can also be analyzed. In infertile obese males, the semen analysis is likely to yield oligozoospermia (Jensen et al., 2004).

An important part of the evaluation is the assessment of comorbidities such as cardiovascular diseases including coronary artery disease and stroke, gallbladder diseases, liver diseases and diabetes. The comorbidities can aggravate infertility or affect treatments options (National Task Force on the Prevention and Treatment of Obesity, 2000).

Therapeutic Interventions

It is important to realize that the relation between obesity and male infertility has not been studied extensively in the literature, as opposed to obesity and female infertility. As a consequence, there is a paucity of studies targeted specifically at the treatment of infertility in obese males as an outcome. The therapeutic interventions discussed in the following section are targeted to reserve the previously discussed abnormalities associated with obesity. Treatment of some of these reversible conditions would improve male fertility and might allow conception to occur with intercourse.

a. Weight Loss and Exercise

Epidemiological studies suggest that modifying health behaviors, including physical activity and leanness, is associated with a reduced risk for sexual dysfunction (Bacon et al., 2003). In a randomized study, obese men who received detailed advice about how to achieve a loss of 10% or more in their total body weight by reducing caloric intake and increasing their level of physical activity, had a higher rate of weight loss and improvement in erectile dysfunction than control (Esposito et al., 2004). Patients with obstructive sleep apnea that loose weight increase their testosterone levels (Semple et al., 1984). Reports stating that bariatric surgery can result in male infertility are worth mentioning. In a recent study, six severely obese men with at least one previous child were found to be azoospermic after Roux-en-Y gastric bypass operation (di Frega et al., 2005). More studies are needed, however, empiric evidence suggests that modification of life style and weight loss may improve semen quality.

b. Aromatase Inhibitors

In oligospermic men with obesity, aromatase inhibitors were proposed to improve spermatogenesis. Serum estrogens of six healthy obese men (body mass index of 38 to 73) were lowered by administering the aromatase inhibitor testolactone (1 g daily for six weeks). Twenty-four-hour mean serum estradiol decreased, and 24-hour mean serum testosterone and luteinizing hormone (LH) increased in all six patients (Zumoff et al., 2003). In a more recent report, Raman evaluated the effect of anastrozole, a newer and more selective aromatase inhibitor, on the hormonal and semen profiles of infertile men with decreased baseline testosterone-to-estradiol ratios. During anastrozole treatment, increases in the testosterone-to-estradiol ratios were seen. This improvement of hormonal parameters was noted for all subgroups. In this report, a total of 25 oligospermic men had an increase in semen volume, sperm concentration and motility index after anastrozole treatment. These changes were similar to those observed in men treated with testolactone. The anastrozole treatment group did show greater decreases in estradiol concentration and increases in testosterone-to-estradiol ratios (Raman and Schlegel, 2002).

c. Testosterone Therapy

In severely hypogonadic men that have low testosterone levels, androgen substitution therapy is indicated to avoid symptoms and sequelae of androgen deficiency. It also helps treat the sexual dysfunction, particularly the decreased in libido (Snyder et al., 2000). However, testosterone treatment in infertile men can have a deleterious effect on spermatogenesis and fertility. Testosterone can block gonadotropin secretion through a negative feedback at hypothalamic-pituitary level. In fact, androgens are used for male contraception (Kamischke et al., 2004)

d. Phosphodiesterase Type 5 Inhibitor

For patients with erectile dysfunction, Sildenafil (Viagra®) was the first oral phosphodiesterase (PDE)-inhibitor to become commercially available. It is considered the first line treatment of erectile dysfunction. The major contra-indications of Sildenafil are concomitant treatment with nitrates or nitric oxide-donating drugs (including amyl nitrite poppers), patients in whom sexual activity is inadvisable (those with unstable angina, severe heart failure, recent infarction), and patients who are allergic or intolerant to the drug. Tadalafil and vardenafil are two newer oral agents used for erectile dysfunction (Brotons et al., 2004).

e. Artificial Reproductive Techniques

If the previous treatments were unsuccessful in reversing semen abnormalities, patients with mild oligospermia can be offered uterine insemination or in vitro fertilization. For severely oligospermic patients, intracytoplasmic sperm injection (ICSI) is a viable option. The criteria for ICSI include severe oligozoospermia (<2–10×10^6 sperm/mL), severe asthenozoospermia (<5–10% motile spermatozoa), poor sperm morphology (<4% normal oval forms) and use of surgically retrieved spermatozoa e.g., for azoospermic patients (Khorram et al., 2001).

CONCLUSION

The parallel increase in obesity and male infertility suggest a link between the two diseases. Obesity is associated with hypogonadism, hyperestrogemia and erectile dysfunction; all these factors affect the ability of a male to participate in the conception of a child. In the absence of clinical trials, treatment options of obese infertile males are essentially empiric and targeted towards reversing the hypogonadism and erectile dysfunction.

REFERENCES

Akingbemi, B.T. (2005). Estrogen regulation of testicular function. *Reprod. Biol. Endocrinol.* 3: 51.

Bacon, C.G., Mittleman, M.A., Kawachi, I., Giovannucci, E., Glasser, D.B. and Rimm, E.B. (2003). Sexual function in men older than 50 years of age: results from the health professionals follow-up study. *Ann. Intern. Med.* **139**(3): 161-168.

Brotons, F.B., Campos, J.C., Gonzalez-Correales, R., Martin-Morales, A., Moncada, I. and Pomerol, J.M. (2004). Core document on erectile dysfunction: key aspects in the care of a patient with erectile dysfunction. *Int J. Impot. Res.* (2): S26-S39.

Bonde, J.P., Ernst, E., Jensen, T.K., Hjollund, N.H., Kolstad, H., Henriksen, T.B., Scheike, T., Giwercman, A., Olsen, J. and Skakkebaek, N.E. (1998). Relation between semen quality and fertility: a population-based study of 430 first-pregnancy planners. *Lancet.* **352**(9135): 1172-1177.

Chung, W.S., Sohn, J.H. and Park, Y.Y. (1999). Is obesity an underlying factor in erectile dysfunction? *Eur. Urol.* **36**(1): 68-70.

De Boer, H., Verschoor, L., Ruinemans-Koerts, J. and Jansen, M. (2005). Letrozole normalizes serum testosterone in severely obese men with hypogonadotropic hypogonadism. *Diabetes Obes. Metab.* **7**(3): 211-215.

Dhindsa, S., Prabhakar, S., Sethi, M., Bandyopadhyay, A., Chaudhuri, A. and Dandona, P. (2004). Frequent occurrence of hypogonadotropic hypogonadism in type 2 diabetes. *J. Clin. Endocrinol. Metab.* **89**: 5462-5468.

di Frega, A.S., Dale, B., Di Matteo, L. and Wilding, M. (2005). Secondary male factor infertility after Roux-en-Y gastric bypass for morbid obesity: case report. *Hum. Reprod.* **20**(4): 997-998.

Diemer, T. and Desjardins, C. (1999). Developmental and genetic disorders in spermatogenesis. *Hum. Reprod. Update.* **5**(2): 120-140.

Esposito, K., Giugliano, F., Di Palo, C., Giugliano, G., Marfella, R., D'Andrea, F.D., Armiento, M. and Giugliano, D. (2004). Effect of lifestyle changes on erectile dysfunction in obese men: a randomized controlled trial. *JAMA* **291**(24): 2978-2984.

Farooqi, I.S. (2005). Genetic and hereditary aspects of childhood obesity. *Best Pract. Res. Clin. Endocrinol. Metab.* **19** (3): 359-374.

Feldman, H.A., Johannes, C.B., Derby, C.A., Kleinman, K.P., Mohr, B.A., Araujo, A.B. and McKinlay, J.B. (2000). Erectile dysfunction and coronary risk factors: prospective results from the Massachusetts male aging study. *Prev. Med.* **30**(4): 328-338.

Giagulli, V.A., Kaufman, J.M. and Vermeulen, A. (1994). Pathogenesis of the decreased androgen levels in obese men. *J. Clin. Endocrinol. Metab.* **79**: 997-1000.

Goldstone, A.P. (2004). Prader-Willi syndrome: advances in genetics, pathophysiology and treatment. *Trends Endocrinol. Metab.* **15**(1): 12-20.

Goyal, H.O., Robateau, A., Braden, T.D., Williams, C.S., Srivastava, K.K. and Ali, K. (2003). Neonatal estrogen exposure of male rats alters reproductive functions at adulthood. *Biol. Reprod.* **68**(6): 2081-2091.

Hedley, A.A., Ogden, C.L., Johnson, C.L., Carroll, M.D., Curtin, L.R. and Flegal, K.M. (2004). Prevalence of overweight and obesity among US children, adolescents, and adults. *JAMA* **291**(23): 2847-2850.

Jarow, J.P., Sharlip, I.D., Belker, A.M., Lipshultz, L.I., Sigman, M., Thomas, A.J., Schlegel, P.N., Howards, S.S., Nehra, A., Damewood, M.D., Overstreet, J.W. and Sadovsky, R. (2002). Male Infertility Best Practice Policy Committee of the American Urological Association Inc. Best practice policies for male nfertility. *J. Urol.* **167**(5): 2138-2144.

Jensen, T.K., Andersson, A.M., Jorgensen, N., Andersen, A.G., Carlsen, E., Petersen, J.H. and Skakkebaek, N.E. (2004). Body mass index in relation to semen quality and reproductive hormones among 1,558 Danish men. *Fertil. Steril.* **82**(4): 863-870.

Kamischke, A., Baumgardt, A., Horst, J. and Nieschlag, E. (2003). Clinical and diagnostic features of patients with suspected Klinefelter syndrome. *J. Androl.* **24**(1): 41-48.

Kamischke, A. and Nieschlag, E. (2004). Progress towards hormonal male contraception. *Trends. Pharmacol. Sci.* **25**(1): 49-57.

Khorram, O., Patrizio, P., Wang, C. and Swerdloff, R. (2001). Reproductive technologies for male infertility. *J. Clin. Endocrinol. Metab.* **86**(6): 2373-2379.

Lanfranco, F., Kamischke, A., Zitzmann, M. and Nieschlag, E. (2004). Klinefelter's syndrome. *Lancet.* **364**(9430): 273-283.

Luboshitzky, R., Zabari, Z., Shen-Orr, Z., Herer, P. and Lavie, P. (2001). Disruption of the nocturnal testosterone rhythm by sleep fragmentation in normal men. *J. Clin. Endocrinol. Metab.* **86**(3): 1134-1139.

Luboshitzky, R., Lavie, L., Shen-Orr, Z. and Herer, P. (2005). Altered luteinizing hormone and testosterone secretion in middle-aged obese men with obstructive sleep apnea. *Obes. Res.* **13**(4): 780-786.

Magnusdottir, E.V., Thorsteinsson, T., Thorsteinsdottir, S., Heimisdottir, M. and Olafsdottir, K. (2005). Persistent organochlorines, sedentary occupation, obesity and human male subfertility. *Hum. Reprod.* **20**(1): 208-215.

Mohsin, N., Marhuby, H., Maimani, Y., Aghanishankar, P., Al-Hassani, M., Seth, M. and Daar, A.S. (2003). Development of morbid obesity after transplantation in Laurence Moon Biedle syndrome. *Transplant Proc.* **35**(7): 2619.

Mokdad, A.H., Serdula, M.K., Dietz, W.H., Bowman, B.A., Marks, J.S. and Koplan, J.P. (1999). The spread of the obesity epidemic in the United States, 1991-1998. *JAMA* **282**(16): 1519-1522.

National Center for Chronic Prevention and Health Promotion. (2005). Defining overweight and obesity. Available at: http://www.cdc.gov/nccdphp/dnpa/obesity/defining.htm Accessed November, 2005.

National Task Force on the Prevention and Treatment of Obesity. (2000). Overweight, obesity, and health risk. *Arch. Intern. Med.* **160**(7): 898-904.

O'Brien, J.H., Lazarou, S., Deane, L., Jarvi, K. and Zini, A. Erectile dysfunction and andropause symptoms in infertile men. *J. Urol.* **174**(5): 1932-1934.

Oehninger, S. (2000). Clinical and laboratory management of male infertility: an opinion on its current status. *J. Androl.* **6**: 814-821.

Oliva, A., Spira, A. and Multigner, L. (2001). Contribution of environmental factors to the risk of male infertility. *Hum. Reprod.* **16**(8): 1768-1776.

Pavlovich, C.P., King, P., Goldstein, M. and Schlegel, P.N. (2001). Evidence of a treatable endocrinopathy in infertile men. *J. Urol.* **165**: 837-841.

Raman, J.D. and Schlegel, P.N. (2002). Aromatase inhibitors for male infertility. *J. Urol.* **167**: 624-629.

Schneider, G., Kirschner, M.A., Berkowitz, R. and Ertel, N.H. (1979). Increased estrogen production in obese men. *J. Clin. Endocrinol. Metab.* **48**: 633-638.

Seftel, A. (2005). Male hypogonadism. Part II: etiology, pathophysiology, and diagnosis. *Int. J. Impot. Res.*

Semple, P.A., Graham, A., Malcolm, Y., Beastall, G.H. and Watson, W.S. (1984). Hypoxia, depression of testosterone, and impotence in pickwickian syndrome reversed by weight reduction. *Br. Med. J. (Clin. Res. Ed.).* **289**(6448): 801-802.

Simpson, E., Rubin, G., Clyne, C., Robertson, K., O'Donnell, L., Davis, S. and Jones, M. (1999). Local estrogen biosynthesis in males and females. *Endocr. Relat. Cancer* **6**(2): 131-137.

Slama, R., Eustache, F., Ducot, B., Jensen, T.K., Jorgensen, N., Horte, A., Irvine, S., Suominen, J., Andersen, A.G., Auger, J., Vierula, M., Toppari, J., Andersen, A.N., Keiding, N., Skakkebaek, N.E., Spira, A. and Jouannet, P. (2002). Time to pregnancy and semen parameters: a cross-sectional study among fertile couples from four European cities. *Hum. Reprod.* **17**(2): 503-515.

Snyder, P.J., Peachey, H., Berlin, J.A., Hannoush, P., Haddad, G., Dlewati, A., Santanna, J., Loh, L., Lenrow, D.A., Holmes, J.H., Kapoor, S.C., Atkinson, L.E. and Strom, B.L. (2000). Effects of testosterone replacement in hypogonadal men. *J. Clin. Endocrinol. Metab.* **85**(8): 2670-2677.

Stellato, R.K., Feldman, H.A., Hamdy, O., Horton, E.S. and McKinlay, J.B. (2000). Testosterone, sex hormone-binding globulin, and the development of type 2 diabetes in middle-aged men: prospective results from the Massachusetts male aging study. *Diabetes Care* **23**: 490-494.

Sullivan, M.E., Thompson, C.S. and Dashwood, M.R. (1999). Nitric oxide and penile erections: is erectile dysfunction another manifestation of vascular disease? *Cardiovasc. Res.* **43**: 658-665.

Suzuki, Y., Sasagawa, I., Tateno, T., Yazawa, H., Ashida, J. and Nakada, T. (2002). Absence of microdeletions in the Y chromosome in patients with Prader-Willi syndrome with cryptorchidism. *Int. J. Androl.* **25**(1): 1-5.

Swan, S.H., Elkin, E.P. and Fenster, L. (2000). The question of declining sperm density revisited: an analysis of 101 studies published 1934-1996. *Environ. Health Perspect.* **108**: 961-966.

Tchernof, A., Després, J.P. and Belanger, A. et al. (1995). Reduced testosterone and adrenal C19 steroid levels in obese men. *Metabolism* **44**: 513-519.

Tsai, E.C., Matsumoto, A.M., Fujimoto, W.Y. and Boyko, E.J. (2004). Association of bioavailable, free, and total testosterone with insulin resistance: influence of sex hormone-binding globulin and body fat. *Diabetes Care* **27**(4): 861-868.

World Health Organization Laboratory. (1999). Manual for the examination of human semen and sperm-cervical mucus interaction, 4th edition. Cambridge University Press, Cambridge.

Yamada, K., Miura, M., Miyayama, H., Sakashita, N., Kochi, M. and Ushio, Y. (2000). Diffuse brainstem glioma in a patient with Laurence-Moon-(Bardet-)Biedl syndrome. *Pediatr. Neurosurg.* **33**(6): 323-327.

Young, T., Skatrud, J. and Peppard, P.E. (2004). Risk factors for obstructive sleep apnea in adults. *JAMA* **291**(16): 2013-2016.

Zumoff, B., Miller, L.K. and Strain, G.W. (2003). Reversal of the hypogonadotropic hypogonadism of obese men by administration of the aromatase inhibitor testolactone. *Metabolism* **52**: 1126-1128.

CHAPTER 5

Melatonin and Induction and/or Growth of Experimental Tumors

Michal Karasek

Abstract

The majority of data linking the pineal gland and tumorigenesis point to its oncostatic action, exerted mainly by its hormone, melatonin, although, there are also reports indicating that the pineal gland has neither any direct effect nor any stimulatory effect on the development and/or growth of some tumors.

This chapter concentrates on the role of melatonin in the induction and/or growth of experimentally-induced animal tumors in vivo and both animal and human tumor cells in vitro.

Melatonin administration resulted in decrease of the incidence and the multiplicity of malignant tumors, as well as in tumor size. It also exerts direct antiproliferative and pro-apoptotic effects and protects DNA and cellular membranes from the oxidative stress abuse caused by carcinogens.

The mechanisms underlying oncostatic action of melatonin seems to be complex, and several well-known activities of melatonin could be responsible for the indirect and direct inhibition of tumor promotion and progression. These mechanisms may include modulation of the endocrine system system, modulation of the immune system, antioxidant action, direct action on tumor cells proliferation and/or apoptosis, inhibition of angiogenesis, signal transduction events, and modulation of hormone receptor expression and regulation.

INTRODUCTION

Although investigations on relationship between the pineal gland and neoplastic disease have a long history, because the first possible association between the pineal and tumors was made in 1929 (Georgiou, 1929), real progress in this area of research has only been made in the last two decades. The majority of data linking the pineal gland and tumorigenesis point to its

Department of Neuroendocrinology, Chair of Endocrinology, Medical University of Lodz, 92-216 Lodz, Czechoslowacka 8/10, Poland. Tel.:/Fax: +48 42 675 7613. E-mail: karasek@umed.lodz.pl oncostatic action, exerted mainly by its hormone, melatonin. However, there are also reports indicating that the pineal gland has neither any direct effect nor any stimulatory effect on the development and/or growth of some tumors (Blask, 1984, 1993; Karasek and Fraschini, 1991; Karasek, 1994, 1997; Karasek and Pawlikowski, 1999; Pawlikowski et al., 2002).

This review concentrates on the role of melatonin in the induction and/or growth of experimentally-induced animal tumors *in vivo* and both animal and human tumor cells *in vitro*.

INFLUENCE OF MELATONIN ON EXPERIMENTALLY-INDUCED ANIMAL TUMORS *IN VIVO* AND ON ANIMAL AND HUMAN TUMOR CELLS *IN VITRO*

Numerous studies dealing with the effects of melatonin on the development and/or growth of various types of experimentally-induced malignant tumors, both *in vivo* and *in vitro*, have been performed in the last three decades. The results of these studies are summarized below.

Melatonin administration resulted in decrease of the incidence and the multiplicity of malignant tumors, as well as in tumor size (Blask, 1993; Musatov et al., 1999). It also exerts direct antiproliferative and pro-apoptotic effects (Eck et al., 1998; Melen-Mucha et al., 1998; Anisimov et al., 2000a, b; Winczyk et al., 2001) and protects DNA and cellular membranes from the oxidative stress abuse caused by carcinogens (Karbownik, 2002).

Tumors of the Reproductive System

Mammary Cancer

Various types of mammary cancer have been the most extensively used for studying the effect of melatonin on tumorigenesis. It appears from many studies that *in vivo* melatonin has a potent inhibitory influence on the development and/or growth of mammary cancer induced by 7,12-dimethylbenzanthracene (DMBA) (Aubert et al., 1980; Tamarkin et al., 1981; Shah et al., 1984; Blask et al., 1986), or by N-methylnitrosourea (MNU) (Blask et al., 1988, 1991; Kothari et al., 1995), transplantable R3290AC mammary cancer in the rat (Karmali et al., 1978), transplantable mammary tumor in the mouse (Anisimov et al., 1973), spontaneous C3H/Jax mammary cancer in the mouse (Subramanian and Kothari, 1991), mammary cancer with c-neu oncogene in transgenic mouse (Rao et al., 2000), and spontaneous mammary cancer in HER-2/neu transgenic mouse (Anisimov et al., 2003). Interestingly, the response of tumor growth to melatonin administration was different in early (slow-growing) and advanced (fast growing) passage of DMBA-induced mammary cancer; melatonin inhibited the early passage growth but had absolutely no effect on advanced passage (Bartsch et al., 1994b). Moreover, serum melatonin concentrations were slightly elevated in animals bearing slow-growing passage, and significantly depressed in fast-growing passage of the tumor (Bartsch et al., 1994a).

In vitro melatonin inhibited proliferation of estrogen receptor-positive human MCF-7 breast cancer line (at concentrations close to physiological – 10^{-9} to 10^{-11} M) (Hill and Blask, 1988; Cos and Blask, 1994; Crespo et al., 1994; Cos et al., 1996a; Blask et al., 1997c; Mediavilla et al., 1999), although some authors detected cytostatic or cytotoxic effects of melatonin only in high pharmacological concentrations (10^{-3} M) (Shellard et al., 1989; Bartsch et al., 1992; Papazisis et al., 1998). Melatonin also inhibited proliferation of EMF-19 (at concentration 10^{-3} M) (Bartsch et al., 1992), and ZR75-1 (at concentration 10^{-9} M) (Blask et al., 1997c) breast cancer cell lines. Moreover, melatonin-treated MCF-7 cells presented greater differentiation, in keeping with their epithelial origin, and displayed ultrastructural features suggestive for cellular injury (Hill and Blask, 1988; Crespo et al., 1994). Estrogen receptor –negative HS578T breast cancer line was not affected by melatonin (at concentration 10^{-9} M) (Blask et al., 1997c). In human T47D breast cancer line, melatonin did not influence cancer growth at concentrations 10^{-3} to 10^{-12} M (Papazisis et al., 1998), whereas in primary human cell culture lines, effects of melatonin depended on the case (inhibition, stimulation, or no effect were observed) (Bartsch et al., 2000). Melatonin exerted inhibitory effect on the proliferation of 16/C mammary cancer cell line, comparable to typical oncostatic agent methotrexate (Winczyk et al., 2006).

Interestingly, capacity of melatonin to augment the inhibitory action of tamoxifen on cultured human MCF-7 breast cancer cells was demonstrated. A comparison of IC_{50} values (i.e., the concentration required to inhibit tumor cell growth by 50%) indicates that tamoxifen, an inhibitor of the growth of these cells, is approximately 100 times more potent, following the pretreatment with a physiological concentration of melatonin (Wilson et al., 1992). Also, the combined use of melatonin and 9-cis-retinoic acid produces additive or synergistic anti-tumor effects on NMU-induced model of mammary carcinogenesis (Nowfar et al., 2002).

It was demonstrated that melatonin inhibits MCF-7 cell proliferation by inducing an arrest of cycle dependent on an increased expression of p21WAF1 protein, which is mediated by the p53 pathway (Mediavilla et al., 1999). Observation that the addition of melatonin to the cultures of MCF-7 cells with conditioned medium from estradiol-treated cells significantly inhibited the growth stimulatory activity of conditioned medium, suggests that melatonin may inhibit the action and/or release of growth stimulatory factors, as well as may stimulate release of growth inhibitory factors in culture (Cos and Blask, 1994). Down-regulation of estrogen receptors expression by melatonin (Molis et al., 1993,1995) suggests that inhibition of breast cancer cell proliferation by melatonin may occur via suppression of the tumor cell's estrogen response pathway (Hill et al., 2001). It was also demonstrated that glutathione is required for melatonin action on estrogen receptor-positive (MCF-7 and ZR75-1) breast cancer cells, but also resulted in 63% decrease in the number of cancer cells of estrogen receptor-negative HS578T breast cancer line, which is normally unresponsive to melatonin (Blask et al., 1997c).

Uterine and Vagina Cancer

Melatonin administration inhibited *in vivo* growth of uterine and vagina cancer induced by DMBA in the rat (Anisimov et al., 2000b) as well as *in vitro* growth of ME-180 cervical cancer cell line at concentration 2×10^{-3} M) (Chen et al., 1995), and uterine endometrial Ishikawa adenocarcinoma cell line (at concentrations 10^{-8} to 10^{-12} M) (Kanishi et al., 2000). However, it was ineffective in human SNG uterine endometrial cancer cell line (Kanishi et al., 2000).

Ovarian Cancer

Melatonin inhibited *in vitro* the growth (judged by IC_{50} value) of human JA-1 and SK-OV-3 ovarian cancer cell lines (at concentrations 505 and 520 µg of melatonin/mL, and 505 and 600 µg/mL, respectively) (Leone et al., 1988; Shellard et al., 1989). Melatonin (at concentrations 10^{-7} to 10^{-9} M) also caused a dose-dependent reduction in cell number in BG-1 ovarian adenocarcinoma cell line *in vitro* (Petranka et al., 1999). In human EFO-27 ovarian carcinoma cell line melatonin (at concentrations 10^{-11} to 10^{-3} M) did not influence cell growth *in vitro* (Bartsch et al., 1992), whereas in primary cell cultures of human ovarian carcinoma effects of melatonin depended on the case (no influence on the growth of four out of seven tumors, inhibition in two tumors, and stimulation in one tumor) (Bartsch et al., 2000). Melatonin (at concentrations 10^{-12} to 10^{-6} M) did not influence the proliferation of OVCAR-3 ovarian cell cancer cell line, although it enhanced sensitivity of these cells to cis-diamminedichloroplatimum (Futagami et al., 2001).

Choriocarcinoma

Transplantable JEG-3 and Jar choriocarcinomas growth *in vivo* was inhibited by melatonin administration (Shiu et al., 2000). Melatonin also inhibited growth of human JA-1 (at concentration 1.61×10^{-4} M) (Sze et al., 1993) and human JEG-3 (at concentrations 10^{-5} to 10^{-10} M) (Shiu et al., 2000) chorioncarcinoma cell lines.

Prostatic Cancer

Melatonin administration reduced the tumor weight and volume, as well as increased tumor doubling time of transplantable, androgen-dependent 3327H Dunning prostate adenocarcinoma in the rat (Philo and Berkowitz, 1988). In the slow-growing, well-differentiated, androgen-dependent W-2 Dunning prostate cancer melatonin did not influence tumor growth, but in the fast-growing, poorly-differentiated, androgen-independent R3327HIF Dunning cancer subline it stimulated tumor growth (Buzzell et al., 1988).

Siu et al. (2002) did not observe *in vivo* the effects of melatonin on proliferation or growth of androgen-independent DU 145 and PC-3 prostate cancer inoculated to athymic nude mice, although melatonin decreased incidence and growth rate of androgen-dependent LNCaP prostate cancer in these animals.

Inhibition of cell proliferation by melatonin (at concentrations 5×10^{-5} to 5×10^{-9} M) was found *in vitro* in androgen-dependent LNCaP prostate cancer cell line (Lupowitz and Zisapel, 1999; Xi et al., 2000). Melatonin *in vitro* (at concentrations $0.5\text{-}5 \times 10^{-9}$ M) also significantly inhibited cell proliferation and induced cell cycle withdrawal by accumulating cells in G0/G1 phase in human androgen-independent DU 145 prostate cancer cell line (Montagnani Marelli et al., 2000). Differential effects of melatonin on proliferation of androgen-independent PC3 prostate cancer cells *in vitro* according to the cell concentration were reported; at low cell density increase in ^{3}H-thymidine incorporation was observed, whereas at high cell density attenuation of thymidine incorporation was found (Gilad et al., 1999).

Tumors of the Digestive System

Hepatoma

A highly significant inhibition of the proliferation of transplanted Kirkman-Robbins hepatoma cells in the Syrian hamster was observed *in vivo* following melatonin administration, whereas *in vitro* melatonin (at concentrations 10^{-5} to 10^{-11} M) was ineffective in these cells (Karasek et al., 1992).

Melatonin treatment was effective in delaying the appearance, suppressing the growth rate, and inhibiting ^{3}H-thymidine incorporation in 7288CTC hepatoma in the rat (Blask et al., 1997a, b, 1998, 1999c). Antiproliferative and pro-apoptotic effects of melatonin (at concentrations 6.4×10^{-7} to 3.2×10^{-3} M) were demonstrated in mouse HEPA 1-6 hepatoma cell line *in vitro*. Moreover antiproliferative influence of tamoxifen and ethanol in these cells were enhanced by melatonin (Hermann et al., 2002).

In the ascites AH 130 hepatoma in the rat melatonin administration inhibited cellular proliferation, doubled mean lifetime, and increased survival, without changes in the apoptotic index. Melatonin slowed cell cycle progression in this tumor type by increasing the number of cells in phase G0/G1 (Cini et al., 1998).

Melatonin also exhibited an anti-tumor-promoting ability in diethylnitrosamine-initiated and phenobarbital-promoted hepatocarcino-genesis in the rat (Rahman et al., 2003).

Intestinal Cancer

Anisimov et al. (1997, 2000) found that *in vivo* melatonin inhibited the development, cell proliferation, multiplicity, and also enhanced the number of apoptotic cells in 1,2-dimethylhydrazine-induced intestinal tumors. In both *in vivo* and *in vitro*, the oncostatic action of melatonin was observed on transplantable murine Colon 38 adenocarcinoma which depends on both antiproliferative and pro-apoptotic effects (Karasek et al., 1998; Pawlikowski et al., 1999; Winczyk et al., 2001, 2002a). *In vitro* antiproliferative action of high, pharmacological concentrations of melatonin ($1\text{-}3 \times 10^{-3}$ M) but not of low, physiological concentrations (10^{-7} to 10^{-9} M) was observed in murine

CT-26 colon carcinoma cell line (Farriol et al., 2000). On the contrary, melatonin (at concentrations 10^{-4} to 10^{-13} M) was ineffective in human DLD-1 adenocarcinoma cell line (Leone and Wilkinson, 1991).

Tumors of the Endocrine System

Antiproliferative effect of melatonin was demonstrated both *in vivo* (Pawlikowski et al., 1999) and *in vitro* (at concentrations 10^{-7} to 10^{-11} M) (Karasek et al., 1988, 2003) in prolactin-secreting pituitary tumor induced in rats by diethylstilbestrol. The influence of melatonin administration on the growth of MtT/F4 anterior pituitary tumor (secreting ACTH, prolactin, and growth hormone) in rats depended on the time of administration. Although latency period of the tumor was significantly increased irrespectively from the time of administration, absolute and relative tumor weights were reduced only in animals in which melatonin was administered in the afternoon, but not following morning injections (Chatterjee and Banerji, 1989).

Melatonin suppressed cell growth of PC12 rat pheochromocytoma cell line *in vitro* (at concentrations 10^{-8} and 10^{-9} M) (Roth et al., 1997, 2001).

Melanoma

Melatonin administration significantly reduced the size and weight of transplantable B16 melanoma (Narita and Kudo, 1985; Narita, 1988) and S-91 melanoma in Mice (Kadekaro et al., 2004). In transplantable MM1 melanoma in the hamster, melatonin administration did not influence tumor growth in intact animals, but inhibited pinealectomy-induced growth and metastases of these tumors (El-Domieri and Das Gupta, 1973, 1976).

In early *in vitro* studies melatonin inhibited cell growth in hamster B7 melanoma cell line at very high concentration (10^{-3} M), whereas at concentrations 10^{-5} to 10^{-15} M either did not influence or even stimulate the tumor growth (Walker et al., 1978). Inhibitory effect of high melatonin concentration (10^{-3} M) on cell growth *in vitro* was also observed in human melanoma cell line (Bartsch et al., 1992). On the contrary, in hamster AbC1 and murine S91 melanoma cell lines, melatonin inhibited cell growth at low concentrations (10^{-8} to 10^{-10} M), whereas concentrations of 10^{-5} to 10^{-7} M were ineffective, and high concentration (10^{-4} M) stimulated cell growth (Slonimski and Pruski, 1993). Melatonin also inhibited growth of: human M-6 (at concentrations 10^{-4} to 10^{-9} M) (Ying et al., 1993)

- mouse M-2R (at concentrations 10^{-5} to 10^{-9} M) (Zisapel and Bubis, 1994),
- human SK-Mel 28 (at concentrations 10^{-5} to 10^{-8} M) (de Souza et al., 2003),
- human S-91 (at concentrations 10^{-4} to 10^{-7} M) (Kadekaro et al., 2004)
- melanoma cell lines as well as of human uveal melanoma cell line (at concentrations 10^{-8} to 10^{-10} M) (Hu and Roberts, 1997; Roberts et al., 2000).

Concentrations of melatonin that correspond to physiological levels (10^{-9} to 10^{-11} M) also inhibited proliferation of mouse melanotic B16BL6 and amelanotic PG19 melanoma cell lines (Cos et al., 2001). The growth of mouse B16 melanoma cell line was not affected by melatonin (at concentrations 5×10^{-4} to 4×10^{-6} M) (Sze et al., 1993). In cultured human melanoma cells from biopsy specimens of metastatic nodules, melatonin (at concentrations 10^{-5} to 10^{-15} M) exerted differential effects (inhibition, no effect, or stimulation), depending on the case (Meyskens and Salmon, 1981). Differential effects of high concentrations of melatonin were also found in slow (4×10^{-4} and 4×10^{-5} M stimulated cell growth) and fast growing (2×10^{-3} and 2×10^{-4} M inhibited cell growth) passages of human melanoma cells (Bartsch et al., 1986).

Other Tumors

In a transplantable LSTRA leukemia in the mouse, both the number of animals developing tumors and the mean tumor weights, were significantly smaller in melatonin-treated group than in control animals (Busswell, 1975). *In vitro* melatonin (at concentration 2×10^{-6} M) inhibited proliferation of human Jurkat T-lymphoma cell line, and did not influence proliferation of human HSB-2 T-lymphoma cell line, but stimulated proliferation of mouse P3X63Ag8U.1 myeloma cell line (B-lineage cell) (Persengiev and Kyurkchiew, 1993). Melatonin also exhibited *in vitro* strong pro-oxidant activity in human Jurkat lymphoma cells leading to the promotion of fast-induced apoptosis (Wölfler et al., 2001). However, cell growth in murine P388 lymphoma cell line (Leone and Wilkinson, 1991) and human K562 erythroleukemia cell line (Bartsch et al., 1992) was not influenced by melatonin (at concentrations 10^{-4} and 10^{-6} M, and 10^{-3} to 10^{-11} M, respectively).

Melatonin administration reduced the viability and volume of Ehrlich ascites carcinoma in the mouse and increased survival of the animals. Melatonin not only delayed the progression of cells from G0/G1 phase to S-phase of the cell cycle, but also reduced DNA synthesis during cell cycle (El-Missiry and Abd El-Aziz, 2000). However, the effect of melatonin on the growth Ehrlich tumor varied depending on the time of administration. Late afternoon melatonin injections inhibited tumor growth, midafternoon injections were without effect, whereas morning injections stimulated tumor growth (Bartsch and Bartsch, 1981).

In a highly malignant, transplantable Guerin epithelioma in the rat, melatonin administration decreased mitotic activity of tumor cells without affecting the life span of the animals (Lewinski et al., 1993). During the early stages, Melatonin also suppressed tumorigenesis of methylcholantrene (MCA)-induced fibrosarcoma in the mouse (Lapin and Ebels, 1976).

In skin papillomas induced by benzo(a)pyrene in the mouse, melatonin treatment significantly decreased the number of animals bearing tumors and the number of tumors per animal, both in the initiation and promotion stages of skin carcinogenesis, as well as inhibited skin cell proliferation (Kumar and Das, 2000).

Melatonin at physiological concentrations (10^{-9} and 10^{-11} M) inhibited the proliferation of human SK-N-SH neurobalstoma cell line, whereas subphysiological (10^{-13} M), and supraphysiological (10^{-5} and 10^{-7} M) concentrations lacked this effects (Cos et al., 1996b)

The concentration of melatonin required to inhibit tumor cell growth by 50% (IC_{50}) was high in human RT112 bladder transitional carcinoma (525 µg/mL) (Shellard et al., 1989), human HEp-2 larynx carcinoma (5×10^{-4} M) (Bartsch et al., 1992), and mouse S-180 sarcoma (2.8×10^{-4} M) (Sze et al., 1993) cell lines.

Melatonin was found to be without effect on the growth of transplantable Walker 256 carcinoma in the rat (Bostelmann et al., 1971). In Yoshida tumor in the rat, melatonin administration had no effect in intact animals, although it prevented tumor growth-promoting effects of pinealectomy and increased the survival time of the pinealectomized animals (Lapin and Ebels, 1976; Lapin and Frowein, 1981).

POSSIBLE REASONS FOR SOME CONTRADICTIONS IN THE RESULTS

As has been presented above, although in majority of the studies oncostatic action of melatonin was demonstrated, there are also results pointing to an exacerbation of malignancy by melatonin, and also those in which no effect of melatonin have been reported. There are several possible reasons causing differences in the results of the studies on the influence of melatonin on the growth and/or development of malignant experimentally-induced tumors. Contradictions in the results may depend on the following factors:

- different species used in the experiments
- various tumor types
- various stages of tumor development
- differences in mode and timing of melatonin administration
- *in vivo* or *in vitro* studies
- various photoperiodic environments used in the experiments
- different methods of tumor growth measurements:
  - neoplastic cell proliferation
  - tumor volume
  - tumor size
  - tumor diameter
  - tumor weight
  - tumor incidence
  - tumor latency
  - survival time of the animals

Different responses to melatonin have been found in slow-growing and fast-growing tumors; in human melanoma cells (Bartsch et al., 1986), Dunning prostate cancer (Buzzell et al., 1988), and early and advanced passages of DMBA-induced mammary cancer (Bartsch et al. 1994a), all point to the importance of stage of tumor development.

On the basis of results obtained in some tumor types (transplantable Ehrlich tumor – Bartsch and Bartsch, 1981; MCA-induced fibrosarcoma – Bartsch and Bartsch, 1981; transplantable MtT/F4 pituitary tumor – Chatterjee and Benerji, 1989; NMU-induced mammary cancer – Blask et al., 1991) it appears that *in vivo* melatonin can exert either oncostatic or oncogenic activity, depending on the circadian stage during which it is administered (late afternoon versus morning, respectively). Moreover, morning intraperitoneal melatonin injections stimulated tumorigenesis in MCA-induced fibrosarcoma, and afternoon injections had an inhibitory influence on this tumor, whereas subcutaneous injections were ineffective at any time (Bartsch and Bartsch, 1981).

In transplanted Kirkman-Robbins hepatoma, inhibition of the cell proliferation was observed *in vivo* following melatonin administration, whereas *in vitro* melatonin was ineffective in these cells (Karasek et al., 1992) suggesting that some tumors differently respond to *in vivo* and *in vitro* conditions.

In the studies on the effects of melatonin administration or pinealectomy in rats bearing transplantable Guerin epithelioma, no direct relationship between tumor cell proliferation and survival time of the animals was found, since pinealectomy shortened the life span of the animals, but did not affect the mitotic activity of tumor cells, whereas melatonin administration decreased mitotic activity of tumor cells without affecting the life span (Lewinski et al., 1993), showing that differences might depend on the method of tumor growth estimation.

THE POSSIBLE MECHANISMS OF ONCOSTATIC ACTION OF MELATONIN

The mechanisms by which melatonin restrains tumor growth seems to be complex, and several well-known activities of melatonin could be responsible for the indirect inhibition of tumor promotion and progression, as well as for the direct action on tumor cells proliferation and/or apoptosis. The possible mechanisms of the oncostatic action of melatonin are presented in Fig. 1 and discussed below.

Melatonin may exert an antitumor activity via different ways, either indirectly or directly. Indirect oncostatic action of melatonin include its modulation of the endocrine and immune systems, antioxidant action, inhibition of angiogenesis, signal transduction events, and modulation of hormone receptor expression and regulation, whereas directly melatonin influences tumor cells proliferation and apoptosis.

It is well-known that in some species melatonin exerts an anti-gonadotropic activity leading to the suppression of gonadal steroids (Reiter, 1980), which may subsequently, influence sex hormone-dependent cancers. However, such a mechanism of action, possible in some animal species, is rather unlikely in human subjects.

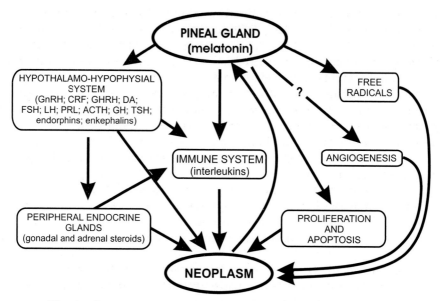

Fig. 1. Possible pathways of the oncostatic action of melatonin.

The modulation of the immune system resulting in the increased production of interleukins 2, 6, and 12, and interferon-γ, as well as the increased activity of interleukin 2, and the enhanced function of natural killer (NK) cells (Angeli et al., 1988; Lissoni et al., 1992; Morrey et al., 1994; Garcia-Maurino et al., 1997; Guerrero et al., 2001; Lissoni, 2001; Maestroni, 2001) may lead to the enhancement of the anti-cancer immune defense.

The recent observation that melatonin decreases the levels of vascular endothelial growth factor (VEGF) in cancer patients (Lissoni et al., 2001) suggests the further indirect mechanism of the oncostatic action of melatonin, namely, the inhibition of tumor angiogenesis.

Toxic-free radicals play an important role at the stage of cancer initiation (because of DNA damage) as well as in tumor progression (because of activation of signal transduction pathways and alteration of the expression of growth and differentiation-related genes) (Reiter, 2004). Melatonin may exert a protective role in cancer development and/or growth because as a potent antioxidant and free radical scavenger (Reiter et al., 2000, 2002), it was shown to protect the DNA from the damage connected with radiation challenge or exposure to chemical carcinogens (Karbownik and Reiter, 2000; Karbownik, 2002).

The effectiveness of melatonin in inhibiting the growth of tumor cells *in vitro* indicates that this hormone can also act directly. Indeed, melatonin was shown to exert direct antiproliferative and pro-apoptotic effects on tumor cells.

The antiproliferative effects of melatonin on normal and neoplastic cells are very well documented (Pawlikowski, 1986; Pawlikowski and Karasek,

1990; Blask, 1993; Karasek and Pawlikowski, 1999). Elegant studies of Blask and his associates (Blask et al., 1999a, b, c; Sauer et al., 1999, 2001) performed on tissue-isolated 7288CTC hepatoma model system clarified, at least partially, the intracellular mechanisms of the antiproliferative action of melatonin. The conclusion from these studies is that melatonin exerts its effect through signal transduction events involving melatonin membrane receptors, which are coupled to the adelylate cyclase via pertussis toxin-sensitive G proteins. The inhibition of the adenylate cyclase leads to the suppression of intracellular cAMP levels, and subsequently, to decreased transport of linoleic acid into the cells, and inhibition of its conversion to 13-hydroxy-actadecadienoic acid (13-HODE), which promotes the cell growth by enhancing epidermal growth factor (EGF)-responsible mitogenesis (Fig. 2). Other possible mechanisms of the antiproliferative action of melatonin include: delayed progression of cells from G0/G1 phase to S-phase of the cell cycle, and reduced DNA synthesis during cell cycle. (Cos et al., 1991; Cini et al., 1998; El-Missiry and Abd El-Aziz, 2000), and biding to calmodulin and blocking microtubule-associated proteins/calmodulin and tubulin/calmodulin complexes formation, and thus inhibiting cytoskeletal disruption (Benitez-King and Anton-Tay, 1993).

Pro-apoptotic effects of melatonin are less recognized (Sainz et al., 2003). As an antioxidant, melatonin could be rather considered as an anti-apoptotic agent, and actually exhibits such activity in normal cells (Sainz et al., 2003).

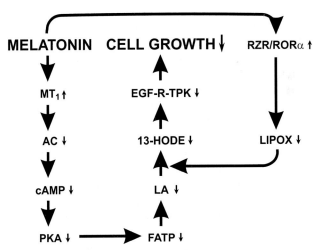

Fig. 2. The putative mechanisms of antiproliferative action of melatonin. MT_1 – membrane MT_1 melatonin receptor, AC – adenylate cyclase, cAMP – cyclic adenosine 3'5'-monophosphate, PKA – cAMP-dependent protein kinase, FATP – fatty acids transporting protein, LA – linoleic acid, 13-HODE – 13-hydroxyoctadecadienoic acid, EGF-R-TPK – epidermal growth factor receptor tyrosine protein kinase, RZR/RORα – nuclear rerinoid Z receptor/retinoid acid receptor related orphan receptor α, LIPOX – 5-lipoxygenase; ↑ – stimulation, ↓ – inhibition (according to Karasek and Pawlikowski, 1999, modified).

However, it has been demonstrated that melatonin stimulates apoptosis in some cancer cells, including murine Colon 38 cancer (Melen-Mucha et al., 1998; Winczyk et al., 2001), rat 1,2-dimethylhydrazine-induced colon cancer (Anisimov et al., 2000), human MCF-7 breast cancer cell line (Eck et al., 1998), human leukemic Jurkat cell line (Wolfer et al., 2001), H22 hepatocarcinoma cells (Qin et al., 2004), and probably murine Ehrlich ascites carcinoma (El-Missiry and Abd El-Aziz, 2000).

Recently, another a novel mechanism whereby melatonin inhibits neoplastic cell growth, was suggested by Leon-Blanco et al. (2004). They demonstrated that melatonin not only inhibited the growth of MCF-7 breast cancer xenographs in BALB/cByJ nude mouse, but also lowered telomerase activity and mRNA expression of the telomerase reverse transcriptase (the catalytic subunit of telomerase), as well as the mRNA of RNA telomerase subunit. It should be stressed that telomerase activity has been proposed as a target for cancer chemotherapy, and pharmacological-induced death of tumor cells is associated with a decline in detectable telomerase activity (Leon-Blanco et al., 2004; Reiter, 2004).

A question arises as to what type of melatonin receptors are involved in the direct oncostatic effects of this hormone. Both membrane (MT_1 and MT_2) and putative nuclear (RZR/ROR) melatonin receptors are considered to be involved in oncostatic action of melatonin. Expression of MT_1 melatonin membrane receptors was found in some experimentally-induced cancers, including:

- human MCF-7 breast cancer (Ram et al., 1998, 2000; Roth et al., 2001),
- rat PC12 pheochromocytoma cell line (Ram et al., 1998),
- human DU-145 and LNCaP prostate cancer cell lines (Montagnani Marelli et al., 2000; Xi et al., 2001),
- mouse N1E-115 neuroblastoma cell line (Bordt et al., 2001),
- mouse Colon 38 cancer (Karasek et al., 2002),
- prostate cancer from a patient with bone metastases (Shiu et al., 2003), whereas expression of MT_2 melatonin receptors was demonstrated in:
- human JAr choriocarcinoma cell line (Shiu et al., 1999)
- mouse Colon 38 cancer (Karasek et al., 2002), human uveal melanoma cell lines (Roberts et al., 2000). On the other hand, $ROR\alpha$ receptors were expressed in:
- human MCF-7 breast cancer line (Dai et al., 2001),
- DU-145 prostate cancer cell line (Moretti et al., 2001b),
- mouse Colon 38 cancer (Karasek et al., 2002).

Many data suggest the involvement of the membrane melatonin receptors. The fact that antiproliferative effect of melatonin on rat 7288CTC hepatoma was reversed by pertussis toxin, forskolin, 8-bromo-cAMP, and melatonin membrane receptor antagonist S-20928, indicates that melatonin-induced suppression of hepatoma growth is mediated via G-protein connected membrane receptors (Fig. 2) (Blask et al., 1999c). Melatonin inhibited the proliferation of human choriocarcinoma cell lines expressing MT_2 receptors, but not in a cell line devoid of melatonin membrane receptors (Shiu et al., 2000). Recently, it has been found that the melatonin membrane receptor ligand UCM 386 reproduces both antiproliferative and pro-apoptotic effects of melatonin on Colon 38 cancer cells (Winczyk et al., 2002b). Melatonin and 6-chloromelatonin (membrane receptor agonist), but not CGP-52608 (the putative nuclear receptor agonist), inhibit the growth of human uveal melanoma cells in vitro (Hu and Roberts, 1997). Moreover, melatonin administration in a 67-year-old man slowed the early biochemical progression of hormone-refractory prostate cancer expressing melatonin MT_1 membrane receptors (Shiu et al., 2003). Additionally, melatonin pretreatment decreased cell proliferation and transformation in NIH3T3 cells transfected with human MT_1 and MT_2 receptors, indicating that an attenuation of receptor-mediated processes are involved in the antiproliferative and anti-transformation capabilities (Jones et al., 2000). Ram et al. (2000) on the basis of the observation of differential responsiveness of human MCF breast cancer cell line stocks to melatonin, suggest that the primary growth-inhibitory effects of melatonin are transducted through the MT_1 melatonin membrane receptor. Differential expression of MT_1 receptors was found in normal and malignant human breast cancer tissue; 68% of nonneoplastic samples were negative or weakly positive, whereas moderate or strong reactivity was seen in 75% in most cancer samples (Dillon et al., 2002).

On the other hand, some data support the involvement of the RZR/ROR receptors in antiproliferative (Fig. 2) and pro-apoptotic action of melatonin on tumor cells. It has been shown that both melatonin and CGP 52608, the RZR/ROR receptor ligand, similarly inhibit the proliferation of Colon 38 cancer cells in vitro and in vivo (Karasek et al., 1998; Pawlikowski et al., 1999). The same effect has been observed in the case of ovarian cancer cell line (Petranka et al., 1999). Moreover, both melatonin and CGP 52608 inhibit the proliferation of prostate cancer cell lines DU-145 and LNCaP, expressing RZR/RORα receptors but not the membrane melatonin receptors (Moretti et al., 2001a, b). Recently, it was also demonstrated that both melatonin and CGP 52608 induce apoptosis in Colon 38 cells (Karasek et al., 1998; Pawlikowski et al., 1999) and the pro-apoptotic action of melatonin was blocked by CGP 54644, assumed as RZR/ROR receptor antagonist (Winczyk et al., 2002a).

Taking together these partially controversial data, it seems that both membrane and nuclear receptors are involved in the oncostatic action of melatonin.

REFERENCES

Angeli, A., Gatti, G., Sartori, M.L., Del Ponte, D. and Carignola, R. (1988). Effect of exogenous melatonin on human natural killer (NK) cell activity. An approach to the immunoregulatory role of the pineal gland. In: *The Pineal Gland and Cancer.* (Gupta, D., Attanasio, A. and Reiter, R.J., eds.). Brain Res. Promotion, London, 145-156.

Anisimov, V.N., Morozov, V.G., Khavinson, V.K. and Dilman, V.M. (1973). Comparison of the anti-tumor activity of extracts of the epiphysis and hypothalamus, melatonin and sigetin in mice with transplanted mammary gland cancer. *Vopr. Onkol.* **19**: 99-101 (in Russian).

Anisimov, V.N., Popovich, I.G. and Zabezhinski, M.A. (1997). Melatonin and colon carcinogenesis: I. Inhibitory effect of melatonin on development of intestinal tumors induced by 1,2-dimethylhydrazine in rats. *Carcinogenesis* **18**: 1549-1553.

Anisimov, V.N., Popovich, I.G., Shtylik, A.V., Zabezhinki, M.A., Ben-Huh, H., Gurevich, R., Berman, V., Tendler, Y. and Zusman, I. (2000a). Melatonin and colon carcinogenesis. III. Effect of melatonin on proliferative activity and apoptosis in colon mucosa and colon tumors induced by 1,2-dimethylhydrazine in rats. *Exp. Toxic. Pathol.* **52**: 71-76.

Anisimov, V.N., Zabezhinski, M.A., Popovich, I.G., Zaripova, E.A., Musatov, S.A., Andre, V., Vigreux, C., Godard, T. and Sichel, F. (2000b). Inhibitory effect of melatonin on 7,12-dimethylbenz[a]anthracene-induced carcinogenesis of the uterine cervix and vagina in mice and mutagenesis in vitro. *Cancer Lett.* **156**: 199-205.

Anisimov, V.N., Alimova, I.N., Batyrin, D.A., Popovich, I.G., Zabezhinski, M.A., Manton, K.G., Semenchenko, A.V. and Yashin, A.I. (2003). The effect of melatonin treatment regimen on mammary adenocarcinoma development in HER-2/neu transgenic mice. *Int. J. Cancer* **103**: 300-305.

Aubert, C., Janiaud, P. and Lacavez, J. (1980). Effect of pinealectomy and melatonin on mammary tumor growth in Sprague-Dawley rats under different conditions of lighting. *J. Neural Transm.* **47**: 121-130.

Bartsch, H. and Bartsch, C. (1981). Effect of melatonin on experimental tumors under different photoperiods and times of administration. *J. Neural Transm.* **52**: 269-279.

Bartsch, H., Bartsch, C. and Fleming, B. (1986). Differential effect of melatonin on slow and fast growing passages of a human melanoma cell line. *Neuroendocrinol. Lett.* **8**: 289-293.

Bartsch, H., Bartsch, C., Simon, W.E., Flehmig, B., Ebels, I. and Lippert, T.H. (1992). Antitumor activity of the pineal gland: effect of unidentified substances versus the effect of melatonin. *Oncology* **49**: 27-30.

Bartsch, C., Bartsch, H., Buchberger, A., Stieglitz, A, Mecke, D. and Lippert, T.H. (1994a). Serial transplants of DMBA-induced mammary tumors in Fisher rats as model system for human breast cancer. I: Effect of melatonin and pineal extracts on slow- and fast-growing passages – in vivo and in vitro studies. *Adv. Pineal Res.* **8**: 479-484.

Bartsch, H., Bartsch, C., Mecke, D. and Lippert, T.H. (1994b). Differential effect of melatonin on early and advanced passages of a DMBA-induced mammary carcinoma in the female rat. *Adv. Pineal Res.* **7**: 247-252.

Bartsch, H., Buchberger, A., Franz, H., Bartsch, C., Maidonis, I., Mecke, D. and Bayer, E. (2000). Effect of melatonin and pineal extracts on human ovarian and mammary tumor cells in a chemosensitivity assay. *Life Sci.* **67**: 2953-2960.

Benitez-King, G. and Anton-Tay, F. (1993). Calmodulin mediates melatonin cytoskeletal effects. *Experientia* **49**: 635-641.

Blask, D.E. (1984). The pineal: an oncostatic gland. In: *The Pineal Gland* (Reiter, R.J., ed.). Raven, New York, 253-284.

Blask, D.E., Hill, S.M., Orstead, K.M. and Mass, J.S. (1986). Inhibitory effects of the pineal hormone melatonin and underfeeding during the promotional phase of

7,12-dimethylbenzanthracene(DMBA)-induced mammary tumorigenesis. *J. Neural Transm.* **67**: 125-138.

Blask, D.E., Hill, S.M. and Pelletier, D.B. (1988). Oncostatic signaling by the pineal gland and melatonin in the control of breast cancer. In: *The Pineal Gland and Cancer* (Gupta, D., Attanasio, A. and Reiter, R.J., eds.). Brain Res. Promotion, London, 195-206.

Blask, D.E., Pelletier, D.B., Hill, S.M., Lemus-Wilson, A., Grosso, D.S., Wilson, S.T. and Wise, M.E. (1991). Pineal melatonin inhibition of tumor promotion in the N-nitroso-N-methylurea model of mammary carcinogenesis: potential involvement of antiestrogenic mechanisms in vivo. *J. Cancer Res. Clin. Oncol.* **117**: 526-532.

Blask, D.E. (1993). Melatonin in oncology. In: *Melatonin – Biosynthesis, Physiological Effects and Perspectives* (Yu, H.S. and Reiter, R.J., eds.). CRC Press, Boca Raton, 447-475.

Blask, D.E., Sauer, L.A. and Dauchy, R.T. (1997a). Melatonin suppression of tumor growth in vitro: a novel mechanism involving inhibition of fatty acid uptake and metabolism. *Front. Horm. Res.* **23**: 107-114.

Blask, D.E., Sauer, L.A. and Dauchy, R.T. (1997b). Melatonin regulation of tumor growth and the role of fatty acid uptake and metabolism. *Neuroendocrinol. Lett.* **18**: 59-62.

Blask, D.E., Wilson, S.T. and Zalatan, F. (1997c). Physiological melatonin inhibition of human breast cancer cell growth in vitro: evidence for a glutathione-mediated pathway. *Cancer Res.* **57**: 1909-1914.

Blask, D.E., Sauer, L.A., Dauchy, R.T., Holowachuk, E.W. and Ruchoff, M.S. (1998). Circadian rhythm in experimental tumor carcinogenesis and progression and the role of melatonin. In: *Biological Clocks – Mechanisms and Applications* (Youitou, Y., ed.). Elsevier, Amsterdam, 469-474.

Blask, D.E., Sauer, L.A., Dauchy, R.T., Holowachuk, E.W. and Ruchoff, M.S. (1999a). New insights into melatonin regulation of cancer growth. *Adv. Exp. Med. Biol.* **460**: 337-343.

Blask, D.E., Sauer, L.A., Dauchy, R.T., Holowachuk, E.W. and Ruchoff, M.S. (1999b). New actions of melatonin on tumor metabolism and growth. *Biol. Signals Recept.* **8**: 49-55.

Blask, D.E., Sauer, L.A., Dauchy, R.T., Holowachuk, E.W., Ruchoff, M.S. and Kopff, H.S. (1999c). Melatonin inhibition of cancer growth in vitro involves suppression of tumor fatty acid metabolism via melatonin-receptor-mediated signal transduction events. *Cancer Res.* **59**: 4693-4701.

Bordt, S.L., McKeon, R.M., Li, P.K., Witt-Enderby, P.A. and Melan, M.A. (2001). N1E-115 mouse neuroblastoma cells express MT1 melatonin receptors and produce neurites in response to melatonin. *Biochim. Biophys. Acta* **1499**: 257-264.

Bostelmann, W., Gocke, H., Ernst, B. and Tesmann, D. (1971). Der Einfluss einer Melatonin Behandlung auf des Wachstum des Walker Carcinosarcoma der Ratte. *A. Aug. Path.* **114**: 289-291.

Buswell, R.S. (1975). The pineal gland and neoplasia. *Lancet* I: 34-35.

Buzzell, G.R., Amerongen, H.M. and Toma, J.G. (1988). Melatonin and the growth of the Dunning R 3327 rat prostatic adenocarcinoma. In: *The Pineal Gland and Cancer* (Gupta, D., Attanasio, A. and Reiter, R.J., eds.). Brain Res. Promotion, London, 295-306.

Chatterjee, S. and Banerji, T.K. (1989). Effects of melatonin on the growth of MtT/F4 anterior pituitary tumor: evidence for inhibition of tumor growth dependent on the time of administration. *J. Pineal Res.* **7**: 381-391.

Chen, L.D., Leal, B.Z., Reiter, R.J., Abe, M., Sewerynek, E., Melchiori, D., Meltz, M.L. and Poeggler, B. (1995). Melatonin's inhibitory effect on growth of ME-180 human cervical cancer cells is not related to intracellular glutathione concentrations. *Cancer Lett.* **91**: 153-159.

Cini, G., Coronnello, M., Mini, E. and Neri, B. (1998). Melatonin's growth-inhibitory effect on hepatoma AH 130 in the rat. *Cancer Lett.* **125**: 51-59.

Cos, S., Blask, D.E., Lemus-Wilson, A. and Hill, A.B. (1991). Effects of melatonin on the cell cycle kinetics and "estrogen-rescue" of MCF-7 human breast cancer cells in culture. *J. Pineal Res.* **10**: 36-42.

Cos, S. and Blask, D.E. (1994). Melatonin's growth-inhibitory effect on hepatoma AH 130 in the rat. *J. Pineal Res.* **17**: 25-32.

Cos, S., Fernandez, F. and Sanchez-Barcelo, E.J. (1996a). Melatonin inhibits DNA synthesis in MCF-7 human breast cancer cells in vitro. *Life Sci.* **58**: 2447-2453.

Cos, S., Verduga, R., Fernandez, F., Megias, M. and Crespo, D. (1996b). Effects of melatonin on the proliferation and differentiation of human neuroblastoma cells in culture. *Neurosci. Lett.* **216**: 113-116.

Cos, S., Garcia-Bolado, A. and Sanchez-Barcelo, E.J. (2001). Direct antiproliferative effects of melatonin on two metastatic cell sublines of mouse melanoma (B16BL6 and PG19). *Melanoma Res.* **11**: 197-201.

Crespo, D., Fernandez-Viadero, R., Verduga, R., Ovejero, V. and Cos, S. (1994). Interaction between melatonin and estradiol on morphological and morphometric features of MCF-7 human breast cancer cells. *J. Pineal Res.* **16**: 215-222.

Dai, J., Ram, P.T., Yuan, L., Spriggs, L.L. and Hill, S.M. (2001). Transcriptional repression of RORalpha activity in human breast cancer cells by melatonin. *Mol. Cell. Endocrinol.* **176**: 111-120.

de Souza, A.V., Visconti, M.A. and Castrucci, A.M.L. (2003). Melatonin biological activity and binding sites in human melanoma cells. *J. Pineal Res.* **34**: 242-248.

Dillon, D.C., Easley, S.E., Asch, B.B., Cheney, R.T., Brydon, L., Jockers, R., Winston, J.S., Brooks, J.S., Hurd, T. and Asch, H.L. (2002). Differential expression of high-affinity melatonin receptors (MT1) in normal and malignant human breast tissue. *Am. J. Clin. Pathol.* **118**: 451-458.

Eck, K.M., Yan, L., Duffy, L., Ram, P.T., Ayetty, S., Chen, I., Cohn, C.S., Reed, J.C. and Hill, S.M. (1998). A sequential treatment regimen with melatonin and all-trans retinoic acid induces apoptosis in MCF-7 tumour cells. *Br. J. Cancer* **77**: 2129-2137.

El-Domeri, A.A.H. and Das Gupta, T.K. (1973). Reversal by melatonin of the effect of pinealectomy on tumor growth. *Cancer Res.* **33**: 2830-2833.

El-Domeri, A.A.H. and Das Gupta, T.K. (1976). The influence of pineal ablation and administration of melatonin on growth and spread of hamster melanoma. *J. Surg. Oncol.* **8**: 197-205.

El-Missiry, M.A. and Abd El-Aziz, A.F. (2000). Influence of melatonin on proliferation and antioxidant system in Ehrlich ascites carcinoma cells. *Cancer Lett.* **151**: 119-125.

Farriol, M., Venereo, Y., Orta, X., Castellanos, J.M. and Segovia-Silvestre, T. (2000). In vitro effects of melatonin on cell proliferation in a colon adenocarcinoma line. *J. Appl. Toxicol.* **20**: 21-24.

Futagami, M., Sato, S., Sakatomo, T., Yokoyama, Y. and Saito, Y. (2001). Effects of melatonin on the proliferation and cis-diamminedichloroplatinum (CDDP) sensitivity of cultured human ovarian cancer cells. *Gynecol. Oncol.* **82**: 544-549.

Garcia-Maurino, S., Gonzalez-Haba, M.G., Calvo, J.R., Raffi-El-Idrissi, M., Sanchez-Margalet, V., Goberna, R. and Guerrero, J.M. (1997). Melatonin enhances IL-2, IL-6, and IFN-gamma production by human circulating CD4+ cells: a possible nuclear receptor-mediated mechanism involving T helper type 1 lymphocytes and monocytes. *J. Immunol.* **159**: 574-581.

Gilad, E., Laufer, M., Matzkin, H. and Zisapel, N. (1999). Melatonin receptors in PC3 human prostate tumor cells. *J. Pineal Res.* **26**: 211-220.

Georgiou, E. (1929). Uber die Natur ind Pathogenese der Krebstumoren. *Z. Krebsforch.* **28**: 562-572.

Guerrero, J.M., Garcia-Maurino, S., Pozo, D., Garcia-Perganeda, A., Carrillo-Vico, A., Molinero, P., Osuna, C. and Calvo, J.R. (2001). Mechanisms involved in the immunomodulatory effects of melatonin on the human immune system. In: *The Pineal and Cancer* (Bartsch, C., Bartsch, H., Blask, D.E., Cardinali, D.P., Hrushesky, W.J.M. and Mecke, D., eds.). Springer, Berlin, 408-416.

Hermann, R., Podhajsky, S., Jungnickel, S. and Lerchl, A. (2002). Potentiation of antiproliferative effects of tamoxifen and ethanol on mouse hepatoma cells by melatonin: possible involvement of mitogen-activated protein kinase and induction of apoptosis. *J. Pineal Res.* **33**: 8-13.

Hill, S.M. and Blask, D.E. (1988). Potentiation of antiproliferative effects of tamoxifen and ethanol on mouse hepatoma cells by melatonin: possible involvement of mitogen-activated protein kinase and induction of apoptosis. *Cancer Res.* **48**: 6121-6126.

Hill, S.M., Kiefer, T., Teplitzky, S., Spriggs, L.L. and Ram, P. (2001). Modulation of the estrogen response pathway in human breast cancer cells by melatonin. In: *The Pineal and Cancer* (Bartsch, C., Bartsch, H., Blask, D.E., Cardinali, D.P., Hrushesky, W.J.M. and Mecke, D., eds.). Springer, Berlin, 343-358.

Hu, D.N. and Roberts, J.E. (1997). Melatonin inhibits growth of cultured human uveal melanocytes and melanoma cells. *Melanoma Res.* **7**: 27-31.

Jones, M.P., Melan, M.A. and Witt-Enderby, P.A. (2000). Melatonin decreases cell proliferation and transformation in a melatonin receptor-dependent manner. *Cancer Lett.* **151**: 133-143.

Kadekaro, A.L., Andrade, L.N.S., Floeter-Winter, L.M., Rollag, M.D., Virador, V., Vieira, W. and Castrucci, A.M.L. (2004). MT-1 melatonin receptor expression increases the antiproliferative effect of melatonin on S-91 murine melanoma cells. *J. Pineal Res.* **36**: 204-211.

Kanishi, Y., Kobayashi, Y., Noda, S., Ishizuka, B. and Saito, K. (2000). Differential growth inhibitory effect of melatonin on two endometrial cancer cell lines. *J. Pineal Res.* **28**: 227-233.

Karasek, M., Kunert-Radek, J., Stepien, H. and Pawlikowski, M. (1988). Melatonin inhibits the proliferation of estrogen-induced rat pituitary tumor *in vitro*. *Neuroendocrinol. Lett.* **10**: 135-140.

Karasek, M. and Fraschini, F. (1991). Is there a role for the pineal gland in neoplastic growth? In: *Role of the Pineal Gland and Melatonin in Neuroimmunomodulation* (Fraschini, F. and Reiter, R.J., eds). Plenum, New York, 243-251.

Karasek, M., Liberski, P., Kunert-Radek, J. and Bartkowiak, J. (1992). Influence of melatonin on the proliferation of hepatoma cells in the Syrian hamster: in vivo and in vitro study. *J. Pineal Res.* **13**: 107-110.

Karasek, M. (1994). Malignant tumors and the pineal gland. In: *Pathophysiology of Immune-Neuroendocrine Communication Circuit* (Gupta, D., Wollman, H.A. and Fedor-Freybergh, P., eds.). Mattes, Heidelberg, 225-244.

Karasek, M. (1997). Relationship between the pineal gland and experimentally induced malignant tumors. *Front. Horm. Res.* **23**: 99-106.

Karasek, M., Winczyk, K., Kunert-Radek, J., Wiesenberg, I. and Pawlikowski, M. (1998). Antiproliferative effects of melatonin and CGP 52608 on the murine Colon 38 adenocarcinoma in vitro and in vivo. *Neuroendocrinol. Lett.* **19**: 71-78.

Karasek, M. and Pawlikowski, M. (1999). Pineal gland, melatonin and cancer. *Neuroendocrinol. Lett.* **20**: 139-144.

Karasek, M., Carrillo-Vico, A., Guerrero, J.M., Winczyk, K. and Pawlikowski, M. (2002). Oncostatic action of melatonin: facts and question marks. *Neuroendocrinol. Lett.* **23**(1): 55-60.

Karasek, M., Gruszka, A., Lawnicka, H., Kunert-Radek, J. and Pawlikowski, M. (2003). Melatonin inhibits growth of diethylstilbestrol-induced prolactin-secreting pituitary tumor in vitro: possible involvement of nuclear RZR/ROR receptors. *J. Pineal Res.* **34**: 294-296.

Karbownik, M. and Reiter, R.J. (2000). Antioxidative effects of melatonin in protection against cellular damage induced by ionizing radiation. *Proc. Soc. Exp. Biol. Med.* **225**: 9-22.

Karbownik, M. (2002). Potential anticarcinogenic action of melatonin and other antioxidants mediated by antioxidative mechanisms. *Neuroendocrinol. Lett.* **23**(1): 39-44.

Karmali, R.A., Horrobin, D.F. and Ghayur, T. (1978). Role of pineal gland in aetiology and treatment of breast cancer. *Lancet* **II**: 1002.

Kothari, A., Borges, A. and Kothari, L. (1995). Chemoprevention by melatonin and combined melatonin-tamoxifen therapy of second generation nitroso-methylurea-induced mammary tumours in rats. *Eur. J. Cancer Prev.* **4**: 497-500.

Kumar, C.A. and Das, U.N. (2000). Effect of melatonin on two stage skin carcinogenesis in Swiss mice. *Med. Sci. Monit.* **6**: 471-475.

Lapin, V. and Ebels, I. (1976). Effects of some low molecular weight sheep pineal fractions and melatonin on different tumors in rats and mice. *Oncology* **33**: 110-113.

Lapin, V. and Frowein, A. (1981). Effects of growing tumours on pineal melatonin levels in male rats. *J. Neural Transm.* **52**: 123-136.

Leon-Blanco, M.M., Guerrero, J.M., Reiter, R.J. and Pozo, D. (2004). RNA expression of human telomerase subunits and TERT is differentially affected by melatonin receptor agonists in the MCF-7 tumor cell line. *Cancer Lett.* **216**: 73-80.

Leone, A.M., Silman, R.E., Hill, B.T., Whelom, R.D.H. and Shellard, S.A. (1988). Growth inhibitory effects of melatonin and its metabolites against ovarian tumor cell lines in vitro. In: *The Pineal Gland and Cancer* (Gupta, D., Attanasio, A. and Reiter, R.J., eds.). Brain Research Promotion, London, 273-281.

Leone, A. and Wilkinson, J. (1991). The effects of melatonin and melatonin analogues on the P388, DLD-1 and MCF-7 tumor cell lines *in vitro*. In: *Role of Melatonin and Pineal Peptides in Neuroimmunomodulation* (Fraschini, F. and Reiter, R.J., eds.). Plenum Press, New York, 241-242.

Lewinski, A., Sewerynek, E., Wajs, E., Lopaczynski, W. and Sporny, S. (1993). Effects of the pineal gland on the growth processes of Guerin epithelioma in male Wistar rats. *Cytobios* **73**: 89-94.

Lissoni, P., Barni, S., Ardizzoia, A., Brivio, F., Tancini, G., Conti, A. and Maestroni, G.J. (1992). Immunological effects of a single evening subcutaneous injection of low-dose interleukin-2 in association with the pineal hormone melatonin in advanced cancer patients. *J. Biol. Regul. Homeost. Agents* **6**: 132-136.

Lissoni, P. (2001). Efficacy of melatonin in the immunotherapy of cancer using interleukin-2. In: *The Pineal and Cancer* (Bartsch, C., Bartsch, H., Blask, D.E., Cardinali, D.P., Hrushesky, W.J.M., Mecke, D., eds.). Springer, Berlin, 465-475.

Lissoni, P., Rovelli, F., Malugani, F., Bukovec, R., Conti, A. and Maestroni, G.J.M. (2001). Anti-angiogenic activity of melatonin in advanced cancer patients. *Neuroendocrinol. Lett.* **22**: 45-47.

Lupowitz, Z. and Zisapel, N. (1999). Hormonal interactions in human prostate tumor LNCaP cells. *J. Steroid Biochem. Molec. Biol.* **68**: 83-88.

Maestroni, G.J.M. (2001). Melatonin and the immune system: therapeutic potential in cancer, viral diseases, and immunodeficiency states. In: *The Pineal and Cancer* (Bartsch, C., Bartsch, H., Blask, D.E., Cardinali, D.P., Hrushesky, W.J.M., Mecke, D., eds.). Springer, Berlin, 384-394.

Mediavilla, M.D., Cos, S. and Sanchez-Barcelo, E.J. (1999). Melatonin increases p53 and p21WAF1 expression in MCF-7 human breast cancer cells in vitro. *Life Sci.* **65**: 415-420.

Melen-Mucha, G., Winczyk, K. and Pawlikowski, M. (1998). Somatostatin analogue Octreotide and melatonin inhibit bromodeoxyuridine incorporation into cell nuclei and enhance apoptosis in the transplantable murine colon 38 cancer. *Anticancer Res.* **18**: 3615-3620.

Meyskens, F.I. and Salmon, S.F. (1981). Modulation of elogenic human melanoma cells by follicle-stimulating hormone, melatonin and nerve growth factor. *Br. J. Cancer* **43**: 111-115.

Molis, T.M., Walters, M.R. and Hill, S.M. (1993). Melatonin modulation of estrogen-receptor expression in MCG-7 human breast cancer cells. *Int. J. Oncol.* **3**: 687-694.

Molis, T.M., Spriggs, L.L., Jupiter, Y. and Hill, S.M. (1995). Melatonin modulation of estrogen-regulated proteins, growth factors, and proto-oncogenes in human breast cancer. *J. Pineal Res.* **18**: 93-103.

Montagnani Marelli, M., Limonta, P., Maggi, R., Motta, M. and Moretti, R.M. (2000). Growth-inhibitory activity of melatonin on human androgen-independent DU 145 prostate cancer cells. *Prostate* **45**: 238-244.

Moretti, R.M., Montagnani Marelli, M., Motta, M. and Limonta, P. (2001a). Oncostatic activity of a thiazolinedione derivative on human androgen-dependent prostate cancer cells. *Int. J. Cancer* **92**: 733-737.

Moretti, R.M., Montagnani Marelli, M., Motta, M., Polizzi, D., Monestiroli, S. and Pratesi, G. (2001b). Activation of the orphan nuclear receptor RORα induces growth arrest in androgen-independent DU 145 prostate cancer cells. *Prostate* **46**: 327-335.

Morrey, K.M., McLachlan, J.A., Serkin, C.D. and Bakouche, O. (1994). Activation of human monocytes by the pineal hormone melatonin. *J. Immunol.* **153**: 2671-2680.

Musatov, S.A., Anisimov, V.N., Andre, V., Vigreux, C., Godard, T. and Sichel, F. (1999). Effects of melatonin on N-nitroso-N-methylurea-induced carcinogenesis in rats and mutagenesis in vitro (Ames test and COMET assay). *Cancer Lett.* **138**: 37-44.

Narita, T. and Kudo, H. (1985). Effect of melatonin on B16 melanoma growth in athymic mice. *Cancer Res.* **45**: 4175-4177.

Narita, T. (1988). Effects of melatonin on B16 melanoma growth. In: *The Pineal Gland and Cancer* (Gupta, D., Attanasio, A. and Reiter, R.J., eds.). Brain Res. Promotion, London, 345-354.

Nowfar, S., Teplitzky, S.R., Melancon, K., Kiefer, T.L., Cheng, Q., Dwived, P.D., Bischoff, E.D., Moro, K., Anderson, M.B., Dai, J., Lai, L., Yuan, L. and Hill, S.M. (2002). Tumor prevention by 9-cis-retinoic acid in the N-nitroso-N-methylurea model of mammary carcinogenesis is potentiated by the pineal hormone melatonin. *Breast Cancer Res. Treat.* **72**: 33-43.

Papazisis, K.T., Kouretas, D., Geromichalos, G.D., Sivridis, E., Tsekreli, O.K., Dimitriadis, K.A. and Kortsaris, A.H. (1998). Effects of melatonin on proliferation of cancer cell lines. *J. Pineal Res.* **25**: 211-218.

Pawlikowski, M. (1986). The pineal gland and cell proliferation. *Adv. Pineal Res.* **1**: 27-30.

Pawlikowski, M. and Karasek, M. (1990). The pineal gland and experimentally-induced tumors. *Adv. Pineal Res.* **4**: 259-265.

Pawlikowski, M., Kunert-Radek, J., Winczyk, K., Melen-Mucha, G., Gruszka, A. and Karasek, M. (1999). The antiproliferative effects of melatonin on experimental pituitary and colonic tumors: possible involvement of the putative nuclear biding site? *Adv. Exp. Med. Biol.* **460**: 369-372.

Pawlikowski, M., Winczyk, K. and Karasek, M. (2002). Oncostatic action of melatonin: facts and question marks. *Neuroendocrinol. Lett.* **23**(1): 24-29.

Persengiew, S.P. and Kyurkchiew, S. (1993). Selective effects of melatonin on the proliferation of lymphoid cells. *Int. J. Biochem.* **25**: 441-444.

Petranka, J., Baldwin, W., Biermann, J., Jayadev, S., Barret, J.C. and Murphy, E. (1999). The oncostatic action of melatonin in an ovarian carcinoma cell line. *J. Pineal Res.* **26**: 129-136.

Philo, R. and Berkowitz, A.S. (1988). Inhibition of Dunning tumor growth by melatonin. *J. Urol.* **139**: 1099-1102.

Qin, L., Wang, X., Duan, Q., Chen, B. and He, S. (2004). Inhibitory effect of melatonin on the growth of H22 hepatocarcinoma cells by inducing apoptosis. *J. Huazhong Univ. Sci. Technolog. Med. Sci.* **24**: 19-21, 31.

Rahman, K.M., Sugie, S., Watanabe, T., Tanaka, T. and Mori, H. (2003). Chemoprotective effects of melatonin on diethylnitrosamine and phenobarbital-induced hepatocarcinogenesis in male F344 rats. *Nutr. Cancer* **47**: 148-155.

Ram, P.T., Kiefer, T., Silverman, M., Song, Y., Brown, G.M. and Hill, S. (1998). Estrogen receptor transactivation in MCF-7 breast cancer cells by melatonin and growth factors. *Mol. Cell Endocrinol.* **141**: 53-64.

Ram, P.T., Yuan, L., Dai, J., Kierer, T., Klotz, D.M., Spriggs, L.L. and Hill, S.M. (2000). Differential responsiveness of MCF-7 human breast cancer cell line stocks to the pineal hormone, melatonin. *J. Pineal Res.* **28**: 210-218.

Rao, G.N., Ney, E. and Herbert, R.A. (2000). Effect of melatonin and linolenic acid on mammary cancer in transgenic mice with c-neu breast cancer oncogene. *Breast Cancer Res. Treat.* **64**: 287-296.

Reiter, R.J. (1980). The pineal gland and its hormone in the control of reproduction in mammals. *Endocr. Rev.* **1**: 109-131.

Reiter, R.J., Tan., D.X., Osuna, C. and Gitto, E. (2000). Actions of melatonin in the reduction of oxidative stress: a review. *J. Biomed. Sci.* **7**: 444-458.

Reiter, R.J., Tan, D.X. and Allegra, M. (2002). Melatonin: reducing molecular pathology and dysfunction due to free radical and associated reactants. *Neuroendocrinol. Lett.* **23**(1): 3-8.

Reiter, R.J. (2004). Mechanisms of cancer inhibition by melatonin. *J. Pineal Res.* **37**: 213-214.

Roberts, J.E., Wiechmann, A.F. and Hu, D.N. (2000). Melatonin receptors in human uveal melanocytes and melanoma cells. *J. Pineal Res.* **28**: 165-171.

Roth, J.A., Rabin, R. and Agnello, K. (1997). Melatonin suppression of PC12 cell growth and death. *Brain Res.* **768**: 63-70.

Roth, J.A., Rosenblatt, T., Lis, A. and Bucelli, R. (2001). Melatonin-induced suppression of PC12 cell growth is mediated by its Gi coupled transmembrane receptors. *Brain Res.* **919**: 139-146.

Sainz, R.M., Mayo, J.C., Rodriguez, C., Tan, D.X., Lopez-Burillo, S. and Reiter, R.J. (2003). Melatonin and cell death: differential actions on apoptosis in normal and cancer cells. *Cell. Mol. Life Sci.* **60**: 1407-1426.

Sauer, L.A., Dauchy, R.T., Blask, D.E., Armstrong, B.J. and Scalici, S. (1999). 13-Hydroxy-octadecadienoic acid is the mitogenic signal for linoleic acid-dependent growth in rat hepatoma 7288CTC *in vivo*. *Cancer Res.* **59**: 4688-4692.

Sauer, L.A., Dauchy, R.T. and Blask, D.E. (2001). Polysaturated fatty acids, melatonin, and cancer prevention. *Biochem. Pharmacol.* **61**: 1455-1462.

Shah, P.N., Mhatre, M.C. and Kothari, L.S. (1984). Effect of melatonin on mammary carcinogenesis in intact and pinealectomized rats in varying photoperiods. *Cancer Res.* **44**: 3403-3407.

Shellard, S.A., Whelan, R.D.H. and Hill, B.T. (1989). Growth inhibitory and cytotoxic effects of melatonin and its metabolites on human tomour cell lines in vitro. *Br. J. Cancer* **60**: 288-290.

Shiu, S.Y.W., Li, L., Xu, J.N., Pang, C.S., Wong, J.T.Y. and Pang, S.F. (1999). Melatonin-induced inhibition of proliferation and G_1/S cell cycle transition delay of human choriocarcinoma Jar cells: possible involvement of MT_2 (MEL_{1B}) receptor. *J. Pineal Res.* **27**: 183-192.

Shiu, S.Y.W., Xi, S.C., Xu, J.N., Mei, L., Pang, S.F., Yao, K.M. and Wong, J.T. (2000). Inhibition of malignant trophoblastic cell proliferation in vitro and in vivo by melatonin. *Life Sci.* **67**: 2059-2074.

Shiu, S.Y.W., Law, I.C., Lau, K.W., Tam, P.C., Yip, A.W.C. and Ng, W.T. (2003). Melatonin slowed the early biochemical progression of hormone-refractory prostate cancer in a patient whose prostate tumor tissue expressed MT1 receptor subtype. *J. Pineal Res.* **35**: 177-182.

Siu, S.W.F., Lau, K.W., Tam, P.C. and Siu, S.Y.W. (2002). Melatonin and prostate cancer cell proliferation: interplay with castration, epidermal growth factor, and androgen sensitivity. *Prostate* **52**: 106-122.

Slonimski, A. and Pruski, D. (1993). Melatonin inhibits proliferation and melonogenesis in rodent melanoma cells. *Exp. Cell Res.* **206**: 189-194.

Subramanian, A. and Kothari, L. (1991). Melatonin, a suppressor of spontaneous murine mammary tumors. *J. Pineal Res.* **10**: 136-140.

Sze, S.F., Ng, T.B. and Liu, W.K. (1993). Antiproliferative effect of pineal indoles on cultured tumor cell lines. *J. Pineal Res.* **14**: 27-33.

Tamarkin, L., Cohen, M., Roselle, D., Reichert, C., Lippman, M. and Chabner, B. (1981). Melatonin inhibition and pinealectomy enhancement of 7,12-dimethyl-benz(a)anthracene-induced mammary tumor in the rat. *Cancer Res.* **41**: 4432-4436.

Walker, M.J., Chanduri, P.K., Beatti, C.W., Tito, W.A. and Das Gupta, T.K. (1978). Neuroendocrine and endocrine correlates to hamster melanoma growth in vitro. *Surg. Forum* **29**: 151-152.

Wilson, S.T., Blask, D.E. and Lemus-Wilson, A.M. (1992). Melatonin augments the sensitivity of MCF-7 human breast cancer cells to tamoxifen in vitro. *J. Clin. Endocrinol. Metab.* **75**: 669-670.

Winczyk, K., Pawlikowski, M. and Karasek, M. (2001). Melatonin and RZR/ROR receptor ligand CGP 52608 induce apoptosis in the murine colonic cancer. *J. Pineal Res.* **31**: 179-182.

Winczyk, K., Lawnicka, H., Pawlikowski, M., Kunert-Radek, J. and Karasek, M. (2006). Growth-inhibitory action of melatonin and thiazolidinedione derivative CGP 52608 on murine 16/C breast cancer cells. *Neuroendocrinol. Lett.* 27: 351-354.

Winczyk, K., Pawlikowski, M., Guerrero, J.M. and Karasek, M. (2002a). Possible involvement of the nuclear RZR/ROR-alpha receptor in the antitumor action of melatonin on murine Colon 38 cancer. *Tumor Biol.* **23**: 298-302.

Winczyk, K., Pawlikowski, M., Lawnicka, H., Kunert-Radek, J., Spadoni, G., Tarzia, G. and Karasek, M. (2002b). Effect of melatonin and melatonin receptor ligand N –[(4-methoxy-1H-indol-2-yl)methyl}propamide on murine Colon 38 cancer growth in vitro and in vivo. *Neuroendocrinol. Lett.* **23**(1): 50-54.

Wölfler, A., Caluba, H.C., Abuja, P.M., Dohr, G., Schauenstein, K. and Liebmann, P.M. (2001). Prooxidant activity of melatonin promotes fas-induced cell death in human leukemic Jurkat cells. *FEBS Lett.* **502**: 127-131.

Xi, S.C., Tam, P.C., Brown, G.M., Pang, S.F. and Shiu, S.Y.W. (2000). Potential involvement of mt_1 receptor and attenuated sex steroid-induced calcium influx in the direct anti-proliferative action of melatonin on androgen responsive LNCaP human prostate cancer cells. *Pineal Res.* **29**: 172-183.

Xi, S.C., Siu, S.W., Fong, S.W. and Shiu, S.Y. (2001). Inhibition of androgen-sensitive LNCaP prostate cancer growth in vivo by melatonin: association of antiproliferative action of the pineal hormone with mt1 receptor protein expression. *Prostate* **46**: 52-61.

Ying, S.W., Niles, L.P. and Crocker, C. (1993).Human malignant melanoma cells express high-affinity receptors for melatonin: antiproliferative effects of mealtonin and 6-chloromelatonin. *Eur. J. Pharmacol.* **246**: 89-96.

Zisapel, N. and Bubis, M. (1994). Inhibition by melatonin of protein synthesis and growth of melanoma cells in culture. *Adv. Pineal Res.* **7**: 259-267.

Developmental Endocrinology

CHAPTER 1

Melatonin and its Role in Avian Embryogenesis

Skwarlo-Sonta Krystyna[1,], Oblap Ruslan[2,3], Majewski Pawel[1], Stepinska Urszula[2], Baranska Anna[1] and Olszanska Bozenna[2]*

Abstract

Melatonin, a multifunctional neurohormone synthesized and secreted by the pineal gland in circadian manner with an elevation during darkness, acts for an organism as a source of information on external lighting conditions. In non-mammalian vertebrates including birds the pineal gland plays an important role as one of the oscillators of the circadian system, as it possesses functions of photoreceptor, circadian clock and a site of melatonin synthesis. Melatonin originates also from extra-pineal tissues (e.g. bone marrow and gastrointestinal tract) where its synthesis is not related with lighting conditions. Besides, melatonin exhibits also a potent free-radical scavenging and antioxidant activity suggesting its important role in the protection against the oxidative stress. Avian embryo, developing in the conditions creating oxidative stress, has to be protected against oxidative damages by the presence of antioxidants and free radical scavengers deposed in the yolk. Melatonin function in the avian embryo antioxidative protection was not examined to date but there are some indications that it is produced/deposed in the freshly laid eggs of Japanese quail. Namely, melatonin receptor gene expression along with that of AA-NAT, a key enzyme in melatonin biosynthesis, has been demonstrated in the developing embryo of this species. Possible role played by melatonin from the egg yolk and early embryo is discussed.

INTRODUCTION

Organization of the Vertebrate Circadian System

For all living organisms, including vertebrate animals, the most important physiological challenges are the measurement of time and recognition of

---

[1] Department of Animal Physiology, Faculty of Biology, Warsaw University, Miecznikowa 1, 02-096 Warszawa, Poland.
[2] Institute of Genetics and Animal Breeding, Polish Academy of Sciences, Jastrzebiec n/Warsaw, 05-552 Wolka Kossowska, Poland.
[3] Institute of Agriecology and Biotechnology, Ukrainian Academy of Agricultural Sciences, 12 Metrologicheskaya St., Kyiv 03 143, Ukraine.
Corresponding author: E-mail: kss@biol.uw.edu.pl

changes taking place in the external environment. Most, if not all, physiological processes exhibit precise temporal organization, so that the phases of high and low intensity of a given process occur in a well coordinated sequence. This circadian and seasonal rhythmicity is generated by an endogenous clock and synchronized with changes in environmental factors, of which lighting conditions are the most important. The endogenous clock structures in mammals are located within the principal pacemaker in the suprachiasmatic nucleus (SCN) which receive photic information from the eyes only, via a direct retinohypothalamic tract. The SCN activity controls the function of the pineal gland, which, in turn, via melatonin (MEL) synthesis and release, influences other rhythmical processes, coupling them to the external lighting conditions (Ganguly et al., 2002). In non-mammalian species, including birds, the circadian organization is much more complex, and comprises also the pineal and extraretinal photoreceptors as well as the eyes. Among avian species there is, however, significant variation in the relative importance of these sites in the coupling of all components of the circadian system (Underwood et al., 2001).

THE PINEAL GLAND AS THE NEUROENDOCRINE TRANSDUCER

MEL Biosynthesis and Degradation

The pineal gland is a neuroendocrine organ, located within the brain, present in most vertebrate species, which acts as a transducer of information on external lighting conditions. This role involves the translation of the lighting conditions into a biochemical message, contained in the concentration of the principal pineal indole hormone, MEL, that is synthesized and released into the blood almost exclusively during the dark phase of the day, because these processes are blocked by the light (Reiter, 1989). MEL (N-acetyl-5-methoxy-tryptamine) is synthesized from tryptophan in a multistep pathway (Fig. 1) in which a key role is played by the penultimate enzyme, arylalkylamine-N-acetyl-transferase (serotonin N-acetyltransferase, AA-NAT, EC 2.3.1.87). The function of this enzyme, converting serotonin into N-acetyl derivative, exhibits the same circadian pattern as MEL concentration, therefore, the measurement of its activity offers an excellent mean to examine the pineal gland function (Axelrod, 1974). The last step in MEL biosynthesis, O-methylation of N-acetylserotonin by hydroxyindole-O-methyltransferase (HIOMT) is not affected by the light conditions (Cassone and Natesan, 1997).

MEL is catabolized in the liver, where it undergoes 6-hydroxylation followed by conjugation mainly to sulphate and, to a lesser extent, glucuronate (Reiter, 2003). The major MEL metabolite, 6-hydroxyMEL sulphate, is excreted in the urine in greater quantities at night than during the day, reflecting the circadian pattern of MEL synthesis and release. Another metabolite, cyclic 3-hydroxy-MEL, is most probably formed in target cells when MEL directly scavenges two hydroxyl radicals (referred to in subsequent paragraphs), and its level is proportional to the oxidative stress experienced by an individual.

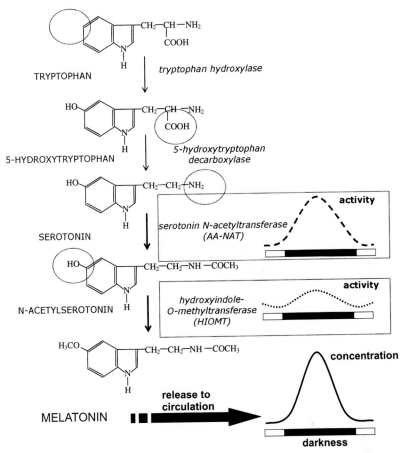

Fig. 1. Multistep pathway of MEL biosynthesis from tryptophan.

The most characteristic feature of the MEL diurnal rhythm is an increase in its concentration during darkness, regardless of the circadian pattern of locomotor activity, i.e., it is the same in both diurnal and nocturnal species. Moreover, since the same pattern of MEL rhythm has been demonstrated in invertebrates and microorganisms, this suggests that the MEL level indicates the time of darkness, and not the time of sleep (rest). MEL has, therefore, been named a "hormone or biochemical substrate of darkness" and is considered to be "a clock" and "a calendar" (Reiter, 1993). The last role comes from the fact that the period of MEL increase depends on the duration of the night, which, in turn, changes seasonally (Fig. 2). An organism

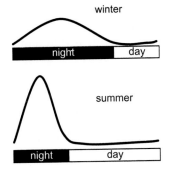

Fig. 2. Seasonal changes in the nocturnal increase in blood MEL concentration.

able to recognize the length of day and night and their reciprocal proportions, is also able to perceive the seasonal changes occurring in the environment and to interpret them in terms of the appropriate physiological adaptations to these changing conditions.

Phylogenesis of the Pineal Gland

In lower vertebrates, including avian species, pinealocytes are directly photosensitive, i.e. light penetrates into the pineal gland, located within the brain beneath the transparent skull and causes inhibition of AA-NAT activity (Collin et al., 1989). In contrast, mammalian pinealocytes, although clearly homologous to the photoreceptive cells comprising the pineal glands of other vertebrate species, receive information on the external lighting conditions detected by the retinal photoreceptors, solely via a multisynaptic neural pathway. Noradrenaline, delivered from postganglionic sympathetic terminals regulates MEL biosynthesis in the pinealocytes. Pinealocytes of birds and reptiles retain several morphological and functional features of those from lower vertebrates (fish and amphibians), but in comparison to the poikilotherms, their sympathetic innervation is increased and they receive photic information from circadian oscillators in the brain structures homologous to the mammalian SCN (Underwood et al., 2001).

Extrapineal Sites of MEL Biosynthesis

The main sources of MEL synthesis in vertebrates are the pineal gland and the photo-sensitive rods and cones of the retina, both being components of the biological clock oscillator. There are also several other tissues where MEL is synthesized, e.g., the Harderian gland, liver, kidney, airway epithelium, cells and glands of the immune system etc., but only in the pineal gland and retina this process occur in the same rhythmical manner (Conti et al., 2002). Among the most important extrapineal sources of MEL are the mucosa of the gastrointestinal tract (Bubenik, 2002) and the bone marrow (Conti et al., 2000), where MEL is particularly abundant. As the level of extrapineal MEL correlates neither with the lighting conditions nor with the quantity or quality of food within the gut, it has been assumed that the role of MEL present in these locations is different from that in the pineal gland. In particular, MEL from the extrapineal sources is considered to be part of a "diffuse neuroendocrine system" (Kvetnoy, 1999) functioning as a universal system of response, control and protection of organisms. In these tissues MEL might act as a paracrine signal molecule, coordinating intercellular communication as well as being a potent protection factor due to its activity as a free radical scavenger (Hardeland and Poeggeler, 2002). The role of free radicals in the pathogenesis of many diseases, including cancer and ageing, is widely accepted, therefore, the presence of MEL in tissues where free radical generation is particularly active, suggests a role as a local paracrine protection factor (Allegra et al., 2003).

MEL Binding Sites

To exert its physiological activities, every messenger molecule needs to be recognized by the target cells, equipped with the appropriate receptor structures. To date, the molecular mechanisms of MEL actions are not fully understood, because they can influence the physiological processes in a variety of ways. Some MEL actions are receptor-mediated, while others are receptor-independent. MEL receptors are widely distributed and belong to both the membrane-bound and nuclear classes (Dubocovich, 1995; Kokkola and Laitinen, 1998). Membrane-bound MEL receptors have seven transmembrane domains and belong to the superfamily of G-protein-coupled receptors. MEL receptors have been cloned and classified into three pharmacologically distinct types. Two of them, the high affinity subtypes mel 1a and mel 1b (MT1 and MT2 according to IUPHAR classification), are remarkably widespread in mammalian tissues. They are particularly abundant in the SCN and pars tuberalis of the adenohypophysis, and it is believed that few issues are totally devoid of MEL receptors. These two receptor subtypes have been found in all vertebrates examined to date, whereas a third, the mel 1c subtype, appears to be limited to *Xenopus laevis* oocytes and various tissues of avian species (Dubocovich et al., 1999; Skwarlo-Sonta et al., 2003). Postreceptoral MEL signal transduction involves several intracellular second messengers; the best known is MEL-induced decrease in the level of cyclic adenosine monophosphate (cAMP), which was previously raised under the influence of factors [like vasointestinal peptide (VIP) or forskolin] stimulating its generation, i.e., activating adenyl cyclase (Lopez-Gonzalez et al., 1992; Vanecek 1998; Markowska et al., 2002).

Due to its high lipophilicity, MEL can easily enter cells and is able to exert receptor-independent effects. These are the results of high affinity binding of MEL to calmodulin (Benitez-King and Anton-Tay, 1996), thereby influencing intracellular calcium levels and cytoskeletal arrangement, and also by its action as a direct and indirect antioxidant (Reiter et al., 2002). This last function is discussed in detail below.

Recently, MEL binding to the orphan ROR/RZR receptors present in the nuclear membrane was demonstrated (Raffi-El-Idrissi et al., 1998). In this case MEL acts as a very efficient transcription factor regulating several cellular function, e.g., synthesis and secretion of cytokines by immune cells (Garcia-Maurino et al., 2000).

Physiological Functions of MEL

MEL acts at the SCN to modulate its activity and influence circadian rhythms. In some avian species, the pineal gland plays an important role as one of the oscillators in the circadian system, since it possesses functions of the photoreceptor, circadian clock, and is a site of MEL-synthesis (Underwood et al., 2001).

The versatile MEL molecule is considered to be one of the ubiquitous molecules, involved in the coordination of several basic physiological processes. Its primary and most important function is to convey circadian signals as a chemical mediator of darkness. Among the physiological processes closely connected with the seasonal changes in external lighting conditions is reproduction, and the first regulatory function attributed to MEL was participation in the regulation of these processes, particularly in wild animals (Goldman and Darrow, 1983). Subsequently, the role of MEL in the regulation of thyroid gland function (Lewinski and Karbownik, 2002), modulation of immunity (Maestroni, 1999; Skwarlo-Sonta et al., 2002), protection against cancer (Barstch et al., 2002) and regulation of gastro-intestinal tract function (Bubenik, 2002) were demonstrated. Protection against oxidative stress, one of the universal MEL functions, is the result of its activity as a potent free radical scavenger.

MEL as a Free Radical Scavenger

In vertebrates, about 4-5% of the inhaled oxygen escapes from the energy-producing mitochondrial reactions and forms partially reduced O_2 molecules, called "reactive oxygen species" (ROS; Reiter, 1998). Some of these possess an unpaired electron and are referred to as "free radicals", that are highly toxic for important biological molecules, including nucleic acids, proteins as well as membrane lipids. Free radicals have been implicated in the development of many degenerative diseases, mutations and as a ageing factor promoting (Allegra et al., 2003).

One of the best known and most highly studied actions of MEL is its effective anti-oxidative potential. This is expressed by the very efficient scavenging of free radicals (MEL interacts with various free radicals and ROS) as well as its role in stimulating the synthesis of another important antioxidant, glutathione, and its influence on anti-oxidant enzymes, such as superoxide peroxidase, catalase, etc. (Reiter et al., 2002). One MEL molecule is able to inactivate up to six free radicals since the derivatives formed during its reaction with free radicals (3-hydroxy-MEL and N^1-acetyl-N^2-formyl-5-metoxykynuramine, AFMK) exhibit this anti-oxidant ability as well (Hardeland and Poeggeler, 2002).

AVIAN PINEAL GLAND STRUCTURE AND FUNCTION

Morphology of Avian Pinealocytes

As previously mentioned, morphologically and functionally the avian pineal gland represents an intermediate step between the fully photoreceptive organ of poikilotherms and the non-photoreceptive mammalian glands (Collin et al., 1989). There are also other special features of avian pinealocytes, one of them being the ability to carry out the rhythmical synthesis of MEL in organ culture, not only by whole pineal glands but also by the isolated pinealocytes. In dispersed pineal cell cultures of different avian species (e.g., chicken, sparrow and quail) MEL synthesis occurs rhythmically for several L:D (light: dark) cycles, and in some species, persists in constant darkness for a few cycles (in chicken but not in Japanese quail, Underwood et al., 2001). This finding confirms that the avian pineal gland possesses not only photoreceptive structures, but also contains self-sustaining circadian pacemakers. The pineal pacemakers of different avian species show variable dependence on the neural inputs from the SCN. In some species, e.g., in the house sparrow, the pineal gland plays the predominant role in circadian organization.

Adrenergic Regulation of MEL Synthesis in Pinealocytes

Information on external lighting conditions reaches the mammalian and avian pineal gland via the multisynaptic adrenergic pathway, starting at the eye photoreceptors (Fig. 3). In both phyla, light causes a decrease in MEL synthesis, but the molecular mechanisms involved are quite different (Fig. 4). In mammals, postganglionic sympathetic nerve terminals release noradrenaline in the dark. Noradrenaline binds to β- and α1-adrenergic receptors on pinealocytes and thereby stimulates cAMP formation, which in turn activates AA-NAT, leading to the increase in MEL level (Vacas et al.,

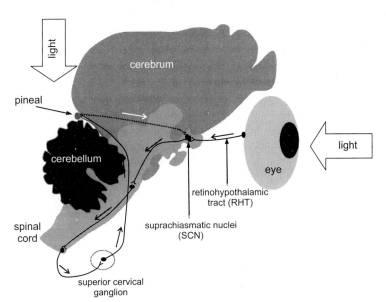

Fig. 3. Multisynaptic tract conveying information of lighting conditions from the eye via the oscillator in the suprachiasmatic nuclei (SCN) to the avian pineal gland, which also receives photic information by its own photoreception. Pineal-derived MEL influences, in turn, SCN function.

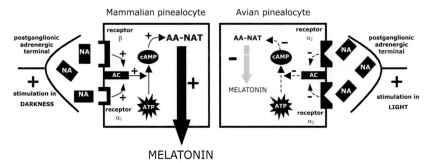

MELATONIN

Fig. 4. Adrenergic regulation of MEL biosynthesis in mammalian and avian pinealocyte. The same regulatory effects (nocturnal increase, depression by light) are achieved by different regulatory mechanisms (NA – noradrenaline, AC – adenylate cyclase).

1985). In birds, noradrenaline is released in the light, and its binding to α2-adrenergic receptors on pinealocytes inhibits cAMP formation, blocks AA-NAT activity and, finally, causes a decrease in MEL level (Zatz and Mullen, 1988; Collin et al., 1989).

SPECIAL CONDITIONS OF AVIAN EMBRYOGENESIS

In contrast to the mammalian embryo, which from the very beginning to the end of its development experiences the nutritive and regulatory influence of the maternal circulation, the avian embryo forms, from the moment of fertilization, an individual unit disconnected from the maternal organism. This means that the avian embryo must be self-sufficient with respect to the delivery of all substances necessary for its development. The main source of such substances is the yolk, which accumulates not only nutrients and energy resources, but also numerous biologically active molecules, such as hormones and neurotransmitters (Emmanuelsson et al., 1988; Mougdal et al., 1990; Petrie et al., 2001), antibodies and enzymes (Stepinska et al., 1996; Hartmann and Wilhelmson, 2001).

The avian embryo develops in conditions creating oxidative stress. Its rapid growth is accompanied by high rates of oxidative metabolism (Blount et al., 2000). The main source of energy for the developing avian embryo is the oxidation of fatty acids from the lipid-rich yolk. This easy oxidizable substrate, rich in unsaturated fatty acids is highly vulnerable to peroxidation by free radicals, produced as normal by-products of metabolism. The risk of peroxidation increases as development proceeds and becomes much higher at hatching, when the embryo is suddenly exposed to the high concentration of atmospheric oxygen. Moreover, the newly hatched chick begins pulmonary respiration and, therefore, enters a new phase of metabolism associated with high risk of oxidative stress. For all these reasons, it is obvious that the developing avian embryo has to be protected against oxidative stress by means of a sufficient content of antioxidant and free radical scavengers in the yolk itself. Recently it was demonstrated that this anti-free radical protection of the developing avian embryo and young chick is afforded by yolk-derived carotenoids and other anti-oxidants such as vitamin E (Surai and Speake, 1998). As per available data, the role of MEL as a free radical scavenger in the yolk and developing avian embryo has so far not been considered.

MEL in the Developing Avian Embryo

It may be assumed, that if MEL is necessary at the beginning of avian development, it should either be stored in the yolk, or its synthesis should be initiated very early in development, before the start of pineal activity, or both events might occur simultaneously. Recently, the presence of MEL, independent of light/dark conditions, was demonstrated in the yolk of rainbow trout (Yamada et al., 2002), but there have been no reports concerning the presence of MEL in the avian yolk or its synthesis in the early avian embryo, before differentiation of the pinealocytes.

The presence of MEL was detected in pineal cell cultures of 11-13-day-old chick embryos (Moller and Moller, 1990) and in cultures of 13-day-old embryo pineal glands (Akasaka et al., 1995). Organization of the chicken embryo pineal gland starts, however, much earlier, when in the 5-day-old embryo the cells aggregate forming rosettes, which are subsequently transformed into vesicles in 10- and 15-day-old embryos (Jove et al., 1999). The appearance of the pineal vesicles is considered to be an indication of the onset of its functional activity.

MEL Receptor Gene Expression in the Developing Avian Embryo

The presence of MEL receptor transcripts in both maternal RNA from oocytes and in RNA from early embryos in freshly laid eggs of Japanese quail has been demonstrated for the first time, using RT-PCR (Revers Transcription Polymerase Chain Reaction). (Oblap and Olszanska, 2001). The quantity of the transcripts was low, but discrimination between the three receptor subtypes was possible. The most abundant subtype of the MEL receptor transcripts in the maternal RNA was that encoding mel 1c, whereas all three transcripts were found in the RNA from blastoderms of eggs, indicating their *de novo* synthesis during uterine embryo development. These results do not necessarily indicate that functional MEL receptors operate in the laid eggs of Japanese quail, instead they suggest that in this species at least, MEL might play a role in the earliest steps of avian embryogenesis. This finding raises the question whether MEL might be present and functioning during early embryonic development, before the pineal gland is formed, or is it produced *de novo* in early embryos. In the latter scenario, the enzymes responsible for MEL synthesis should be expressed in the avian oocytes and embryo, and further experiments have been undertaken to check the presence of the relevant transcripts.

AA-NAT Gene Expression in the Developing Avian Embryo

The presence of transcripts encoding AA-NAT, a key enzyme in MEL biosynthesis, in the oocytes and blastoderms from freshly laid eggs of Japanese quail has been detected by using the same method (Oblap and Olszanska, 2003). Furthermore, in these developmental stages (oocytes and blastoderms) two transcripts for AA-NAT (323- and 238-bp RT-PCR products) have been identified, whereas in the pineal gland of adult females only one (238-bp) was present. The two products differed by an intron of 85-bp present in the larger and absent in the smaller band. In small 3-mm oocytes, the 328-bp band was predominant (323-bp: 238-bp transcript ratio was ~17:1), in immature vitellogenic oocytes this ratio was ~4:1, and in mature, preovulatory oocytes, the transcripts showed equal abundance (ratio 1:1). This ratio was reversed to ~0.2: 1 (1:5) in blastoderms from fertile freshly laid eggs, corresponding to an embryo of ~40,000 cells. Developmental changes in the transcription patterns of AA-NAT and MEL receptor genes during ontogenesis of the Japanese quail pineal gland (Oblap and Olszanska, 2004) were examined in subsequent experiments. The findings provide evidence for the existence of two developmental phases of AA-NAT expression in the quail embryo pineal gland. The first, embryonic-type phase, lasted from the beginning until seven-ten days of incubation and was marked by the presence of both the aforementioned transcripts for AA-NAT, whereas in the second, adult-type phase (starting at seven-ten days of incubation and continuing through hatching and development of the adult pineal gland), only one, intron-less transcript was present.

It is too early to conclude about the physiological meaning of the presence the two AA-NAT transcripts in the early embryos and the switch towards the one, adult form in the later stages of development. Probably it reflects another, still unknown, mechanism of regulating AA-NAT activity during ontogenesis, at the level of mRNA processing, whose specificity (or not) for embryonic development is yet to be established in the future.

The presence of a transcript does not necessarily imply the existence of a functional protein, although it is a prerequisite for such an event. The presence of serotonin, a substrate for the action of AA-NAT, has been conclusively proved in hen yolk (Emmanuelsson et al., 1988), as it has the presence of MEL in the yolk of trout (Yamada et al., 2002). Moreover, it has been proposed that the longer form of the AA-NAT transcript, containing an unprocessed intron, represents its translationally inactive form, present in maternal RNA (Oblap and Olszanska, 2003). Collectively, these data provide the basis for the assumption that MEL can be stored in oocytes and/or synthesized in early embryonic cells, albeit at a low level. This notion was verified in subsequent experiments, where surprisingly, AA-NAT activity in the yolk and early embryos of Japanese quails was found to be relatively high and decreased with time during embryogenesis (unpublished data).

The Role of the MEL during Avian Embryonic Development

The role played by MEL stored or synthesized in the yolk and early avian embryos has yet to be examined in detail, but taking into consideration available data, some suggestions could be made in this context.

The primary and most well recognized role of MEL seems to be the circadian organization of embryonic physiology, especially because the rhythmical pineal gland function appears relatively early in avian ontogenesis, i.e., during the last third phase of embryogenesis. This may reflect an adaptation to the absence of maternal MEL, which in mammals continuously influences the foetus where the circadian pineal function develops only postnatally. This rhythmicity was measured by AA-NAT activity (Binkley and Geller, 1977; Zeman and Illnerova 1990), MEL level (Akasaka et al., 1995; Zeman et al., 1999) or by the accumulation of transcripts of AA-NAT gene (Herichova et al., 2001). A recent study (Adachi et al., 2002) has demonstrated that 17-day-old embryonic chick glial cell cultures express high-affinity MEL receptors (mel 1a and mel 1c subtypes) and adapt their metabolic rhythms (2-deoxyglucose uptake, release of lactate and pyruvate) to the rhythms imposed by MEL administered to the culture at physiological concentrations. Moreover, the initial heart rate response of Muscovy duck and chick embryos to exogenous MEL coincides with the start of rhythmical MEL production (Hochel and Nichelman, 2001).

Another possible function of the yolk and embryonic MEL biosynthesis, directly connected with its aforementioned role in the circadian organization of embryo physiology, could be in the memorization of external lighting conditions. In sparrow it has been demonstrated that the circadian patterns of the pineal gland biosynthetic activity imposed during *in vivo* synchronization to the distinct photoperiods could be maintained as an internal representation of time within the isolated pineal glands (Brandstetter et. al., 2000). This was demonstrated by the significant difference in the circadian rhythm of *in vitro* MEL synthesis by pineal glands maintained in constant darkness but isolated from donors exposed to the different photoperiods. Information of maternal origin, such as the mRNA encoding AA-NAT and MEL receptors identified in the blastoderm of laid eggs of Japanese quail (Oblap and Olszanska, 2001, 2003) might also be considered as a source of the "photoperiodic memory". In the recent experiments (unpublished data), it has been found that young male chickens kept from the time of hatching in precisely controlled L:D:=12:12 conditions exhibited a highly pronounced circadian rhythm of AA-NAT activity, strictly correlated, however, with the season in which the chickens were hatched and reared. Surprisingly, the season influenced not only the time of the nocturnal peak of AA-NAT activity, but also the development of an experimentally induced inflammatory reaction. In a previous study it was demonstrated that exogenous MEL exerted an anti-inflammatory effect on developing peritonitis (Skwarlo-Sonta et al., 2002), and this experiment indicates that the hormone coming from endogenous sources might play a similar role. It should be pointed out that the chickens used for the experiment did not experience natural lighting conditions and their only information on the season most probably originated from maternal sources.

Indirect evidence that the pineal gland is functionally active in very early chick embryos was obtained by Jankovic and co-workers (1994). By using embryos pinealectomized at 96 hours of incubation, thus developing without any influence of the pineal gland, they found retarded development of the lymphoid glands and immune response, accompanied by a significant change in the concentration of biogenic amines (serotonin, dopamine and noradrenaline) both in the immune organs and in the brain. This finding strongly supports the notion of reciprocal dependence of the development of the neuroendocrine and immune systems, and highlights the important role played by the pineal gland in these interrelationships (Skwarlo-Sonta, 2002).

MEL from the yolk may easily penetrate into the developing embryo, as was demonstrated recently by Kovacikova and co-workers (2003) in Japanese quail. *In ovo* MEL treatment had no effect on the pineal MEL level or on hatchability and organ development. On the contrary, it might participate in intercellular communication, as suggested by Kvetnoy (1999) for mammalian tissues. Moreover, as already mentioned above (see p. 179), the avian embryo develops in conditions in which free radical generation is particularly high and, therefore, the anti-oxidative protection has to be very efficient (Blount et al., 2000). Until now, this role of MEL as a particularly potent free radical scavenger, present/synthesized in the yolk and early embryos, was not considered and anti-oxidative protection was attributed to other antioxidant substances (carotenoids and vitamin E) present in the yolk of avian species. It is believed that future studies will elucidate this putative role of MEL.

CONCLUSION

While the role of MEL in the circadian organization of the avian organism is well recognized, far less is known about its possible role in embryogenesis of binds. Experimental data on pineal gland development and function in the avian embryo are rather limited. Nevertheless, there are some indications that MEL itself and/or the key enzyme in its biosynthetic pathway (AA-NAT), as well as the substrate for this enzyme, are present in the yolk and the early embryo, thus MEL is available to fulfill its functions before the embryonic pineal gland starts its own biosynthetic activity. A semi-processed transcript encoding AA-NAT of maternal origin seems to be deposited in eggs and thereafter undergoes posttranscriptional modification leading to the synthesis of the functionally active enzyme. Consideration of the available evidence has led to some suggestions regarding the role played by MEL from the yolk and early embryo. In particular, it seems to be involved in the circadian organization of the developing embryo along with its putative roles in the memorization of external lighting conditions, as a member of the "diffuse neuroendocrine system", involved in intercellular communication and as a free radical scavenger. This last function might protect the developing embryo against the oxidative stress accompanying the metabolic processes occurring in the very special conditions created in the avian egg.

ACKNOWLEDGEMENTS

This work was supported in part by a grant of the Committee for Scientific Research (KBN) No. 3 P06D 38 24. RO was the recipient of a NATO Science Fellowship (Poland).

REFERENCES

Adachi, A., Natesan, A.K., Whitfield-Rucker, M.G., Weigum, S.E. and Cassone, V.M. (2002). Functional melatonin receptors and metabolic coupling in cultured chick astrocytes. *Glia* **39**: 268-278.

Akasaka, K., Nasu, T., Katayama, T. and Murakami, N. (1995). Development of regulation of melatonin release in pineal cells in chick embryo. *Brain. Res.* **692**: 283-286.

Allegra, M., Reiter, R.J., Tan, D.-X., Gentile, C., Tesoriere, L. and Livrea, M.A. (2003). Minireview: The chemistry of melatonin's interaction with reactive species. *J. Pineal Res.* **34**: 1-10.

Axelrod, J. (1974). The pineal gland: a neurochemical transducer. *Science* **184**: 1341-1348.

Bartsch, C., Bartsch, H. and Karsek, M. (2002). Melatonin in clinical oncology. *Neuroendocrinol. Lett.* **23**(1): 30-38.

Benitez-King, G. and Anton-Tay, F. (1996). Role of melatonin in cytoskeletal remodeling is mediated by calmodulin and protein kinase C. In: *Melatonin: A Universal Photoperiodic Signal with Diverse Action.* (Reiter, R.J., Pang, S.F. and Tang, P.L., eds.), Karger, Basel, Switzerland, 154-159.

Binkley, S. and Geller, E.B. (1977). Pineal enzymes in chickens. Development of daily rhythmicity. *Gen. Comp. Endocrinol.* **27**: 424-429.

Blount, J.D., Houston, D.C. and Moller, A.P. (2000). Why egg yolk is yellow. *TREE* **15**: 47-49.

Brandstatter, R., Kumar, V., Abraham, U. and Gwinner, E. (2000). Photoperiodic information acquired and stored *in vivo* is retained *in vitro* by a circadian oscillator, the avian pineal gland. *PNAS* **97**: 12324-12328.

Bubenik, G.A. (2002). Gastrointestinal melatonin: a Cinderella story of melatonin research. In: *Treatise on Pineal Gland and Melatonin.* (Haldar, C., Singaravel, M., Maitra, S.K., eds.), Science Publishers, Enfield, (NH) USA, Plymouth, UK, 145-156.

Cassone, V.M. and Natesan, A.K. (1997). Time and time again: The phylogeny of melatonin as a transducer of biological time. *J. Biol. Rhythms* **12**: 489-497.

Collin, J.P., Voisin, P., Falcon, J., Faure, J.P., Brisson, P. and Defaye, J.R. (1989). Pineal transducers in the course of evolution: Molecular organization, rhythmic metabolic activity and role. *Arch. Histol. Cytol.* **52**(1): 441-449.

Conti, A., Conconi, S., Hertens, S., Skwar³o-Soñta, K., Markowska, M. and Maestroni, G.J.M. (2000). Evidence for melatonin synthesis in mice and human bone marrow. *J. Pineal Res.* **28**: 193-202.

Conti, A., Tettamanti, C., Singaravel, M., Haldar, C., Pandi-Perumal, S.R. and Maestroni, G.J.M. (2002). Melatonin: an ubiquitous and evolutionary hormone. In: *Treatise on Pineal Gland and Melatonin* (Haldar, C., Singaravel, M. and Maitra, S.K., eds.), Science Publishers, Enfield (NH) USA, Plymouth, UK, 105-143.

Dubocovich, M.L. (1995). Melatonin receptors: are there multiple subtypes? *Trends Pharmacol. Sci.* **16**: 50-56.

Dubocovich, M.L., Masana, M.I. and Benloucif, S. (1999). Molecular pharmacology and melatonin receptor subtypes. *Adv. Exp. Med. Biol.* **460**: 181-190.

Emmanuelsson, H., Carlberg, M. and Lowkvist, B. (1988). Presence of serotonin in early chick embryos. *Cell. Differ.* **24**: 191-200.

Ganguly, S., Coon, S.L. and Klein, D.C. (2002). Control of melatonin synthesis in the mammalian pineal gland: the critical role of serotonin acetylation. *Cell Tissue Res.* **309**: 127-137.

Gracia-Maurino, S., Pozo, D., Calvo, J.R. and Guerrero, J.M. (2000). Correlation between nuclear melatonin receptor expression and enhanced cytokine production in human lymphocytic and monocytic cell lines. *J. Pineal Res.* **29**: 120-137.

Goldman, B.D. and Darrow, J.M. (1983). The pineal gland and mammalian photoperiodism. *Progr. Neuroendocrinol.* **37**: 386-396.

Hardeland, R. and Poeggeler, B. (2002). Chemistry and biology of melatonin oxidation. In: *Treatise on Pineal Gland and Melatonin*. (Haldar, C., Singaravel, M. and Maitra, S.K., eds.), Science Publishers, Enfield, (NH) USA, Plymouth, UK, 407-422.

Hartmann, C. and Wilhelmson, M. (2001). The hen's egg yolk: A source of biologically active substances. *The World's Poultry Sci.* **57**: 13-28.

Herichova, I., Zeman, M., Mackova, M. and Griac, P. (2001). Rhythms of the pineal N-acetyltransferase mRNA and melatonin concentration during embryonic and post-embryonic development in chicken. *Neurosci. Lett.* **298**: 123-126.

Hochel, J. and Nichelman, M. (2001). Ontogeny of heart rate responses to exogenous melatonin in Muscovy duck and chicken embryos. *Life Sci.* **69**: 2295-2309.

Jankovic, B.D., Knezevic, Z., Koji, L. and Nikoli, V. (1994). Pineal gland and immune system. Immune function in the chick embryo pinealectomized at 96 hours of incubation. *Ann. NY Acad. Sci.* **719**: 398-409.

Jove, M., Cobos, P., Torrente, M., Gilabert, R. and Piera, V. (1999). Embryonic development of pineal gland vesicles: A morphological and morphometric study in chick embryos. *Eur. J. Morph.* **37**: 29-35.

Kokkola, T. and Laitinen, J.T. (1998). Melatonin receptor genes. *Ann. Med.* **30**: 88-94.

Kovacikova, Z., Kuncova, T., Lamosova, D. and Zeman, M. (2003). Embryonic and post-embryonic development of Japanese quail after *in ovo* melatonin treatment. *Acta Vet. Brno.* **73**: 157-161.

Kvetnoy, I.M. (1999). Extrapineal melatonin: location and role within diffuse neuroendocrine system. *Histochem. J.* **31**: 1-12.

Lewiñski, A. and Karbownik, M. (2002). Melatonin and the thyroid gland. *Neuroendocrinol. Lett.* **23**(1): 73-78.

Lopez-Gonzalez, M.A., Calvo, J.R., Osuna, C. and Guerrero, J.M. (1992). Interaction of melatonin with human lymphocytes: evidence for binding sites coupled to potentiation of cyclic AMP stimulated by vasoactive intestinal peptide and activation of cyclic GMP. *J. Pineal Res.* **12**: 97-104.

Lopez-Gonzalez, M.A., Martin-Cacao, A., Calvo, J.R., Reiter, R.J., Osuna, C. and Guerrero, J.M. (1993). Specific binding of 2-[^{125}I]melatonin by partially purified membranes of rat thymus. *J. Neuroimmunol.* **45**: 121-126.

Maestroni, G.J.M. (1999). MLT and the immune-hematopoietic system. *Adv. Exp. Med. Biol.* **460**: 395-405.

Markowska, M., Mrozkowiak, A. and Skwarlo-Soúta, K. (2002). Influence of melatonin on chicken lymphocytes in vitro: involvement of membrane receptors. *Neuroendocrinol. Lett.* **23**(1): 67-72.

Moller, W. and Moller, G. (1990). Structural and functional differentiation of the embryonic chick pineal organ in vivo and in vitro. *Cell Tissue Res.* **260**: 337-342.

Moudgal, R.P., Panda, J.N. and Mohan, J. (1990). Catecholamines are present in egg yolk in fairly stable form: Elevated adrenaline indicates the stress. *Curr. Sci.* **59**: 937-939.

Oblap, R. and Olszanska, B. (2001). Expression of melatonin receptor transcripts (mel-1a, mel-1b and mel-1c) in Japanses quail oocytes and eggs. *Zygote* **9**: 237-244.

Oblap, R. and Olszanska, B. (2003). Presence and developmental regulation of serotonin N-acetyltransferase transcripts in oocytes and early quail embryos (*Coturnix coturnix japonica*). *Mol. Reprod. Develop.* **65**: 132-140.

Oblap, R. and Olszanska, B. (2004). Transition from embryonic to adult transcription pattern of serotonin N-acetyltransferase gene in avian pineal gland. *Mol. Reprod. Develop.* **67**: 145-153.

Petrie, M., Schwabl, H., Brande-Lavridsen, N. and Burke, T. (2001). Sex differences in avian yolk hormone levels. *Nature* **412**: 498.

Raffi-El-Idrissi, M., Calvo, J.R., Harmouch, A., Garcia-Maurino, A. and Guerrero, J.M. (1998). Specific binding melatonin by purified cell nuclei from spleen and thymus of the rat. *J. Neuroimmunol.* **86**: 190-197.

Reiter, R.J. (1989). The pineal gland and its indole products: basic aspects and clinical applications. In: *The Brain as an Endocrine Organ.* (Cohen, M.P. and Foley, P.P., eds.) Springer, Vienna, 96-149.

Reiter, R.J. (1993). The melatonin rhythm: both a clock and a calendar. *Experientia* **49**: 654-664.

Reiter, R.J. (1998). Oxidative damage in the central nervous system: protection by melatonin. *Progr. Neurobiol.* **56**: 359-384.

Reiter, R.J., Tan, D.X. and Allegra, M. (2002). Melatonin: reducing molecular pathology and dysfunction due to free radicals and associated reactants. *Neuroendocrinol. Lett.* **23**(1): 3-8.

Reiter, R.J. (2003). Melatonin: clinical relevance. *Best Practice Res. Clin. Endocrinol. Metab.* **17**: 273-285.

Skwarlo-Sonta, K., Majewski, P., Markowska, M., Jakubowska, A. and Waloch, M. (2002). Bimodal effect of melatonin on the inflammatory reaction in young chickens. In: *Treatise on Pineal Gland and Melatonin.* (Haldar, C., Singaravel, M. and Maitra, S.K., eds.) Science Publishers, Enfield, (NH) USA, Plymouth, UK, 225-238.

Skwarlo-Sonta, K., Majewski, P., Markowska, M., Oblap, R. and Olszanska, B. (2003). Bidirectional communication between the pineal gland and the immune system. *Can. J. Physiol. Pharmacol.* **81**: 342-349.

Stepinska, U., Malewska, A. and Olszanska, B. (1996). RNase activity in Japanese quail oocytes. *Zygote* **4**: 219-227.

Surai, P.F. and Speake, B.K. (1998). Distribution of carotenoids from the yolk to the tissues of the developing embryo. *J. Nutr. Biochem.* **9**: 645-651.

Underwood, H., Steele, C.T. and Zivkovic, B. (2001). Circadian organization and the role of the pineal in birds. *Micro. Res. Techn.* **53**: 48-62.

Vacas, M.I., Sarmiento, I.K. and Cardinali, D.P. (1985). Interaction between β- and α-adrenoceptors in rat pineal adenosine cyclic 3′,5′-monophosphate phosphodiesterase activation. *J. Neural. Trans.* **26**: 295-304.

Vanecek, J. (1998). Cellular mechanism of melatonin action. *Physiol. Rev.* **78**: 687-721.

Yamada, H., Chiba, H., Amano, M., Iigo, M. and Iwata, M. (2002). Rainbow trout eyed-stage embryos demonstrate melatonin rhythms under light-dark conditions as measured by a newly developed time resolved fluoroimmunoassay. *Gen. Comp. Endocrinol.* **125**: 41-46.

Zatz, M. and Mullen, D.A. (1988). Norepinephrine, acting via adenylate cyclase, inhibits melatonin output but does not phaseshift the pacemaker in cultured chick pineal cells. *Brain Res.* **450**: 137-143.

Zeman, M. and Illnerova, H. (1990). Ontogeny of N-acetyltransferase activity rhythm in pineal gland of chick embryo. *Comp. Biochem. Physiol.* A., **97**: 175-178.

Zeman, M., Gwinner, E., Herichowa, I., Lamosowa, D. and Kostal, L. (1999). Perinatal development of circadian melatonin production in domestic chicks. *J. Pineal Res.* **26**: 28-34.

CHAPTER 2

Early Avian Development: Molecular Aspects

Bożenna Olszańska

Abstract

Avian egg/embryo, a well-known object for morphological and embryological studies, was somehow neglected in molecular biology. In this chapter the general characteristics of avian ovary and oocyte is given, specifying the maternal RNA accumulation in oogenesis and its metabolism. The amounts of maternal RNA for different avian species are given in comparison with mammalian oocytes. The term "extra-embryonic RNA" designates the RNA localised in the cytoplasmic layer around a yolk sphere, outside of germinal disc region. This portion of maternal RNA is supposed to participate in oocyte growth and not in embryo development, and is in part delivered into oocyte by granulosa cells. The hypothesis of supernumerary sperm degradation by DNases in avian oocyte cytoplasm is proposed. During physiological polyspermic fertilisation in birds many sperm enter the oocyte cytoplasm due to the lack of the early block to polyspermy existing in mammals, but only a single spermatozoon forms the zygotic nucleus. Other sperm could be degraded by DNase I and II present in the oocyte cytoplasm. The spermatozoon entering a germinal disc at its central point, into karyoplasm at the territory of nucleus, can escape the DNase degradation and participate in embryo development. The segmentation of egg cytoplasm on activation, without concomitant divisions of nuclei, is described, the pattern of the segmentation strictly resembling normal cleavage divisions.

INTRODUCTION

While considering different aspects of avian embryo development, one should bear in mind that an avian embryo represents an individual unit separated from the mother organism since the earliest developmental stages. This is in contrast to mammals, whose development since the very beginning until the time of birth occurs within the mother, and the embryo is exposed to the influence of all substances circulating in the maternal organism.

Institute of Genetics and Animal Breeding of Polish Academy of Sciences, Jastzębiec n/Warsaw, 05-552 Wólka Kossowska, Poland. Tel.: +48 22 756 1711(to 18), Fax: +48 22 756 1417. E-mail: B.Olszanska@ighz.pl

Consequently, the mammalian embryo obtains from the mother all the substances needed for its growth and development, not only the nutrients but also biologically active compounds; while an avian embryo must have all the nutrients and biologically active compounds accumulated in the oocyte, in the yolk, or it should commence their synthesis by itself from the earliest developmental stages. Therefore, the avian embryos should be equipped with all the molecular machinery required for synthesis of cellular components, which are not stored in oocytes in sufficient amounts or are rapidly used during intensive metabolism at the early stages of development (cleavage divisions, migratory movements and protein synthesis). The indispensable constituents of such a machinery include, e.g., a store of maternal ribosomes, tRNAs and mRNA templates, as well as the factors necessary for transcription and translation. With respect to the molecular mechanisms of development, highly studied models have been the organisms in which the embryo develops externally: the *Echinoderms* and *Amphibia* (Davidson, 1986).

For this kind of study, the avian embryo culture under *in vitro* conditions is an invaluable tool allowing direct observations and permitting the precise control of environmental and culture medium factors. Since the early 1970s, techniques have been developed for mammalian embryos, the application of which has resulted in extraordinary breakthroughs including the transfer of an embryo from one female to another, cloning, production of transgenic animals, and, above all, has led to an avalanche of investigations on molecular mechanisms of fertilization and embryo development. The techniques applied to birds lag considerably behind those for mammals, probably because of the difficulty in manipulating large oocytes heavily loaded with yolk, and, possibly, due to the cost of the biological material: from one superovulated mouse or rabbit female it is possible to obtain several tens of oocytes/embryos, while for a single bird oocyte, zygote or cleaving embryo, it is necessary to sacrifice one female. Considering the costs, the most convenient model for investigations seems to be the Japanese quail (*Coturnix coturnix japonica*). Its maintenance is cheaper, the dimensions of its body and of the oocyte/egg are smaller so they are easier to manipulate, the generation time is shorter (maturity at ~6 weeks of age) and egg laying and embryo development are comparable to other domesticated birds, except for the length of the incubation period, which is species specific. Most physiological and molecular mechanisms in birds described here were originally delineated to the hen and quail, thus some individual modifications relevant to other avian groups/species might be required.

The term "egg" in the present chapter will have one of two meanings, depending on the context: 1st – in the common sense of the word meaning the laid egg of a bird, containing yolk and albumen enclosed within a shell, 2nd – the female gamete (*ovum*) equivalent to the yolk sphere after ovulation from an ovarian follicle, traversing an oviduct until the moment when the fertilized egg becomes a zygote, i.e., a 1-cell embryo.

THE STRUCTURE AND CONTENT OF OOCYTE

Among animals, oocytes of birds attain the largest dimensions. It is a remarkable fact that avian oocyte, corresponding to an egg yolk sphere, comprises just one cell. The dimensions of mammalian oocytes are decidedly smaller (~60 µm in diameter for the mouse, ~140 µm for cattle) because of the insignificant amounts of stored substances (alecital oocytes), while the avian oocytes loaded with great amount of yolk (polylecital/megalecithal oocytes) can attain a diameter of 1-2 cm for the quail (F1), ~3-4 cm for the hen, and ~10 -14 cm for the ostrich.

The avian (laying hen) ovary is an ordered structure, containing a hierarchy of several (5-10) large yellow ovarian follicles filled with yolk (LYF), usually designated in descending order as F1, F2, F3....Fn. The oocytes mature and ovulate individually in succession, usually following a circadian rhythm lasting from 22 to 28 hours, depending on the individual female. There are a number (5-7) of small yellow follicles (SYF) of diameter 5-10 mm, about 10 to 20 large white follicles (LWF) of diameter 3-5 mm and usually >1,000 small white follicles (SWF), representing the total pool of oocytes of the ovary (Etches, 1990; Etches and Petitte, 1990). In the Japanese quail, the ovary (Fig. 1) has a similar hierarchical structure: 3 LYF, 2-3 SYF, ~20 LWF and 39-50 SWF (data for young quail laying females, ~10 weeks of age, from a randomly bred flock). The oocytes in large white follicles (LWF) already have, by this stage, a germinal disc with a centrally placed nucleus at the animal pole.

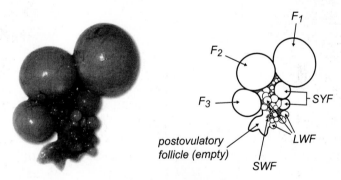

Fig. 1. The Japanese quail ovary.

The bird oocyte is an extremely valuable source of huge amounts of nutritive substances stored in the form of yolk; at its animal pole there is a discoidal clump of cytoplasm, the germinal disc, that appears as a whitish spot on the yellow yolk sphere (Fig. 2). In spite of the great differences in yolk dimensions, the diameter of a germinal disc is surprisingly constant, e.g. 2.5-3.0 mm for quail, 4-5 mm for hen and ~6 mm for ostrich.

The nucleus of an immature oocyte (*germinal vesicle*) is also exceptionally big; for quail its diameter is ~0.3 mm and is readily observed under a stereomicroscope as a dark vesicle in the centre of the germinal disc (Fig. 2a), and sometimes it may even be visible with the naked eye. After excising the germinal disc and clearing it from the yolk, the nucleus is seen as a transparent vesicle (Fig. 2b) which can easily be separated from the germinal disc cytoplasm. The separation procedure should be performed in a coloured medium, otherwise the transparent germinal vesicle may be lost in solution. Considering its large volume and ease of access, the avian oocyte nucleus could be a convenient model for biochemical and molecular investigations, although it has not become so, as yet. The presence/absence of a germinal vesicle in the centre of a germinal disc may serve as a diagnostic marker to estimate the time before ovulation: The beginning of the maturation process (about four hours before oocyte ovulation) is marked by the disappearance of the nuclear membrane and the nucleus is no longer visible on the yolk sphere (Fig. 2c). An oocyte in its ovarian follicle is surrounded by an oocyte membrane (*inner perivitelline layer*) underlaid by a thin cytoplasmic layer and filled with yolk. There is a certain inconsistency about the terminology and some authors use the name "vitelline membrane" for both the membrane surrounding an oocyte in an ovary and for the membrane surrounding it in the infundibulum, after the penetration of spermatozoa. In this chapter, *inner perivitelline membrane* is used to describe the membrane existing in the ovary, and *continuous membrane* and *outer perivitelline membrane*, for the membranes

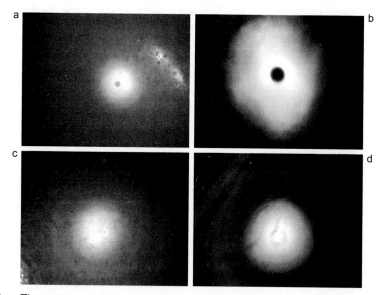

Fig. 2 . The oocyte nucleus (germinal vesicle) as seen under a stereo microscope:
a) in the centre of a germinal disc at a yolk sphere, before maturation;
b) in the centre of a germinal disc which was excised and cleared from the yolk, before maturation;
c) germinal disc at a yolk sphere, after maturation; no germinal vesicle is seen;
d) germinal disc excised and cleared from the yolk, the contour of the nucleus is diffused and only the nuclear territory can be assumed.

formed and added after fertilization in the infundibulum. All three layers together will form the *vitelline membrane* surrounding the yolk in the oviduct (Bellairs et al., 1963; Burley and Vadehra, 1989; Wishart and Staines, 1999; Pan et al., 2000). Their role and origin are described later in this chapter.

MATERNAL RNA

The yolk contains all the substances needed for future embryo development; the oocyte nucleus contains the DNA which is responsible for inheritance, and germinal disc contains cytoplasm, proteins and maternal RNA. The maternal RNA (sometimes also called oocytal RNA), is accumulated in the oocyte during oogenesis and comprises many different species of RNAs (Davidson, 1986). A part of it, after fertilization, assists in protein synthesis during early embryo development, before the commencement of embryonic transcription. This RNA fraction is very stable, being bound to the specific proteins in the form of *ribonucleoprotein particles (RNPs)*. In mature oocytes it is translationally inactive, but undergoes activation during maturation and fertilization of the oocyte, and is then transferred onto polysomes participating in translation. Another part of the RNA assisting oocyte growth and development during oogenesis is translationally active and is rapidly turning over on polysomes (Brevini-Gandolfi and Gandolfi, 2001). The period of dependence on the maternal RNA varies in different groups of animals, e.g., during 2-, 4-, 16- cell embryo stages for mammals to mid-blastula stage for Amphibiae *(Xenopus)*. Maternal RNA is of great importance for normal embryonic development – although frequently neglected in poultry science. Irregularities in its synthesis and accumulation during oogenesis might influence early embryonic development during the uterine period of embryogenesis, and a portion of eggs usually qualified as infertile might result from such irregularities. Investigations of the synthesis and metabolism of maternal RNA during oogenesis and early embryo development have been carried out mostly on model organisms such as sea urchins, *Xenopus* and the fruit fly *(Drosophila melanogaster)*, with very few studies on birds (Davidson, 1986).

The amount of maternal RNA in oocytes varies in different animal species. As a rule, small mammalian oocytes contain much less maternal RNA than the big oocytes of *Amphibia* (Davidson, 1986) and *Aves* (Olszańska and Borgul, 1993). In mammals, the rate of early development is much slower than in birds; the time from fertilization until the 2-cell embryo in mouse takes ~24 hours while over the same period the hen embryo increases to ~60,000 cells. The slow internally-developing mammalian embryos are less dependent on the maternal RNA store than the rapidly developing embryos of Amphibia and birds. As shown in Table 1, the amounts of maternal RNA in birds are generally about three orders of magnitude greater than in mammals.

The RNA contents given in Table 1 for birds are not the total amount of RNA in the oocytes, since they represent only the fraction present in the

Table 1. The amounts of maternal RNA contained in oocytes of mammals and in germinal discs of birds (after Olszańska and Borgul, 1993)

Species	RNA amount
Mammals	**(ng/oocyte)**
Mouse	0.47
Rabbit	15.0
Cattle	0.98
Pig	0.65
Sheep	0.76
Birds	**(μg/germinal disc)**
Hen	2.1
Japanese quail	1.1
Guinea hen	1.9
Turkey	1.8

germinal discs. The fraction is highly stable and is not degraded even after 24 hours of incubation at 41°C (the temperature inside the Japanese quail uterus). There is an additional more labile pool of RNA, outside the germinal disc territory, which is termed as "extra-embryonic RNA" contained in a cytoplasmic layer under the perivitelline membrane around the yolk sphere (Fig. 3), which comprises about three fourths (3-4 μg) of all quail oocyte RNA (Malewska and Olszańska, 1999). This fraction is rapidly degraded within the 24 hours passage of an egg through the oviduct. It is presumed that this is the RNA fraction utilized by the oocyte for its own growth and development, while the RNA stored in the germinal disc (~1 μg) is predestined for future embryo development.

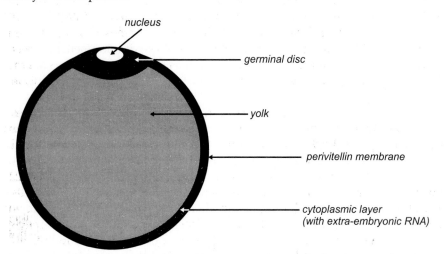

Fig. 3. Schematic representation of localization of maternal RNA in the avian oocyte.

The presence of two such pools of maternal RNA has also been found in other animal species (Davidson, 1986; Brevini-Gandolfi and Gandolfi, 2001), but in birds – probably because of the large oocyte dimensions – the two fractions are differently localized and spatially separated. The role of the extra-embryonic RNA is not yet fully understood, but in this pool (Malewska and Olszańska, 1999) rRNA, tRNA (~90%) and a large amount of poly(A+)RNA representing up to 8-9% of the total RNA has been observed. The percentage of polyadenylated RNA is four to five times higher than in somatic cells thus it comprises a store of mRNA templates. Investigations conducted (by RT-PCR method) on quail maternal RNA from a germinal disc and from extra-embryonic RNA, revealed the presence of 11 individual transcripts in both pools (unpublished observations). It seems that all or most mRNA coding sequences are accumulated in oocytes, independently of whether they are to be translated immediately or at a certain time after fertilization.

The kinetics of maternal RNA accumulation in the quail oocyte during oogenesis is shown in Fig. 4.

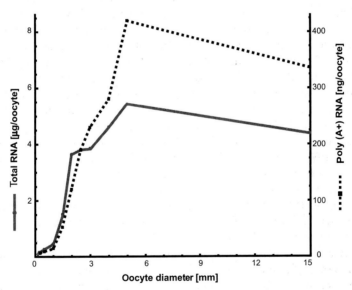

Fig. 4. Kinetics of maternal RNA accumulation during oogenesis in Japanese quail (from Malewska and Olszańska, 1999, by kind permission of Cambridge University Press).

The rapid accumulation of total RNA occurs at the phase of oocyte growth between 1 and 2 mm in diameter and coincides with the presence of the lampbrush chromosomes (Callebaut, 1973; 1974; Kropotova and Gaginskaya, 1984). There is then a plateau in oocytes of 2-3 mm of diameter, which corresponds to the period of chromosome condensation, completed in 2-mm hen and quail oocytes (Callebaut, 1973, 1974; Kropotova and Gaginskaya,

1984), that is accompanied by a cessation of RNA synthesis (Callebaut, 1973). However, RNA accumulation is resumed in oocytes of 3-5 mm in spite of the reported lack of RNA synthesis in the oocyte nucleus. This may be explained by the external delivery of RNA from granulosa cells into the oocyte. Such a possibility was already considered much earlier by the authors who, with an electron microscope, observed RNA-containing vesicles (*transosomes*) traversing from granulosa cells into oocytes (Schjeide et al., 1970; Paulson and Rosenberg, 1972; Gorbik, 1976; Gaginskaya and Hau, 1980; Guraya, 1989). In 5-15-mm oocytes, a plateau or even a decrease of RNA accumulation was once again observed. It is supposed that at that time a certain equilibrium is attained between the processes of RNA delivery from granulosa cells and RNA utilization during translation and its degradation due to RNases present in the cytoplasmic layer (Stępińska and Olszańska, 1996; Stępińska et al., 1996; Malewska and Olszańska, 1999).

BIOLOGICALLY ACTIVE SUBSTANCES OF THE AVIAN OOCYTES

Besides the large amounts of nutritive substances (lipids, proteins, lipoproteins and carbohydrates), avian oocytes also contain a lot of biologically active compounds which are indispensable for the future embryo. These include vitamins, with the exception of vitamin C (Fraser and Emtage, 1976; White, 1985; White and Whitehead, 1987; Burley and Vadehra, 1989; Bush and White, 1989; Squires and Naber, 1992, 1993; Sherwood et al., 1993; Blount et al., 2000; Surai et al., 2001; Grobas et al., 2002), proteolytic and lipolytic enzymes (Burley and Vadehra, 1989; Hartmann and Wilhelmson, 2001), DNases and RNases (Stępińska and Olszańska, 1996; Stępińska et al., 1996; Stępińska and Olszańska, 2001, 2003), steroid hormones (Schwabl, 1996; Lipar et al., 1999; French et al., 2001; Petrie et al., 2001; Elf and Fivizzani, 2002) and thyroid hormones (Wilson and McNabb, 1997; Flamant and Samarut, 1998). The presence of antibodies in chick yolk is well-known and is used for medical and research purposes (Kuronen et al., 1997; Morrison et al., 2001; Mine and Kovacs-Nolan, 2002). Other biologically active substances are also present in the avian yolk, such as: growth factors (Scavo et al., 1989; Sasse et al., 2000; Nakane et al., 2001), transcription factors (Knepper et al., 1999), signalling molecules like serotonin (Emanuelsson et al., 1988; Mougdal et al., 1992) and melatonin (Skwarlo-Sońta et al., 2007, this volume; Yamada et al., 2002 – in the egg yolk of rainbow trout). The presence of neurotransmitters is of special interest because they probably to serve as components of a "diffuse neuroendocrine system" providing a general means of cell communication and control (Kvetnoy, 1999), especially in the early embryos before formation of the nervous system (Buznikov et al., 2001). In addition, melatonin may also serve as a potent antioxidant (Reiter, 1996; Reiter et al., 2000) protecting, together with other compounds like glutathione, carotenoids and vitamin E (Blount et al., 2000), the young embryo from the harmful actions of free radicals produced during intensive cell metabolism and proliferation at the earliest developmental stages.

FERTILIZATION IN BIRDS

Unlike mammals where fertilization is, as a rule, monospermic (only one spermatozoon enters each oocyte), in birds physiological polyspermy occurs. This means that under physiological conditions many spermatozoa enter an oocyte, although only one of them participates in the process of zygotic nucleus formation. Fertilization of an ovulated oocyte occurs in the *infundibulum* within several minutes of ovulation (usually within 15 minutes). In early studies from the beginning of the 20th century some authors also found fertilized eggs in the body cavity, but such observations do not exclude fertilization occurring in the *infundibulum* followed by movement of an egg into the body cavity. Some authors supposed that spermatozoa could penetrate the walls of ovarian follicles to fertilize the unovulated oocyte while still contained in its follicle (Romanoff, 1960). This opinion was refuted by Olsen and Neher (1948) who introduced *in vitro* ovulated oocytes obtained from inseminated hens into the oviducts of virgin hens and were unable to produce fertile eggs in this manner. In the reverse experiment, *in vitro* ovulated oocytes from non-inseminated hens introduced into the oviducts of inseminated hens did produced fertile eggs.

Spermatozoa can penetrate an egg all over its surface, both in the region of the germinal disc and over the remaining surface of the yolk sphere, but they show ~20 times greater preference for the germinal disc territory where several to several hundred spermatozoa can enter (Fofanova, 1965; Bekhtina, 1966; Bramwell and Howarth, 1992; Wishart and Staines, 1999). However, considering the significantly greater surface area of a yolk sphere compared with the germinal disc, about 50 times more spermatozoa penetrate the yolk membrane outside the germinal disc region (Wishart and Staines, 1999). On average, spermatozoa enter the first inner perivitelline layer (IPVL) within 15 minutes of ovulation, leaving tracks of their entrance in the form of small holes made by the acrosomal enzymes. The IPVL membrane is probably produced in an ovarian follicle by granulosa cells and is deposited around an oocyte during its growth phase; it seems to be the equivalent of zona pellucida protein C in mammals and probably, as in mammals, contains receptor sites for sperm binding (Bramwell and Howarth, 1992; Kuroki and Mori, 1995, 1997; Waclawek et al., 1998; Pan et al., 2000). Within a few minutes after fertilization in the *infundibulum*, two other membranes are formed around the egg: a continuous membrane and outer perivitelline layer (OPVL), which together form a barrier impenetrable to the sperm. The spermatozoa entering the IVPL, but not penetrating into the cytoplasm, are retained in the OVPL; which in number are about 10 times more than those successfully entering the cytoplasm. The small holes in the IVPL left after sperm entrance, and the sperm caught in the OPVL are, after staining, visible under a microscope in the yolk membrane taken from a laid egg and they can be counted. There is a linear relationship between the amount of spermatozoa entering an egg at the territory of a germinal disc (holes in IPVL from above a germinal disc), the amount of spermatozoa entering the egg cytoplasm from outside the germinal disc (holes in the IPVL from outside the germinal disc) and the amount of spermatozoa caught in the OPVL (Wishart and Staines, 1999). These authors report a logarithmic relationship between the above parameters and the percentage of fertilized eggs for hen and turkey, and conclude that these parameters could be used for the estimation of insemination efficiency in reproductive flocks of poultry.

The sperm heads penetrating the IVPL to enter the cytoplasm seem to lose their own nuclear envelope; they transform into male pronuclei by swelling, changing their form from elongated to oval, then to spherical, accompanied by progressive chromatin decondensation and reconstruction of the nuclear envelope (Okamura and Nishiyama, 1978; Perry, 1987). In the vicinity of the male pronuclei the remnants of sperm flagella can often be seen.

Since many sperm enter the germinal disc cytoplasm during polyspermic fertilization, many male pronuclei are formed in this territory. Nevertheless, as mentioned above, only one of them – usually the one at the centre of a germinal disc – together with a female pronucleus, will form the zygotic nucleus. The others migrate towards the periphery of the germinal disc, some of them may even divide but all degenerate and disappear at some time during cleavage (Fofanova, 1965; Perry, 1987; Nakanishi et al., 1990; Waddington et al., 1998). Till recently it was not known what happened to the supernumerary sperm entering an egg during polyspermic fertilization, when they apparently "degenerate and disappear". Lately the presence of high DNase I and DNase II activities in the oocyte cytoplasm have been identified (Stępińska and Olszańska, 2001); these enzymes digested naked DNA from phage λ and chromatin DNA in intact quail sperm under *in vitro* conditions. The activity of both DNases was low in previtellogenic oocytes, it increased during oogenesis, was at its highest in ovulated oocytes and decreased again in the blastoderm from a laid egg. The activity was not observed in the yolk of quail oocytes or in mouse oocytes (Stępińska and Olszańska, 2003). On the basis of this finding, it is postulated that in bird oocytes, which lack the early block against polyspermy existing in mammals, a late block against polyspermy has developed, that relies on the degradation of supernumerary sperm by the DNases present in the oocyte cytoplasm. In the mammalian oocyte, where only one spermatozoon penetrates each oocyte, the presence of cytoplasmic DNases would be useless or even harmful.

Nevertheless, the question arises, if all sperm entering the avian oocyte are destroyed by the action of DNases, then how is only one sperm predestined to form a zygotic nucleus saved? What is the mechanism(s) saving this spermatozoon from destruction? It is suggested that the protective mechanism may depend on the point of entry of the spermatozoon into the egg. At the time of fertilization, the oocyte undergoes a maturation process and the nuclear material, devoid of its nuclear envelope, spreads just under the oocyte membrane (Bekhtina, 1966). The chromatin of a sperm entering the egg cytoplasm outside or within the territory of the germinal disc would experience the degrading action of the DNases present there, while the sperm entering the germinal disc at the central point avoids contact with these nucleases because the sperm is located in the karyoplasm and not in the cytoplasm (Stępińska and Olszańska, 2003). Indirect support for this hypothesis is provided by photographs of the perivitelline layer (PL) over the germinal disc territory showing holes made by the entering sperm (Bramwell and Howarth, 1992). The holes are very densely packed over the cytoplasmic part of the germinal disc and much less so over the central part (corresponding to the nuclear territory) so that they form the shape of a doughnut. This result suggests that the sperm receptor proteins contained in the IPVL are much more abundant over the cytoplasm of the germinal disc territory than over that of the germinal vesicle. Thus, the probability of sperm entering at that particular part of a germinal disc is greatly reduced. Other mechanisms that are also possible, although in the author's opinion, less probable, include the presence of DNase inhibitors at a specific location in the germinal disc or some intrinsic properties of the particular spermatozoon.

Birds are the only group of higher vertebrates where the complete parthenogenetic development has been observed. Olsen (1966, 1967, 1974) selected lines of chicken and turkey where such development proceeded up to quite advanced embryonic stages. In the case of chicken, parthenogenetic development led to the formation of degenerated membranous structures and the embryos never hatched; for turkeys, however, healthy adult males were obtained, which were able to reproduce. The parthenogenetic birds had the normal diploid chromosome number, probably due to suppression of the 2nd polar body extrusion (Olsen, 1967, 1974). Recently, it has been observed in *in vitro* activated quail oocytes that the phenomenon of segmentation of the cytoplasm strictly follows the cleavage pattern of normal *in vivo* fertilized embryos, as seen under a stereo microscope, but without the concomitant divisions of the nuclei (Olszańska et al., 2002). In this case "embryos" devoid of nuclei have been obtained (Fig. 5) that are morphologically the same as the

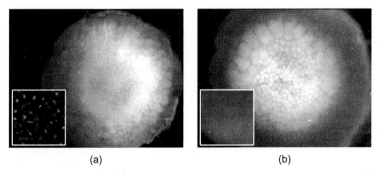

(a)   (b)

Fig. 5. Comparison of cleavage pattern of a normal *in vivo* fertilized egg (a) and an *in vitro* activated germinal disc, without nuclei (b). the cleavage patterns are alike, but in (b) the nuclei are absent. The inserts show the parts of the corresponding embryo (a) and the germinal disc (b) stained with 4,6-Diamidino-2-phenylindole (DAPI) to disclose the presence of nuclei.

normally cleaved embryos at uterine stages IV-V (according to the classification of Eyal-Giladi and Kochav, 1976). Cytoplasmic segmentation in the absence of nuclei was observed previously in sea urchin eggs (Harvey, 1940) and frogs (Briggs et al., 1951) and is considered evidence of maternal information contained in the ooplasm, which is effective during early developmental stages (Davidson, 1986).

THREE MAIN DEVELOPMENTAL PHASES OF THE AVIAN EMBRYO

a. Uterine Development

Fertilization of the avian oocyte takes place in the upper part of the oviduct – the *infundibulum* – usually within 15 minutes after ovulation (Fig. 6). Then,

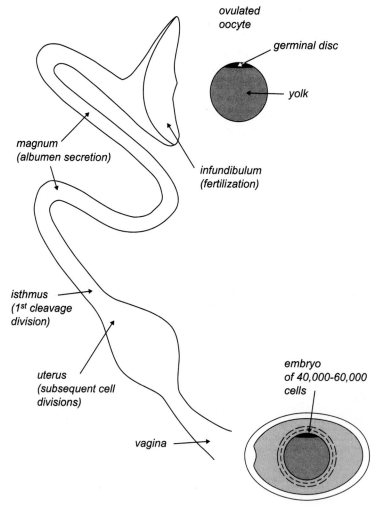

Fig. 6.   Passage of an egg through the avian oviduct (schematic drawing).

within 4-5 hours, the egg (the naked yolk sphere) passes along the longest, secretory part of the oviduct – the *magnum* – where it is surrounded by albumen; at that time no cleavage divisions are observed. The first one or two divisions occur in the shorter, narrow part of the oviduct – the *isthmus* – within 4.5 - 5.5 hours after ovulation and there the two main shell membranes (egg membranes), the inner and outer, are deposited around the albumen (Olsen, 1942; Olsen and Neher, 1948; Stępińska and Olszańska, 1983; Perry, 1987). The inner and outer shell membranes are made up of intertwining proteinaceous fibres. They protect the embryo against the entry of bacteria, offer a support for the developing *chorioallantois* and retain the liquid of the albumen (Burley and Vadehra, 1989). Initially, they are quite thin, delicate and translucent, but they become solid and opaque after entering the next part of the oviduct called the *uterus*, or more correctly, the shell gland. There, the egg spends the major time during its formation (18-20 hours); the albumen is filled with the *plumping fluid* (consisting of water, mineral salts and some proteins), the calcified shell is deposited over the shell membranes and cell divisions continue to give, at oviposition, an embryo of ~40,000 to 60,000 cells (Kochav et al., 1980; Stępińska and Olszańska, 1983). In quail, about 2.5-3 hours before oviposition, the shell becomes pigmented with brownish spots in a pattern characteristic for each individual female (Woodard and Mather, 1964; author's observations). Due to consecutive cleavage, divisions are correlated with the time that the egg remains in the uterus. It is possible to estimate – by palpation – the degree of egg shell calcification and to deduce the developmental stage of the embryo in the forming egg (Stêpińska and Olszańska, 1983). In this manner it is possible to select and manually extract, for experimental purposes, a uterine embryo of the required developmental stage without sacrificing the female.

The intrauterine period of chick embryo development was morphologically described and classified into ten developmental stages by Eyal-Giladi and Kohav (1976) and Kochav et al. (1980). The stages are designated by Roman numerals from I to X with additional stages XI –XIII for the embryo (blastoderm) in the laid egg.

Stages I-IV:	Uterine stages I-VI correspond to individual cleavage stages.
Stages IV-V:	During stages IV/V the subgerminal cavity filled with liquid is formed: a narrow cavity separating the multilayered germ from the yolk.
Stage VI:	At stage VI the formation of nucleoli is observed (Raveh et al., 1976).
Stage VII:	At stage VII the first signs of postero-anterior axis formation in the embryo can be observed, starting with the shedding of cells into the subgerminal cavity from its posterior part. This process is unique to birds and does not occur in any other animal group (Eyal-Giladi, 1984). This period is significant for embryo development

because it represents the process of symmetry axis formation, probably under the effect of gravity (Eyal-Giladi, 1984; Olszańska et al., 1984b) At this stage the nucleoli are already formed in the cell nuclei (Raveh et al., 1976), accompanied by the start of rRNA synthesis (Olszańska et al., 1984a).

Stages VIII-IX: During stages VIII/IX the *area pellucida* in the central part of the blastoderm is formed by further cell shedding from the central part of the blastoderm to the bottom of the subgerminal cavity. As a result, a single-cell layer transparent area is formed in the central part of the blastoderm, surrounded by the *area opaca*: an opaque multicellular whitish ring around the *area pellucida* (Eyal-Giladi and Kochav, 1976; Kochav et al., 1980). It is now accepted that blastulation in avian embryos corresponds to the process of formation of the hypoblast and blastocoele: the narrow space formed between the epiblast and hypoblast (Eyal-Giladi, 1984). Thus, the avian embryo at oviposition corresponds to the pre-blastula (hen) where usually no hypoblast is present yet (stage X), or to the blastula (quail) where the hypoblast formation starts (author's observations).

Stage X: Hen eggs are usually laid at stage X, but in other species the stage of embryo development at oviposition may be earlier (duck, goose) or later (quail) and differs by the extent of hypoblast formation. There are indications that a more advanced stage of the embryo at oviposition is positively correlated with egg hatchability (Coleman and Siegel, 1966).

b. Period of Rest

The period of rest between oviposition and the beginning of incubation may be regarded as a second, separate period in avian embryo development. It is generally known that eggs lose their hatching ability during extended period of storage. This can be partially prevented by periodic warming of the eggs to incubation temperature during storage (Kraszewska-Domańska and Pawluczuk, 1977). In nature too, a female entering her nest to lay a consecutive eggs, warms the former ones at the same time. However, the biochemical processes responsible for the loss of hatchability that occur in eggs during storage are completely unknown. Researchers have somehow neglected this period. Long ago an attempt had been made to clarify the processes going on in the embryo during the storage of Japanese quail eggs (Olszańska et al., 1981). The metabolism of proteins and RNA was gradually quenched with the prolongation of storage time; the decrease in protein metabolism (leucine incorporation) could be somehow reduced by periodic warming of the eggs, but not the fall in RNA metabolism (uridine incorporation). It was also found that cell divisions proceed during the 1st day after oviposition in eggs held at room temperature (Stępińska and Olszańska, 1983). Therefore, it is important to indicate the time after oviposition when eggs are collected for experiments and the designation "freshly laid" should be reserved for eggs collected while they are still warm.

Morphologically, the stage of hen embryos at oviposition was designated by Hamburger and Hamilton (1951, 1992) as stage 1. This was further subdivided by Eyal-Giladi and Kochav (1976) and Kochav et al. (1980) into stages X-XIII corresponding to hypoblast formation (blastulation).

c. Incubation Development

Sometimes avian development is described in terms of the incubation time, e.g., 24-hour embryo, 3-day embryo, etc. It is evident that such a definition is neither exact nor sufficient. Great differences may exist between embryos incubated for the same period and in the same conditions, but originating from separate hen lines or from eggs stored for different lengths of time. Many factors can affect the rate of embryo development during incubation, e.g., the egg size, the embryo stage at the beginning of incubation, the temperature and time of storage, the type of incubator and variations in the incubation temperature. For example, quail embryos of the same chronological age (collected the same day) incubated for 30 hours in the same incubator were classified as between stages 5 to 8 H.H. (Olszańska and Lassota, 1980). In the past such differences led to discrepancies in the interpretation of results from different laboratories or, at least, prevented direct comparison of such data.

The incubation development of hen embryos from oviposition until hatch was classified into 45 developmental stages by Hamburger and Hamilton (1951, 1992), and they are designated by Arabic numerals from 1 to 45; sometimes followed by the letters H.H. to indicate the initials of the authors of this classification (i.e., H.H. stand for Hamburger and Hamilton). Other authors proposed different classification schemes for the turkey uterine embryo (Gupta and Bakst, 1993) and the incubated quail embryo (Zacchei, 1961), but these are not as popular.

Stage 1 H.H. corresponds to the blastoderm of a laid hen egg (stages X-XIII of E-G., i.e., Eyal-Giladi), with a well-developed *area pellucida* and *opaca* and a defined axis of symmetry. The *epiblast* layer is already formed and the *hypoblast* layer is developing at the posterior and sometimes at the central part of the *area pellucida*.

Stages 2-5 H.H. comprise gastrulation and the beginning of the neurulation process. The criteria for classification are the level of advancement of primitive streak formation and the presence of the primitive groove and the Hensen node. The primitive streak is the equivalent of the blastopore in *Amphibia*, the place where, as a result of intensive migratory movement, the cells from the upper layer (*epiblast*) enter the interior of an embryo, giving rise to two other embryonic layers: the mesoderm and endoderm. The proper embryo structures are formed within the territory of the *area pellucida*, while the *area opaca* participates in the formation of the yolk follicle and the *chorioallantois*. Stages 4-5 H.H. are the phases when primordial germ cells (PGCs) can be visualized and start to migrate onto the hypoblast layer towards the anterior part of the embryo. PGCs probably start to differentiate as early as stage 1 H.H. (stages X-XIII of E-G), possibly arising from the epiblastic cells. At stages 5-7 H.H. they accumulate in number to between 100-400 at the anterior crescent-like territory (*the germinal crescent*). There they are easily identified and visualized under a microscope as large, PAS (Periodic Acid Schiffis) positive, round cells, often forming groups of 2-4 (Ginsburg and Eyal-Giladi, 1986, 1987, 1989; Pardanaud et al., 1987). From there they migrate within blood vessels to the germinal ridges, multiply intensively and differentiate into male/female gametes.

For subsequent stages (from 7-14 H.H.) the diagnostic marker is the number of somites, and thereafter, the appearance and development of embryo structures and organs including limb primordia, brain, eyes, feather-germs and visceral arches.

IN VITRO CULTURE OF AVIAN EMBRYOS

Short-term avian embryo culture was practised for many years (New, 1955; Olszańska and Lassota, 1980; Olszańska et al., 1984b) and it was often used for embryological and biochemical purposes. The crucial point in the development of *in vitro* methods was the complete embryo culture from a one-cell zygote to hatch first described by Margaret Perry (1988) from the Roslin Institute. Her system consisted of a triple change in culture conditions. Briefly, the fertile ova surrounded by the albumen capsula were taken from the middle part of the oviduct; first, they were cultured in a glass jar filled with medium, at 41-42°C, for 24 hours, and then they were transferred to egg shells filled with liquid albumen, closed with a plastic sheet (permitting the free flow of air) and incubated for three days in a horizontal position, with periodic rocking. They were then transferred to larger egg shells filled with another medium and with an artificial air chamber, and were incubated at 37°C in a vertical position, with periodic rocking. Initially, the efficiency of the system was rather low: only ~7% of the embryos hatched. After several years the technique was improved and its use spread to quail embryos, reaching efficiencies of up to 35% (Naito and Perry, 1989; Perry and Mather, 1991; Ono et al., 1994, 1996; Naito et al., 1995).

Efforts to perform *in vitro* fertilization of avian ova were also made (Howarth, 1971; Nakanishi et al.,1990, 1991). These attempts used hen oocytes ovulated *in vivo* and then fertilized *in vitro*. Quail oocytes ovulated *in vitro* from ovarian follicles, hanging in a culture medium in glass containers have been used, as shown in Fig. 7 (Olszańska et al., 1996, 2002). After fertilization with freshly prepared quail semen and following 20-22 hours of incubation at

Fig. 7. *In vitro* ovulation of quail oocytes from mature ovarian follicles.

41°C, ~14% of embryos were obtained which were correctly cleaved at stages IV-VI (E-G). At the same time ~25% of "embryos" with a normal cleavage pattern but containing no nuclei (Fig. 5), resulted from cytoplasm segmentation. This phenomenon emphasized the importance of testing early avian embryos for the presence of nuclei to distinguish true cleavage from cytoplasmic segmentation because the correct morphological pattern of the embryo is not an adequate proof.

Thus, at present, it is already possible to link ovulation, fertilization and early cleavage development under *in vitro* conditions. It would be interesting, in the future, to go back to earlier vitellogenic oocytes before maturation and, from another side, to go forward and prolong the *in vitro* embryo culture until hatch. In this manner, it would be feasible to develop an *in vitro* system comprising several consecutive steps, which permitted the complete reproductive process to occur from oocyte maturation until hatching. This is already within reach using the present techniques, but it is uncertain whether it will ever be possible to extend an *in vitro* system to even earlier stages, to vitellogenesis and primordial germ cell differentiation into oocytes and sperm cells.

ACKNOWLEDGEMENTS

This work was performed with support of the Institute of Genetics and Animal Breeding of the Polish Academy of Sciences. The author wishes to express her gratitude to all collaborators, as well as to M. Prokopiuk and P. Przybylska, whose technical expertise and assistance made it possible to successfully carry out several experimental works mentioned in this chapter.

REFERENCES

Bekhtina, W. (1966). Morphologitscheskije osobiennosti polispermnogo opłodo-tworienija u kur. In: *Nasliedstwiennost' i izmientschivost'* sielskokhoziajstwiennoj pticy (Morphological peculiarities of polyspermic fertilisation in hen). In: *Heredity and Variability of Poultry*). (Fomin, A.L., ed.), Kolos, Moskwa, 92-108 (in Russian).

Bellairs, R., Harkness, M. and Harkness, R.D. (1963). The vitelline membrane of the hen's egg: a chemical and electron microscope study. *J. Ultrastruc. Res.* **8**: 339-359.

Blount, J.D., Houston, D.C. and Moller, A.P. (2000). Why yolk is yellow. *Trends Ecol. Evol.* **15**: 47-49.

Bramwell, R.K. and Howarth, B. Jr. (1992). Preferential attachment of cock spermatozoa to the perivitelline layer directly over the germinal disc of the hen's ovum. *Biol. Reprod.* **47**: 1113-1117.

Brevini-Gandolfi, T.A.L. and Gandolfi, L. (2001). The maternal legacy to the embryo: cytoplasmic components and their effects on early development. *Theriogenology* **55**: 1255-1276.

Briggs, R., Green, E.U. and King, T.J. (1951). An investigation of the capacity for cleavage and differentiation in *Rana pipiens* eggs lacking functional chromosomes. *J. Exp. Zool.* **116**: 455-497.

Burley, R.W. and Vadehra, D.V. (1989). *The Avian Egg, Chemistry and Biology.* John Wiley & Sons Inc., New York, 472.

Bush, L. and White, H.B. 3rd. (1989). Avidin traps biotin diffusing out of chicken egg yolk. *Comp. Biochem. Physiol.* B. **93**: 543-547.

Buznikov, G.A., Lambert, H.W. and Lauder, J.M. (2001). Serotonin and serotonin-like substances as regulators of early embryogenesis and morphogenesis. *Cell Tissue Res.* **305**: 177-186.

Callebaut, M. (1973). Correlation between germinal vesicle and oocyte development in the adult Japanese quail (*Coturnix coturnix japonica*): a cytochemical and autoradiographic study. *J. Embryol. Exp. Morph.* **29**: 145-157.

Callebaut, M. (1974). La formation de l'oocyte d'oiseau: Etude radiographique chez la caille japonese (*Coturnix coturnix japonica*) pondeuse a l'aide de la leucine tritiee. *Arch. Biol.* (Bruxelles) **85**: 201-233.

Coleman, J.W. and Siegel, P.B. (1966). Selection for body weight at eight weeks of age. 5. Embryonic stage at oviposition and its relationship to hatchability. *Poultry Sci.* **45**: 1008-1011.

Davidson, E.H. (1986). *Gene Activity in Early Development.* 3rd edition. Academic Press, London, England. 679.

Elf, P.K. and Fivizzani, A.J. (2002). Changes in sex steroid levels in yolks of the Leghorn chicken, *Gallus domesticus*, during embryonic development. *J. Exp. Zool.* **293**: 594-600.

Emanuelsson, H., Carlberg, M. and Lowkvist, B. (1988). Presence of serotonin in early chick embryos. *Cell Differ.* **24**: 191-200.

Etches, R.J. (1990). The ovulatory cycle of the hen. *Crit. Rev. Poultry Biol.* **2**: 293-318.

Etches, R.J. and Petitte, J.N. (1990). Reptilian and avian follicular hierarchies: Models for the study of ovarian development. *J. Exp. Zool.* Suppl. **4**: 112-122.

Eyal-Giladi, H. and Kochav, S. (1976). From cleavage to primitive streak formation: A complementary normal tables and a new look at the first stages of the development of the chick. *Dev. Biol.* **49**: 321-337.

Eyal-Giladi, H. (1984). The gradual establishment of cell commitments during the early stages of chick development. *Cell Differ.* **14**: 245-255.

Flamant, F. and Samarut, J. (1998). Involvement of thyroid hormone and its alpha receptor in avian neurulation. *Dev. Biol.* **197**: 1-11.

Fofanova, K.A. (1965). Morphological data on polyspermy in chickens. *Fed. Proc. Transl.* Suppl. **24**: 239-247.

Fraser, D.R. and Emtage, J.S. (1976). Vitamin D in the avian egg. Its molecular identity and mechanism of incorporation into yolk. *Biochem. J.* **160**: 671-682.

French, J.B. Jr., Nisbet, I.C. and Schwabl, H. (2001). Maternal steroids and contaminants in common tern eggs: a mechanism of endocrine disruption. *Comp. Biochem. Physiol. C Toxicol. Pharmacol.* **128**: 91-98.

Gaginskaya, E.P. and Hau, C-S. (1980). Peculiarities of ontogenesis in the chick. II. Follicular period in oocyte development. *Ontogenez* **11**: 213-221 (in Russian).

Ginsburg, M. and Eyal-Giladi, H. (1986). Temporal and spatial aspects of the gradual migration of primordial germ cells from the epiblast into the germinal crescent in the avian embryo. *J. Embryol. Exp. Morph.* **95**: 53-71.

Ginsburg, M. and Eyal-Giladi, H. (1987). Primordial germ cells of the young chick blastoderm originate from the central zone of the area pellucida irrespective of the embryo-forming process. *Development* **101**: 209-219.

Ginsburg, M. and Eyal-Giladi, H. (1989). Primordial germ cell development in cultures of dispersed central disks of stage X chick blastoderms. *Gamete Res.* **23**: 421-427.

Grobas, S., Mendez, J., Lopez, B.C., De, B.C. and Mateos, G.G. (2002). Effect of vitamin E and A supplementation on egg yolk alpha-tocopherol concentration. *Poultry Sci.* **81**: 376-381.

Gorbik, T.A. (1976). On the ultrastructure and transformation of transosomes – the specific organelles of the avian ovary. *Archiv Anat. Histol. Embryol.* (Leningrad) **70**: 79-86 (in Russian).

Gupta, S.K. and Bakst, M.R. (1993). Turkey embryo staging from cleavage through hypoblast formation. *J. Morph.* **217**: 313-325.

Guraya, S.S. (1989). *Ovarian Follicles in Reptiles and Birds*. Springer-Verlag, Berlin-Heidelberg, 285.

Hamburger, V. and Hamilton, H.L. (1951). A series of normal stages in the development of the chick embryo. *J. Morph.* 88: 49-92 (reprinted in: *Developmental Dynamics* 1992. **195**: 195-231).

Hartmann, C. and Wilhelmson, M. (2001). The hen's egg yolk: A source of biologically active substances. *World's Poultry Sci. J.* **57**: 13-28.

Harvey, E.B. (1940). A comparison of development of nucleate and non-nucleate eggs of *Arbacia punctulata*. *Biol. Bull.* **79**: 166-187.

Howarth, B. (1971). An examination for sperm capacitation in the fowl. *Biol. Repr.* **3**: 338-341.

Knepper, P.A., Chandra, S.K., Mayanil, C.S.K., Hayes, E., Goossens, W., Byrne, R.W. and Mclone, D.G. (1999). The presence of transcription factors in chicken albumin, yolk and blastoderm. *In vitro Cell. Dev. Biol. – Animal* **35**: 357-363.

Kochav, S., Ginsburg, M. and Eyal-Giladi, H. (1980). From cleavage to primitive streak formation: A complementary normal table and a new look at the first stages of the development of the chick. II. Microscopic anatomy and cell population dynamics. *Dev. Biol.* **79**: 296-308.

Kropotova, E.V. and Gaginskaya, E.R. (1984). Lampbrush chromosomes from the Japanese quail oocytes: light and electron microscopy study. *Cytology* **26**: 1008-1015 (in Russian).

Kraszewska-Domańska, B. and Pawluczuk, A. (1977). The effect of periodic warming of stored quail eggs on their hatchability. *British Poultry Sci.* **18**: 531-533.

Kuroki, M. and Mori, M. (1995). Origin of 33 kD protein of vitelline membrane of quail egg: immunological studies. *Dev. Growth Differ.* **37**: 545-550.

Kuroki, M. and Mori, M. (1997). Binding of spermatozoa to the perivitelline layer in the presence of a protease inhibitor. *Poultry Sci.* **76**: 748-752.

Kuronen, I., Kokko, H., Mononen, I. and Parviainen, M. (1997). Hen egg yolk antibodies purified by antigen affinity under highly alkaline conditions provide new tools for diagnostics. Human intact parathyrin as a model antigen. *Eur. J. Clin. Chem. Clin. Biochem.* **35**: 435-440.

Kvetnoy, J.M. (1999). Extrapineal melatonin: Location and role within diffuse neuroendocrine system. *Histochem. J.* **31**: 1-12.

Lipar, J.L., Ketterson, E.D., Nolan, V. Jr. and Casto, J.M. (1999). Egg yolk layers vary in the concentration of steroid hormones in two avian species. *Gen. Comp. Endocrinol.* **115**: 220-227.

Malewska, A. and Olszańska, B. (1999). Accumulation and localisation of maternal RNA in oocytes of Japanese quail. *Zygote* **7**: 551-559.

Mine, Y. and Kovacs-Nolan, J. (2002). Chicken egg yolk antibodies as therapeutics in enteric infectious disease: a review. *J. Med. Food.* **5**: 159-169.

Morrison, S.L., Mohammed, M.S., Wims, L.A., Trinh, R. and Etches, R. (2001). Sequences in antibody molecules important for receptor-mediated transport into the chicken egg yolk. *Mol. Immunol.* **38**: 619-625.

Moudgal, R.P., Panda, J.N. and Mohan, J. (1992). Serotonin in egg yolk and relation to cage density stress and production status. *Indian J. Animal Sci.* **62**: 147-148.

Naito, M. and Perry, M.M. (1989). Development in culture of the chick embryo from cleavage to hatch. *British Poultry Sci.* **30**: 251-256.

Naito, M., Nirasawa, K. and Oishi, T. (1995). An *in vitro* culture method for chick embryos obtained from the anterior portion of the magnum of oviduct. *British Poultry Sci.* **36**: 161-164.

Nakane, S., Tokumura, A., Waku, K. and Sugiura, T. (2001). Hen egg yolk and white contain high amounts of lysophosphatidic acids, growth factor-like lipids: distinct molecular species compositions. *Lipids* **36**: 413-419.

Nakanishi, A., Utsumi, K. and Iritani, A. (1990). Early nuclear events of in vitro fertilization in the domestic fowl (*Gallus domesticus*). *Mol. Repr. Dev.* **26**: 217-221.

Nakanishi, A., Miyake, M., Utsumi, K. and Iritani, A. (1991). Fertilizing competency of multiple ovulated eggs in the domestic fowl (*Gallus domesticus*). *Mol. Repr. Dev.* **28**: 131-135.

New, D.A.T. (1955). A new technique for the cultivation of the chick embryo *in vitro*. *J. Embryol. Exp. Morph.* **3**: 325-331.

Okamura, F. and Nishiyama, H. (1978). Penetration of spermatozoon into the ovum and transformation of the sperm nucleus into the male pronucleus in the domestic fowl (*Gallus gallus*). *Cell Tissue Res.* **190**: 89-98.

Olsen, M.W. (1942). Maturation, fertilization and early cleavage in the hen's egg. *J. Morph.* **70**: 513-533.

Olsen, M.W. and Neher, B.H. (1948). The site of fertilization in the domestic fowl. *J. Exp. Zool.* **109**: 355-366.

Olsen, M.W. (1966). Frequency of parthenogenesis in chicken eggs. *J. Heredity* **57**: 23-25.

Olsen, M.W. (1967). Parthenogenetic development in turkey and chicken eggs and its contribution to genetics. In: Societa Italiana per il Progresso della Zootecnica, 56th Meeting on "Il contributo della ricerca, in campo avicolo, al progresso della genetica mondiale", 25-29 June 1966, Varese, Italy, 95-109.

Olsen, M.W. (1974). Frequency and cytological aspects of diploid partenogenesis in turkey eggs. *Theor. Appl. Genetics* **44**: 216-221.

Olszańska, B. and Lassota, Z. (1980). Simple *in vitro* system for molecular studies of early avian development in the quail. *British Poultry Sci.* **21**: 395-403.

Olszańska, B., Stępińska, U. and Kraszewska-Domańska, B. (1981). Metabolism of proteins and nucleic acids in avian blastoderms during storage and ageing of eggs. *British Poultry Sci.* **22**: 49-51.

Olszańska, B., Kłudkiewicz, B. and Lassota, Z. (1984a). Transcription and polyadenylation processes during early development of quail embryos. *J. Embr. Exp. Morph.* **79**: 11-24.

Olszańska, B., Szołajska, E. and Lassota, Z. (1984b). Effect of spacial position of uterine quail blastoderms cultured *in vitro* on bilateral symmetry formation. *Roux's Arch. Dev. Biol.* **193**: 108-110.

Olszańska, B. and Borgul, A. (1993). Maternal RNA content in oocytes of several mammalian and avian species. *J. Exp. Zool.* **265**: 317-320.

Olszańska, B., Malewska, A. and Stêpińska, U. (1996). Maturation and ovulation of Japanese quail oocytes under *in vitro* conditions. *British Poultry Sci.* **37**: 929-935.

Olszańska, B., Stępińska, U. and Perry, M. (2002). Development of embryos from *in vitro* ovulated and fertilized oocytes of quail (*Coturnix coturnix japonica*). *J. Exp. Zool.* **292**: 580-586.

Ono, T., Murakami, T., Mochii, M., Agata, K., Kino, K., Otsuka, K., Ohta, M., Mizutani, M., Yoshida, M. and Eguchi, G. (1994). A complete culture system for avian transgenesis, supporting quail embryos from the single-cell stage to hatching. *Dev. Biol.* **161**: 126-130.

Ono, T., Murakami, T., Tanabe, Y., Mizutani, M., Mochii, M. and Eguchi, G. (1996). Culture of naked quail eggs (*Coturnix coturnix japonica*) ova *in vitro* for avian transgenesis: culture from the single-cell stage to hatching with pH-adjusted chicken thick albumen. *Comp. Biochem. Physiol.* **113A**: 287-292.

Pardanaud, L., Buck, C. and Dieterlen-Lievre, F. (1987). Early germ cell segregation and distribution in the quail blastodisc. *Cell Differ.* **22**: 47-60.

Pan, J., Sasanami, T., Nakajima, S., Kido, S., Doi, Y. and Mori, M. (2000). Characterisation of progressive changes in ZPC of the vitelline membrane of quail oocyte following oviductal transport. *Mol. Repr. Dev.* **55**: 175-181.

Paulson, J. and Rosenberg, M.D. (1972). The function and transposition of lining bodies in developing avian oocytes. *J. Ultrastruct. Res.* **40**: 25-43.

Perry, M.M. (1987). Nuclear events from fertilisation to the early cleavage stages in the domestic fowl (*Gallus domesticus*). *J. Anat.* **150**: 99-109.

Perry, M.M. (1988). A complete culture system for the chick embryo. *Nature* **331**: 70-72.

Perry, M.M. and Mather, C.M. (1991). Satisfying the needs of the chick embryo in culture, with emphasis on the first week of development. In: *Avian Incubation and Embryology* (Tullet, S., ed.). *Poultry Sci. Symp.* **22**: 91-105.

Petrie, M., Schwabl, H., Brande-Lavridsen, N. and Burke, T. (2001). Sex differences in avian yolk hormone levels. *Nature* **412**: 498-499.

Raveh, D., Friedlander, M. and Eyal-Giladi, H. (1976). Nucleolar ontogenesis in the uterine chick germ correlated with morphological events. *Exp. Cell. Res.* **100**: 195-203.

Reiter, R.J. (1996). Functional aspects of the pineal hormone melatonin in combating cell and tissue damage induced by free radicals. *Eur. J. Endocrin.* **134**: 412-420.

Reiter, R.J., Tan, D.X., Osuna, C. and Gitto, E. (2000). Actions of melatonin in reduction of oxidative stress. *J. Biomed. Sci.* **7**: 444-458.

Romanoff, A.L. (1960). *The Avian Embryo*. The Macmillan Co., New York, 1303.

Sasse, M., Lengwinat, T., Henklein, P., Hlinak, A. and Schade, R. (2000). Replacement of fetal calf serum in cell cultures by an egg yolk factor with cholecystokinin/gastrin-like immunoreactivity. *Altern. Lab. Anim.* **28**: 815-831.

Scavo, L., Alemany, J., Roth, J. and de Pablo, F. (1989). Insulin-like growth factor I activity is stored in the yolk of the avian egg. *Biochem. Biophys. Res. Commun.* **162**: 1167-1173.

Schjeide, O.A., Galey, F., Grellert, E.A., I-San Lin, R., de Vellis, J. and Mead, J.F. (1970). Macromolecules in oocyte maturation. *Biol. Reprod.* Suppl. **2**: 14-43.

Schwabl, H. (1996). Environment modifies the testosterone levels of a female bird and its eggs. *J. Exp. Zool.* **276**: 157-163.

Sherwood, T.A., Alphine, R.L., Saylor, W.W. and White, H.B. III. (1993). Folate metabolism and deposition in eggs by laying hens. *Arch. Biochem. Biophys.* **307**: 66-72.

Squires, M.W. and Naber, E.C. (1992). Vitamin profiles of eggs as indicators of nutritional status in the laying hen: Vitamin B12 study. *Poultry Sci.* **71**: 2075-2082.

Squires, M.W. and Naber, E.C. (1993). Vitamin profiles of eggs as indicators of nutritional status in the laying hen: riboflavin study. *Poultry Sci.* **72**: 483-494.

Stępińska, U. and Olszańska, B. (1983). Cell multiplication and blastoderm development in relation to egg envelope formation during uterine development of quail (*Coturnix coturnix japonica*) embryo. *J. Exp. Zool.* **228**: 505-510.

Stępińska, U. and Olszańska, B. (1996). Characteristics of poly(A)-degrading factor present in avian oocytes and early embryos. *J. Exp. Zool.* **271**: 19-29.

Stępińska, U., Malewska, A. and Olszańska, B. (1996). RNase A activity in Japanese quail oocytes. *Zygote* **4**: 219-227.

Stępińska, U. and Olszańska, B. (2001). Detection of deoxyribonuclease I and II activities in Japanese quail oocytes. *Zygote* **9**: 1-7.

Stępińska, U. and Olszańska, B. (2003). DNase I and II present in avian oocytes: a possible involvement in sperm degradation at polyspermic fertilisation. *Zygote* **11**: 35-42.

Surai, P.F., Speake, B.K., Decrock, F. and Groscolas, R. (2001). Transfer of vitamins E and A from yolk to embryo during development of the king penguin (*Aptenodytes patagonicus*). *Physiol. Biochem. Zool.* **74**: 928-936.

Waclawek, M., Foisner, R., Nimpf, J. and Schneider, W.J. (1998). The chick homologue of zona pellucida protein-3 is synthesized by granulosa cells. *Biol. Repr.* **59**: 1230-1239.

Waddington, D., Gribbin, C., Sterling, R.J., Sang, H.M. and Perry, M.M. (1998). Chronology of events in the first cycle of the polyspermic egg of the domestic fowl (*Gallus domesticus*). *Int. J. Dev. Biol.* **42**: 625-628.

White, H.B. 3rd (1985). Biotin-binding proteins and biotin transport to oocytes. *Ann. NY Acad. Sci.* **447**: 202-211.

White, H.B. 3rd and Whitehead, C.C. (1987). Role of avidin and other biotin-binding proteins in the deposition and distribution of biotin in chicken eggs. Discovery of a new biotin-binding protein. *Biochem. J.* **241**: 677-684.

Wilson, C.M. and McNabb, F.M. (1997). Maternal thyroid hormones in Japanese quail eggs and their influence on embryonic development. *Gen. Comp. Endocrinol.* **107**: 153-165.

Wishart, G.J. and Staines, H.J. (1999). Measuring sperm:egg interaction to assess breeding efficiency in chickens and turkey. *Poultry Sci.* **78**: 428-436.

Woodard, A.E. and Mather, F.B. (1964). The timing of ovulation, movement of the ovum through oviduct, pigmentation and shell deposition in Japanese quail. *Poultry Sci.* **43**: 1427-1432.

Yamada, H., Chiba, H., Amano, M., Iigo, M. and Iwata, M. (2002). Rainbow trout eyed-stage embryos demonstrate melatonin rhythms under light-dark conditions as measured by a newly developed time resolved fluoroimmunoassay. *Gen. Comp. Endocrinol.* **125**: 41-46.

Zacchei, A.M. (1961). Lo sviluppo embrionale della quaglia giapponese (*Coturnix coturnix japonica*). *Arch. Ital. Anat. Embriol.* **66**: 36-62.

CHAPTER 3

Melatonin in Anuran Development and Metamorphosis

M.L. Wright, T.A. Shea, R.F. Visconti, B.L. Ramah,
L.L. Francisco, J.L. Scott and A.L. Wickett*

Abstract

This chapter considers the evidence for a developmental role of melatonin in the anuran amphibian. The profile of whole embryo melatonin has been identified in amphibian embryos but there are no data on a possible role of melatonin during early development. Melatonin levels and rhythms have been determined in the plasma, and in many organs, mainly of the Rana catesbeiana (bullfrog) tadpole. Melatonin is found in high concentration in the pineal gland, eye, and intestine, which are organs that contribute to circulating melatonin. Melatonin is initially high in the thyroid gland, whose hormones induce metamorphosis, but the melatonin concentration decreases as development proceeds. The phase of the diel rhythms of hormones that modulate thyroid hormone activity changes with relation to the thyroxine (T_4) acrophase during larval development, which could influence the efficacy of the hormones as antagonists or synergists, and consequently affect the rate of metamorphosis. Melatonin may have a biphasic effect on the thyroid gland, stimulating it at low concentrations and inhibiting it at high levels. However, melatonin does not affect iodide uptake or deiodination of T_4, although high doses antagonize the response of the thyroid to thyrotropin in vitro. Pineal secretion of melatonin in vitro is stimulated by exogenous T_4 at all developmental stages, while the retina is stimulated by T_4 only at the metamorphic climax. Melatonin significantly decreases in the plasma at climax, and a hypothetical scenario is proposed to account for the decline. Other possible roles for melatonin in amphibian development which need investigation include melatonin's antioxidant capacity its interaction with corticotropin-releasing hormone (a major stimulator of both the thyroid and adrenal axes at climax), and its inhibition of increase in intracellular calcium.

Biology Department, College of Our Lady of the Elms, Chicopee, MA 01013, USA.
Corresponding author: Tel.: 413 265 2298, Fax: 413 592 4871. E-mail: wrightm@elms.edu

INTRODUCTION

The primary role of melatonin is to act as a link between enviromental lighting conditions and the internal state of the organism. As such, the highest concentration is secreted during darkness, and it regulates the circadian rhythmicity of other hormones and their physiological functions. Many other roles have been indicated for melatonin, including modulating immune and hematopoietic events, and scavenging free radicals (Conti et al., 2002). Recently interest has been expressed in identifying possible developmental roles of melatonin. Receptors for melatonin exhibited developmentally-specific expression in fetal sheep brain and pituitary (Helliwell and Williams, 1994), melatonin stimulated cell proliferation and accelerated the development of zebra fish embryos (Danilova et al., 2004), while on the other hand, exogenous melatonin failed to affect the development of quail eggs (Kováčiková et al., 2003). Functions of melatonin in the anuran amphibian may include participation in development and growth, or in the hormonal regulation of metamorphosis (Delgado et al., 1984; Gutierrez et al., 1984; Edwards and Pivorun, 1991; Wright et al., 1991). Metamorphosis is induced by the thyroid hormones, acting synergistically with the corticosteroids and, perhaps, prolactin (Kikuyama et al., 1993), under the influence of pituitary and hypothalamic stimulating hormones (Kaltenbach, 1996). This chapter focuses first on the origin and presence of melatonin in the anuran amphibian embryo and larva, and subsequently discusses the evidence for a role of the hormone in development.

Origin of Circulating Melatonin

Melatonin is secreted in mammals by the pineal gland and the gastrointestinal (GI) tract (Bubenik, 2002). The mammalian retina also synthesizes melatonin, but does not secrete it into the circulation (Tosini and Fukuhara, 2003). In contrast, the eye has been reported to contribute to circulating melatonin in birds, reptiles, and fish, mainly during the scotophase (reviewed in Filadelfi and Castrucci, 1996; Mayer et al., 1997). Information available about the GI tract melatonin secretion in lower vertebrates is insufficient, although the hormone has been identified in the GI tissues (Bubenik and Pang, 1997). Delgado and Vivien-Roels (1989) suggested that melatonin in the frog *Rana perezi* comes mainly from the eyes, based on their observations that the eyes had more melatonin than either the plasma or the pineal gland, and that the diel profiles of melatonin in the eye and the plasma were similar. However, the origin of the melatonin identified in the tissues and plasma of the anuran amphibian has not been studied experimentally until recently. Research carried out in our laboratory showed that over one-half of the circulating melatonin, particularly during the scotophase, comes from the eyes in the *Rana* tadpole (Wright et al., 2006), based on changes in plasma melatonin after ophthalmectomy (enucleation). Current intestinal ablation and pinealectomy studies show that the intestine and the pineal gland also contribute to blood melatonin in the tadpole (unpublished data). Moreover, studies in progress further indicate tentatively that the contributions of various organs to blood melatonin in the adult frog are similar to those of the tadpole.

Melatonin in the Embryo and in Larval Tissues

Melatonin was first localized in the *Xenopus* embryo using a spectrophotofluormetric method (Baker, 1969). Until shortly after hatching, there were very low levels of melatonin with only minor peaks. At the beginning of active swimming, there was a sharp rise in embryo melatonin, which gradually declined. We have recently studied the changes in melatonin levels in the early *Rana pipiens* embryo from the egg through the end of embryonic development using a melatonin radioimmunoassay (Fig. 1). The melatonin level was low and fairly constant in the embryo until it increased about eight-fold at around stage 21 in a transient peak that fell back, on average, to about twice as high as in early embryonic life. Thus, the pattern coincides very well with the findings in *Xenopus*, and the sharp melatonin rise occurs in *Rana* at a corresponding stage of development. It is likely that the low level of melatonin in the embryo prior to this peak represents maternal melatonin. The peak at stage 21, and the subsequent higher level of melatonin

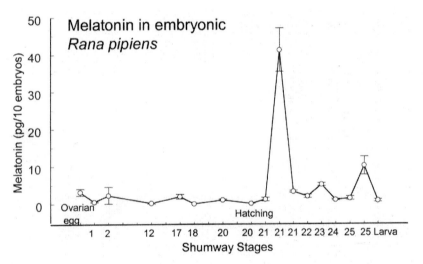

Fig. 1. Melatonin levels in embryonic *Rana pipiens* determined by radioimmunoassay. During development, the embryos were kept on a 12L:12D cycle at 22°C. Each point on the curve represents the mean ± S.E. of five samples, each consisting of ten embryos with jelly coats removed. Repetition of the same stage on the x axis label indicates that several observations were made at successively later intervals in that stage. Shumway (1940) stages 1 and 2 represent unfertilized and fertilized eggs, respectively. The peaks at mid stage 21 (132 hours after fertilization) and late stage 25 (204 hours after fertilization) were significantly different from all the other points on the curve, and from each other (P<0.05).

(Fig. 1) might reflect the beginning of synthesis of embryonic melatonin in the pineal gland and retina since both organs are developing at this time (Hendrickson and Kelly, 1969; Green et al., 1999). In fact, the time of the melatonin peak in both *Xenopus* and *Rana* embryos coincides with detection of N-acetyltransferase (NAT), a key enzyme in melatonin synthesis, in the eye (Alonso-Gómez et al., 1994). In subsequent stages of development, a rhythm of NAT is established (Alonso-Gómez et al., 1994), which could account for the subsequent decline in the embryonic melatonin level. A second, much lower peak of whole embryo melatonin occurs in *Rana pipiens* at stage 25, the last embryonic stage (Fig. 1). There are no data as yet which show that melatonin influences amphibian embryonic development.

Melatonin has also been identified in many tissues of the *Rana catesbeiana* (bullfrog) tadpole, such as the pineal gland, the thyroid gland, the gut, the gills (Wright et al., 2001), the eyes, and the plasma (Wright and Racine, 1997). In premetamorphic tadpoles, melatonin was much higher in the pineal and the thyroid glands, than in the other tissues studied (Table 1). Gills and intestine had the next highest melatonin levels, followed by very low concentrations in the tail and limbs. Intestine, as well as the pineal gland, are sources of melatonin synthesis and secretion, while the gill may be involved in excretion of melatonin into the holding solution (Wright et al., 2001). The source and significance of the high level of melatonin in the tadpole thyroid gland at early stages is not known.

Table 1. Melatonin levels in various tissues of late premetamorphic bullfrog tadpoles

Tissue	Melatonin (pg/g)[1]
Pineal gland	194,218.3 ± 76,571.9
Thyroid[2]	2,439.1 ± 438.7
Gill	186.4 ± 49.4
Intestine	135.6 ± 19.9
Hind limb	20.6 ± 5.2
Tail	8.7 ± 2.9

[1] Mean ± S.E. of data collected at 6 intervals during one 24-hour period. Tissues listed all came from the same animals (n = 5/interval).
[2] Thyroid includes underlying cartilage.

Developmental and Diel Changes in Melatonin

Developmental changes in levels of ocular melatonin varied with the light: dark (LD) cycle in bullfrog tadpoles. Retinal melatonin decreased on 18L:6D (Wright and Racine, 1997), increased on 6L:18D, and remained the same on 12L:12D from prometamorphosis to metamorphic climax (Wright et al., 2003a). Pineal, intestinal, and hind limb concentrations of melatonin did not change significantly during larval development, whereas tail melatonin peaked in mid-larval life and gill melatonin during metamorphic climax (Fig. 2). Interestingly, melatonin levels decreased in the thyroid gland at later stages of metamorphosis (Fig. 2). Plasma melatonin decreased significantly at climax on all the LD cycles studied (Wright and Racine, 1997; Wright et al., 2003a; Wright and Bruni, 2004). Melatonin was also identified in the plasma of *Rana pipiens*, where the concentration did not change significantly up to the end of prometamorphosis (Wright, 2002), but then fell at climax (Wright et al., 2003a).

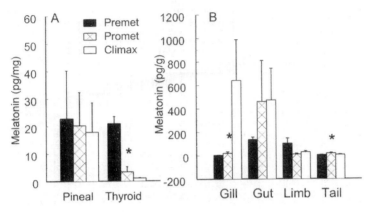

Fig. 2. Changes in the levels of melatonin in various glands and tissues during metamorphosis in the bullfrog tadpole. Melatonin in pineal and thyroid is represented in pg/mg tissue, while the hormone level in the other tissues is scaled in pg/g tissue. Asterisks indicate a significant difference in the levels of melatonin among the developmental periods (P<0.05). (Premet = premetamorphosis, a young tadpole in growth stages; Promet = prometamorphosis, beginning of development of some adult organs; Climax = stage of rapid loss of larval organs and transformation into the frog.)

Melatonin levels in various tissues of the premetamorphic bullfrog tadpole manifested diel fluctuations, as illustrated in Fig. 3 compiled from data collected from tadpoles kept on 18L:6D. The peak of the rhythm occurred during, or near the onset, of the brief dark period in the pineal gland, gill (Fig. 3A), tail and limb, and eye (Wright, 2002), whereas it occurred late in the light in the thyroid gland (Fig. 3B), and early in the photophase in the intestine (Fig. 3C). These findings indicate that the rhythms in the various tissues are not synchronized, and, in some cases (e.g., pineal gland, eye, and intestine) the peak probably represents the time of maximum synthesis and secretion of melatonin into the circulation. The peak of melatonin in the thyroid gland occurred about midway between the peak of plasma T_4 at 1400 hours and the peak of plasma melatonin at around 0500 hours (Wright and Racine, 1997). Plasma melatonin exhibited diel fluctuations, which varied with the LD cycle and changed with the developmental stage (Wright and Racine, 1997; Wright et al., 2003a). Changing diel rhythmicity during metamorphosis has the effect of bringing corticosteroid rhythms in phase at climax (Wright et al., 2003b), and of establishing the typical scotophase peak of melatonin at the end of metamorphosis (Wright et al., 2003a). Since these changes occur in the

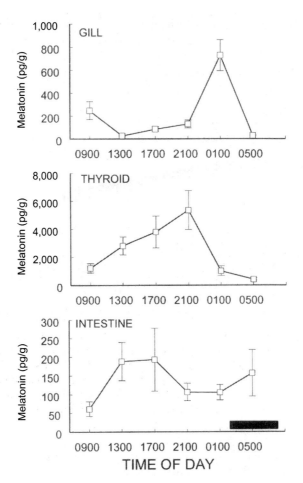

Fig. 3. Diel rhythms of melatonin in gill, thyroid gland, and intestine of the bullfrog tadpole. The bar on the lower graph indicates the scotophase in the 18L:6D lighting regimen. In gill, the peak at 0100 hr was significantly different from all the other points on the curve, and the peak at 0900 was significantly different from 1300 and 0500 hr (p<0.05). In the thyroid data, the peak at 2100 was significantly different (P<0.05) from 0900, 1300, 0100, and 0500 hr whereas in the intestine, the two high points at 1300 and 1700 hr together were significantly different from the three low points at 0900, 2100, and 0100 hr together (P<0.02).

laboratory without an alteration in the LD cycle, they indicate that the larval rhythms are influenced by developmental events, even though they are entrained to each LD cycle (Wright and Bruni, 2004). The developmental changes in the rhythms also alter the coincidence of the phase of the thyroid hormone rhythm with those of a synergist (corticosteroids) and a modulator (melatonin), probably affecting the rate of metamorphosis through this mechanism (Wright et al., 2003a, b).

Influence of Melatonin on the Thyroid Gland and Metamorphosis

Melatonin has been reported to increase, decrease, or have no effect on the progress of anuran metamorphosis, which may be due to differences among the studies in the melatonin concentrations used, the LD cycle, and other factors (Wright, 2002; Wright and Alves, 2001). In any case, the mode of action of melatonin does not include affecting deiodination of the metamorphic hormone T_4 in its conversion to the tissue-active form of the hormone, triiodothyronine (T_3). When bullfrog tadpoles were immersed for five days in T_4 (50 µg/l), or T_4 and low (150 pg/ml), or high (150 ng/ml) melatonin concentrations, the plasma levels of T_4 and T_3 rose over the control, indicating that uptake of T_4 had occurred and that the excess exogenous T_4 was partially deiodinated to T_3 (Fig. 4). The T_3/T_4 ratio was the same in the T_4 alone and the T_4/melatonin groups, showing that melatonin did not affect deiodination of exogenous T_4 to T_3. However, plasma T_4, and the corresponding deiodination to T_3, were less in the group treated with 150pg melatonin and T_4, than in the T_4 alone and T_4/150 ng melatonin groups (Fig. 4), indicating that physiological melatonin concentrations might block T_4 uptake from the holding solution. An influence of melatonin on intestinal uptake of T_4 in vitro had been previously observed (Wright et al., 1997a). Melatonin failed to affect

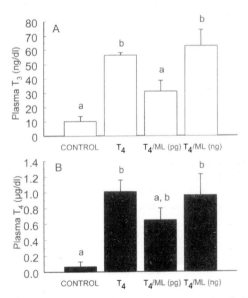

Fig. 4. The antagonistic action of melatonin on metamorphosis is not mediated via an influence on deiodination of T_4 to T_3. Prometamorphic bullfrog tadpoles on 12L:12D were immersed for five days in 10% Holtfreter's solution (control), 50 µg/l T_4, T_4 and 150 pg/ml melatonin (T_4/ML [pg]), or T_4 and 150 ng/ml melatonin (T_4/ML [ng]), after which time blood was collected. Radioimmunoassay was performed for plasma T_3 (A) and T_4 (B). Bars with different letters are significantly different (P<0.05).

iodide uptake by the thyroid gland in larval and adult tiger salamanders (Norris and Platt, 1973).

The most compelling evidence for a possible influence of melatonin on metamorphosis comes from in vivo (Wright et al., 1996) and *in vitro* (Wright et al., 1997b, 2000) studies showing that thyroid cell proliferation, T_4 secretion, and response to thyrotropin were reduced by melatonin. Recent studies have shown that plasma T_4 in early climax tadpoles significantly (P<0.05) increased from 0.021 ± 0.019 to 0.138 ± 0.040 µg/dl after ophthalmectomy, a procedure that reduces endogenous plasma melatonin, indicating that melatonin was inhibiting the thyroid gland secretion at that time. Similar influences of melatonin on the thyroid gland have been observed in mammals, although stimulation of the thyroid gland also occurred under some circumstances, usually involving a specific mode or timing of administration, or concentration of melatonin (review of mammalian studies in Lewiński et al., 2002). Indeed, immersion of prometamorphic tadpoles in physiological (150 pg/ml) melatonin significantly increased plasma T_4 without changing plasma melatonin, however a higher concentration of melatonin (100 ng/ml) failed to significantly affect plasma T_4 while it increased plasma melatonin (Fig. 5). These findings explain why metamorphosis was accelerated at low melatonin concentrations, but retarded at high concentrations in some of the earlier work (e.g. Edwards and Pivorun, 1991). Studies are underway in our laboratory to determine if removing endogenous sources of circulating melatonin has any effect in development or metamorphosis.

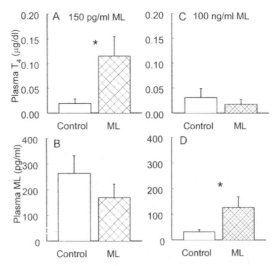

Fig. 5. Plasma T_4 (A, C) and plasma melatonin (B, D) after immersing prometamorphic bullfrog tadpoles on 18L:6D in 150 pg/ml (A, B) or 100 ng/ml melatonin (C, D) for five days. Asterisks indicate a significant difference between the control and treated groups (P=0.03). ML = melatonin.

Effect of Metamorphosis-inducing Hormones on Melatonin

Plasma melatonin decreased markedly at climax during both spontaneous and T_4-induced metamorphosis (Wright and Alves, 2001; Wright et al., 2003a). T_4, directly or indirectly, could stimulate the degradation and elimination of endogenous melatonin, increase uptake into the tissues, or reduce the secretion of melatonin by the pineal gland, eyes, and gut. T_4 increased the uptake of tritiated melatonin into peripheral tissues, but not into neural or endocrine organs such as the pineal gland or thyroid gland when it was administered by injection (Wright et al., 1997a), however, immersion in T_4 did not increase tissue melatonin content (Wright et al., 2001). There is a threshold above which exogenous melatonin in the plasma is rapidly cleared (Wright et al., 2001; Wright, 2002), resulting in a melatonin level lower than in controls. An alternate possibility to explain the climax decline in plasma melatonin is that melatonin-secreting organs increase activity under the influence of the rising T_4 at late metamorphosis, and the resulting higher level of plasma melatonin exceeds the threshold and so is cleared from the blood. This hypothesis is supported by the experimental evidence that administration of T_4 in vitro significantly increased melatonin secretion into the media of pineal glands from tadpoles and froglets, and of retinal tissue from climax tadpoles (Fig. 6). This finding is not surprising in light of the general growth and developmental effects of the thyroid hormones on virtually every organ in the tadpole during metamorphosis (Shi, 2000). Pineal, but not retinal, melatonin was significantly higher at prometamorphosis in the controls, and at this stage the pineal was most stimulated by T_4 (Fig. 6). There was a significant fall in pineal melatonin secretion between prometamorphosis and climax in the controls, whereas the change in the retina was not significant. These results suggest that the climax decline in plasma melatonin may also be linked to inhibition of pineal secretion.

Recent developments in our understanding of the hormonal control of metamorphosis brought the recognition that in larval amphibians corticotropin-releasing hormone (CRH) stimulated both pituitary thyrotropin and pituitary adrenocorticotropin (ACTH), thus controlling both the thyroid and the adrenal axes (Denver, 1993, 1996, 1997), the key regulators of metamorphosis. There are evidences in mammals for an effect of melatonin on CRH, and for the reverse, an influence of CRH on the melatonin level. An *in vitro* study showed that melatonin inhibited the releasing effect of both 5-hydroxytryptamine and acetylcholine on CRH from rat hypothalami (Jones et al., 1976). Administration of melatonin to rats in vivo lowered both the adrenocortical response to stress and hypothalamic CRH content (Konakchieva et al., 1997). ACTH-stimulated cortisol production by the monkey adrenal gland was inhibited by melatonin (Torres-Farfan et al., 2003). In turn, CRH inhibited pineal secretion of melatonin in humans (Kellner et al., 1997). A feedback loop has been proposed that includes melatonin inhibition of CRH and consequent lowering of β-endorphin from the pituitary, which then depresses melatonin secretion (Chamberlain and

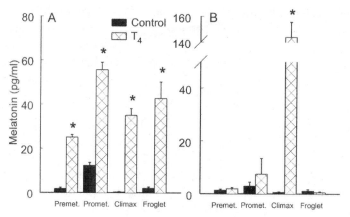

Fig. 6. Effect of 2.5×10^{-7} M T_4 on pineal (A) and retinal (B) secretion of melatonin into culture media during 24 hr of culture in the dark. Tissues were taken from bullfrog tadpoles and froglets previously on a 12L:12D cycle. Melatonin radioimmunoassay of the media showed that T_4 significantly increased pineal melatonin secretion without regard to developmental stage, whereas it only increased melatonin from the retina at climax. Asterisks indicate significant differences between the control and T_4 groups ($P<0.001$). In addition, pineal melatonin secretion (A) in the controls was significantly higher at prometamorphosis (Promet.) than at the other stages ($P<0.006$), while T_4-stimulated secretion was higher at prometamorphosis than at premetamorphosis (Premet.) or climax ($P<0.01$), and higher in the froglet than at premetamorphosis ($P=0.037$). There were no significant differences in retinal melatonin secretion (B) among the controls, while in the T_4-treated, climax was significantly higher than the other stages ($P<0.004$).

Herman, 1991). The climax of metamorphosis is marked by high levels of thyroid hormones and corticosterone, indicating that both the thyroid and the adrenal axes are activated, presumably by a rise in CRH, since CRH receptors and CRH binding protein peak at climax in *Xenopus* (Manzon and Denver, 2004). An increased level of CRH in climax, coupled with the possibility of its inhibitory action on the pineal gland and perhaps other melatonin-secreting sites, might provide an explanation for the decrease in pineal melatonin secretion between prometamorphosis and climax in the controls of Fig. 6 and for the decline in plasma melatonin at climax discussed above. A possible scenario with reference to Fig. 6 and discussion above might be that the rising level of T_4 stimulates the development of the pineal and the retina in prometamorphosis. These organs increase secretion at the onset of climax, melatonin levels then exceed the threshold, and the blood level falls. Inhibition of CRH secretion by melatonin is reduced (other hormone changes also influence CRH (other hormone changes also influence CRH - see Manzon and Denver, 2004) and the CRH level rises to stimulate the high levels of climax T_4 and corticosteroids, which induce the final metamorphic events. The high level of CRH antagonizes secretion of melatonin, which remains low during climax. Verification of this proposed explanation, and of relationships between CRH and melatonin, await experimental investigation.

MELATONIN AND METAMORPHOSIS: NEW DIRECTIONS

There has been little investigation of several other vital roles of melatonin as they apply to anuran metamorphosis, particularly the effect of the hormone on calcium levels in tissues, its antioxidant capacity, and its effect on apoptosis. Treatment with calcium accelerated metamorphosis, possibly through an effect on the thyroid (Menon et al., 2000), while a calcium channel blocker and a calmodulin inhibitor antagonized development and metamorphosis (Kumar et al., 1993). In the habenular nuclei of the frog brain, the calcium binding protein calretinin varied in level during development, peaking during metamorphosis, and declining after its completion (Guglielmotti et al., 2004), suggesting that calcium and its binding proteins may have important functions in brain morphogenesis. Melatonin interacts with calmodulin, and it has been suggested that this interaction allows melatonin to regulate and modulate many aspects of cellular physiology (Benitez-King and Anton-Tay, 1993). Melatonin is also a potent anti-oxidant because of its widespread distribution within tissues, and its ability to cross membranes and function within various cell compartments (Karbownik and Reiter, 2000). Physiological concentrations of melatonin inhibited nitric oxide synthase activity, a process in which calmodulin may be involved (Bettahi et al., 1996). Melatonin also inhibited apoptosis (Karbownik and Reiter, 2000). Future studies need to ascertain if melatonin is involved in calcium influence on metamorphosis or apoptotic events in degenerating larval tissues. Does the antioxidant capacity of melatonin play a role in the various fates of larval tissues during metamorphosis? Clarification of the developmental role of melatonin in the amphibian obviously awaits the results of future research.

ACKNOWLEDGEMENTS

The authors thank Meagan Hathaway, Arielle Noyes, and Nicole Bruni for their technical expertise in some of the experiments presented. The research was funded in part by grants IBN-9723858 and IBN-0236251 from the National Science Foundation.

REFERENCES

Alonso-Gómez, A.L., de Pedro, N., Gancedo, B., Alonso-Bedate, M., Valenciano, A.I. and Delgado, M.J. (1994). Ontogeny of ocular serotonin N-acetyltransferase activity daily rhythm in four anuran species. *Gen. Comp. Endocrinol.* **94**: 357-365.

Baker, P.C. (1969). Melatonin levels in developing *Xenopus laevis*. *Comp. Biochem. Physiol.* **28**: 1387-1393.

Benitez-King, G. and Anton-Tay, F. (1993). Calmodulin mediates melatonin cytoskeletal effects. *Experientia* **49**: 635-641.

Bettahi, I., Pozo, D., Osuna, C., Reiter, R.J., Acuna-Castroviejo, D. and Guerrero, J.M. (1996). Melatonin reduces nitric oxide synthase activity in rat hypothalamus. *J. Pineal Res.* **20:** 205-210.

Bubenik, G.A. and Pang, S.F. (1997). Melatonin levels in the gastrointestinal tissues of fish, amphibians, and a reptile. *Gen. Comp. Endocrinol.* **106:** 415-419.

Bubenik, G.A. (2002). Gastrointestinal melatonin: localization, function, and clinical relevance. *Digestive Diseases and Sciences* **47:** 2336-2348.

Chamberlain, R.S. and Herman, B.H. (1990). A novel biochemical model linking dysfunctions in brain melatonin, proopiomelanocortin peptides, and serotonin in autism. *Biol. Psych.* **28:** 773-793.

Conti, A., Tettamanti, C., Singaravel, M., Haldar, C., Pandi-Perumal, R.S. and Maestroni, G.J.M. (2002). Melatonin: an ubiquitous and evolutionary hormone. In: *Treatise on Pineal Gland and Melatonin* (Haldar, C. Singaravel, M. and Kumar-Maitra, S., eds.), Science Publishers, Enfield NH, USA, 105-143.

Danilova, N., Krupnik, V.E., Sugden, D. and Zhdanova, I.V. (2004). Melatonin stimulates cell proliferation in zebrafish embryo and accelerates its development. *Faseb Journal* Express Article 10.1096/fj.03-0544fje. Published online February 6, 2004.

Delgado, M.J., Gutierrez, P. and Alonso-Bedate, M. (1984). Growth response of premetamorphic *Rana ridibunda* and *Discoglossus pictus* tadpoles to melatonin injections and photoperiod. *Acta Embryol. Morphol. Exper.* **5:** 23-39.

Delgado, M.J. and Vivien-Roels, B. (1989). Effect of environmental temperature and photoperiod on the melatonin levels in the pineal, lateral eye, and plasma of the frog, *Rana perezi*: Importance of ocular melatonin. *Gen. Comp. Endocrinol.* **75:** 46-53.

Denver, R.J. (1993). Acceleration of anuran amphibian metamorphosis by corticotropin-releasing hormone-like peptides. *Gen. Comp. Endocrinol.* **91:** 38-51.

Denver, R.J. (1996). Neuroendocrine control of amphibian metamorphosis. In: *Metamorphosis: Postembryonic Reprogramming of Gene Expression in Amphibian and Insect Cells* (Gilbert, L.I., Tata, J.R. and Atkinson, B.G., eds.), Academic Press, Inc., New York, 433-464.

Denver, R.J. (1997). Environmental stress as a developmental cue: corticotropin-releasing hormone is a proximate mediator of adaptive phenotypic plasticity in amphibian metamorphosis. *Hormones and Behavior* **31:** 169-179.

Edwards, M.L.O. and Pivorun, E.B. (1991). The effects of photoperiod and different dosages of melatonin on metamorphic rate and weight gain in *Xenopus laevis* tadpoles. *Gen. Comp. Endocrinol.* **81:** 23-38.

Filadelfi, A.M.C. and Castrucci, A.M.D.L. (1996). Comparative aspects of the pineal/melatonin system of poikilothermic vertebrates. *J. Pineal Res.* **20:** 175-186.

Green, C.B., Liang, M.Y., Steenhard, B.M. and Besharse, J.C. (1999). Ontogeny of circadian and light regulation of melatonin release in *Xenopus laevis* embryos. *Develop. Brain Res.* **117:** 109-116.

Guglielmotti, V., Cristino, L., Sada, E. and Bentivoglio, M. (2004). The epithalamus of the developing and adult frog: calretinin expression and habenular asymmetry in *Rana esculenta*. *Brain Res.* **999:** 9-19.

Gutierrez, P., Delgado, M.J. and Alonso-Bedate, M. (1984). Influence of photoperiod and melatonin administration on growth and metamorphosis in *Discoglossus pictus* larvae. *Comp. Biochem. Physiol.* **79:** 255-260.

Helliwell, R.J. and Williams, L.M. (1994). The development of melatonin-binding sites in the ovine fetus. *J. Endocrinol.* **142:** 475-484.

Hendrickson, A.E. and Kelly, D.E. (1969). Development of the amphibian pineal organ. *Anat. Rec.* **165:** 211-227.

Jones, M.T., Hillhouse, E.W. and Burden, J. (1976). Effect on various putative neurotransmitters on the secretion of corticotrophin-releasing hormone from the rat hypothalamus in vitro – a model of the neurotransmitters involved. *J. Endocrinol.* **69:** 1-10.

Kaltenbach, J. (1996). Endocrinology of amphibian metamorphosis. In: *Metamorphosis: Postembryonic Reprogramming of Gene Expression in Amphibian and Insect Cells* (Gilbert, L.I., Tata, J.R. and Atkinson, B.G., eds.), Academic Press, Inc., New York, 403-431.

Karbownik, M. and Reiter, R.J. (2000). Antioxidative effects of melatonin in protection against cellular damage caused by ionizing radiation. *Proc. Soc. Exp. Biol. Med.* **225:** 9-22.

Kellner, M., Yassouridis, A., Manz, B., Steiger, A., Holsboer, F. and Wiedemann, K. (1997). Corticotropin-releasing hormone inhibits melatonin secretion in healthy volunteers – a potential link to low-melatonin syndrome in depression? *Neuroendocrinol.* **65:** 284-290.

Kikuyama, S., Kawamura, K., Tanaka, S. and Yamamoto, K. (1993). Aspects of amphibian metamorphosis: hormonal control. *Int. Rev. Cytol.* **145:** 105-148.

Konakchieva, R., Mitev, Y., Almeida, O.F.X. and Patchev, V.K. (1997). Chronic melatonin treatment and the hypothalamo-pituitary-adrenal axis in the rat: Attenuation of the secretory response to stress and effects of hypothalamic neuropeptide content and release. *Biology of the Cell* **89:** 587-596.

Kováčiková, Z., Kuncová, T., Lamošová, D. and Zeman, M. (2003). Embryonic and post-embryonic development of Japanese quail after in ovo melatonin treatment. *Acta Vet. Brno.* **72:** 157-161.

Kumar, B.A., Vinod, K.R., Paul, V.F. and Pilo, B. (1993). Effect of calcium and calmodulin antagonists on metamorphosis in the anuran tadpole, *Rana tigerina*. *Funct. Develop. Morphol.* **3:** 237-242.

Lewiński, A., Sewerynek, E. and Karbownik, M. (2002). Melatonin from the past into the future – our own experience. In: *Treatise on Pineal Gland and Melatonin* (Haldar, C., Singaravel, M. and Kumar-Maitra, S., eds.), Science Publishers, Enfield, NH, USA, 157-175.

Manzon, R.G. and Denver, R.J. (2004). Regulation of pituitary thyrotropin gene expression during *Xenopus* metamorphosis: negative feedback is functional throughout metamorphosis. *J. Endocrinol.* **182:** 273-285.

Mayer, I., Bornestaf, C. and Borg, B. (1997). Melatonin in non-mammalian vertebrates: physiological role in reproduction? *Comp. Biochem. Physiol.* **118**A: 515-531.

Menon, J., Gardner, E.E. and Vail, S. (2000). Developmental implications of differential effects of calcium in tail and body skin of anuran tadpoles. *J. Morphol.* **244:** 31-43.

Norris, D.O. and Platt, J.E. (1973). Effects of pituitary hormones, melatonin, and thyroidal inhibitors on radioiodide uptake by the thyroid glands of larval and adult tiger salamanders, *Ambystoma tigrinum* (Amphibia: Caudata). *Gen. Comp. Endocrinol.* **21:** 368-376.

Shi, Y. (2000). *Amphibian Metamorphosis*. Wiley-Liss, New York.

Shumway, W. (1940). Stages in the normal development of *Rana pipiens* I. External form. *Anat. Rec.* **78:** 139-147.

Torres-Farfan, C., Richter, H.G., Rojas-García, P., Vergara, M., Forcelledo, M.L., Valladares, L.E., Torrealba, F., Valenzuela, G.J. and Serón-Ferré, M. (2003). mt1 melatonin receptor in the primate adrenal gland: inhibition of adrenocorticotropin-stimulated cortisol production by melatonin. *J. Clin. Endocrinol. Met.* **88**: 450-458.

Tosini, G. and Fukuhara, C. (2003). Photic and circadian regulation of retinal melatonin in mammals. *J. Neuroendocrinol.* **15**: 364-369.

Wright, M.L., Cykowski, L.J., Mayrand, S.M., Blanchard, L.S., Kraszewska, A.A., Gonzales, T. M. and Patnaude, M. (1991). Influence of melatonin on the rate of *Rana pipiens* tadpole metamorphosis *in vivo* and regression of thyroxine-treated tail tips *in vitro*. *Dev. Growth & Differ.* **33**: 243-249.

Wright, M.L., Pikula, A., Cykowski, L.J. and Kuliga, K. (1996). Effect of melatonin on the anuran thyroid gland: Follicle cell proliferation, morphometry, and subsequent thyroid hormone secretion *in vitro* after melatonin treatment *in vivo*. *Gen. Comp. Endocrinol.* **103**: 182-191.

Wright, M. and Racine, C. (1997). Inverse relationship between plasma T_4 and plasma and ocular melatonin in prometamorphic and climax bullfrog (*Rana catesbeiana*) tadpoles. In: *Advances in Comparative Endocrinology* (Kawashima, S. and Kikuyama, S., eds.), Monduzzi Editore, Bologna, 403-407.

Wright, M.L., Pikula, A., Babski, A.M. and Kuliga, K. (1997a). Distribution and reciprocal interactions of ^{3}H-melatonin and ^{125}I-thyroxine in peripheral, neural, and endocrine tissues of bullfrog tadpoles. *Comp. Biochem. Physiol.* **118**A: 691-698.

Wright, M.L., Pikula, A., Babski, A.M., Labieniec, K.E. and Wolan, R.B. (1997b). Effect of melatonin on the response of the thyroid to thyrotropin stimulation *in vitro*. *Gen. Comp. Endocrinol.* **108**: 298-305.

Wright, M.L., Cuthbert, K.L., Donohue, M.J., Solano, S.D. and Proctor, K.L. (2000). Direct influence of melatonin on the thyroid and comparison with prolactin. *J. Exp. Zool.* **286**: 625-631.

Wright, M.L. and Alves, C.D. (2001). The decrease in plasma melatonin at metamorphic climax in *Rana catesbeiana* tadpoles is induced by thyroxine. *Comp. Biochem. Physiol.* **129**A: 653-663.

Wright, M., Alves, C., Francisco, L., Shea, T., Visconti, R. and Bruni, N. (2001). Regulation of the plasma melatonin level in anuran tadpoles. In: *Perspectives in Comparative Endocrinology* (Th. Goos, H.J., Rastogi, R.K., Vaudry, H. and Pierantoni, R., eds.), Monduzzi Editore, Bologna, 669-674.

Wright, M.L. (2002). Melatonin in the anuran amphibian larva and its reciprocal interactions with the thyroid gland. In: *Treatise on Pineal Gland and Melatonin* (Haldar, C., Singaravel, M. and Kumar-Maitra, S., eds.), Science Publishers, Enfield, NH, USA, 200-210.

Wright, M.L., Duffy, J.L., Guertin, C.J., Alves, C.D., Szatkowski, M.C. and Visconti, R.F. (2003a). Developmental and diel changes in plasma thyroxine and plasma and ocular melatonin in the larval and juvenile bullfrog. *Gen. Comp. Endocrinol.* **130**: 120-128.

Wright, M.L., Guertin, C.J., Duffy, J.L., Szatkowski, M.C., Visconti, R.F. and Alves, C.A. (2003b). Developmental and diel profiles of plasma corticosteroids in the bullfrog, *Rana catesbeiana*. *Comp. Biochem. Physiol.* **135**A: 585-595.

Wright, M.L. and Bruni, N.K. (2004). Influence of the photocycle and thermocycle on rhythms of plasma thyroxine and plasma and ocular melatonin in late metamorphic stages of the bullfrog tadpole, *Rana catesbeiana*. *Comp. Biochem. Physiol.* **139**A: 33-40.

Wright, M.L., Francisco, L.L., Scott, J.L., Richardson, S.R., Carr, J.A., King, A.B., Noyes, A.G. and Visconti, R.F. (2006). Effects of bilateral and unilateral ophthalmectomy on plasma melatonin in *Rana* tadpoles and froglets under various experimental conditions. *Gen. Comp. Endocrinol.* **147**: 158-166.

Development and Regeneration of the Vertebrate Retina: Role of Tissue Interaction and Signaling Molecules on the Retinal Fate Determination and RPE Transdifferentiation

Masasuke Araki

Abstract

Development of eye tissues has been the most intensively studied organ development among vertebrates, and recent studies are elucidating gene regulatory mechanisms involved in eye development using molecular genetics. Regeneration of eye tissues is also one of the classical subjects of developmental biology, and it is now being vigorously studied to reveal the cellular and molecular mechanisms. Although many experimental animal models have been studied, there may be a common basic mechanism that governs retinal regeneration. This can also control ocular development, indicating the existence of a common principle between the development and regeneration of eye tissues. This notion is now getting more widely accepted by recent studies on the genetic regulation of ocular development. In the present review, tissue interactions which underlie both retinal development and regeneration are described, an area which seems not to have been paid much attention to in comparison with gene regulatory mechanisms. We propose that tissue interaction, particularly mesenchyme-neuroepithelial interaction, is considered to play a fundamental role both in retinal development and regeneration.

Developmental Neurobiology Laboratory, Department of Biological Sciences, Nara Women's University, Nara 630-8506, Japan. Fax: 81 742 20 3411. E-mail: masaaraki@cc.nara-wu.ac.jp

INTRODUCTION

Overview of Eye Development

Recent progress in the genetic analysis of vertebrate eye development has elucidated a number of genes involved in ocular development and their regulated expression patterns (Chow and Lang, 2001; Bailey et al., 2004; Hatakeyama and Kageyama, 2004). Retinal development is one of the most intensively studied subjects among various organogenesis phenomena, and includes many topics such as fate determination, cell proliferation, precursor cell differentiation, organized layer formation, projection mapping etc. (Adler, 2000; Kondoh, 2002). Eye development, however, has not yet been paid much attention to, from the viewpoint of endocrinology. This clearly contrasts with studies on of vertebrate brain development, in which hormonal effects on brain development, particularly from a functional aspect, have been extensively studied. Since the retina is a derivative of the brain vesicle and has various aspects in common with the nervous system, it is a suitable experimental model for the analysis of brain development, and will also provide a good future model for developmental endocrinology.

Recent studies have elucidated numerous signaling molecules that have critical roles both in retinal development and regeneration (Moshiri et al., 2004). In the present paper, we describe how these molecules are involved in the processes of retinal development and regeneration, with particular emphasis on tissue interactions and related signaling molecules.

Dorsal-Ventral Polarity of the Eye Primordium and Its Developmental Significance

The retina initially develops as a lateral protrusion of the forebrain called the optic vesicle, which further develops to form a double-layered optic cup by invagination (Fig. 1). Subsequently, the optic cup develops into two different regions, neural retina (NR) and retinal pigment epithelium (RPE), by inductive signals from neighboring tissues (Saha et al., 1992; Chow and Lang, 2001). At the same time, the invagination of the optic vesicle extends from the distal to the proximal direction at the ventral part to form the future optic fissure in the ventral optic stalk, through which the retinal ganglion axons pass into the optic tectum. Thus, the early optic vesicle appears to consist of at least two discrete dorsal and ventral compartments that show different developmental fates. This is also suggested by the different expression patterns of various transcription factors (Schulte et al., 1999; Koshiba-Takeuchi et al., 2000). The optic vesicle makes contact with various tissues, such as the overlying surface ectoderm and periocular mesenchyme, and it has been assumed that the specification of the NR and RPE is influenced by signals from the surface ectoderm and mesenchyme, respectively (Dragomirov, 1937; Lopashov, 1963; Hyer et al., 1998; Fuhrmann et al., 2000).

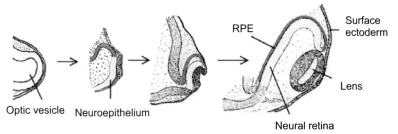

Fig. 1. Schematic drawings of transversal images of the developing eye. The optic vesicle neuroepithelium is subdivided into the dorsal and ventral regions, which later develop into the retinal pigmented epithelium (RPE) and neural retina.

Fate mapping of the chick optic vesicle with fluorescent dye reveals the developmental fates of discrete regions in the optic vesicle according to the dorsal-ventral and proximal-distal directions (Dütting and Thanos, 1995). The dorsal and ventral polarity is well understood for its importance in eye morphogenesis, and several transcription factors, such as Pax2 and Pax6, are known to play essential roles in the fate determination of the optic vesicle regions (Fig. 2) (Schwarz et al., 2000; Baümer et al., 2003). Several signaling molecules have been implicated in the regulation of ventral optic development, and Sonic hedgehog (Shh) is considered to contribute to the bilateral pattern formation of the vertebrate eye, particularly in the development of the ventral structure (Jean et al., 1998; Oliver and Gruss, 1997). Pax2 and Pax6 expressions in the optic vesicle are regulated by Shh,

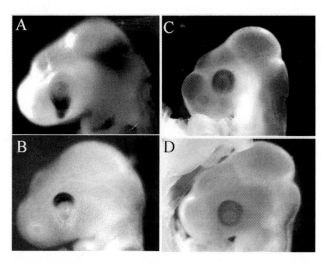

Fig. 2. Localized expression profiles of four transcription factors in three-day chick embryos: (A) Pax2, (B) Tbx5, (C) Pax6 and (D) Chx10. All these genes are expressed in the developing eyes and differ in their localization from each other. Pax6 is expressed in the lens and retina, while Chx10 is expressed only in the retina.

and alterations in Shh activity perturb Pax6 and Pax2 expressions in zebra fish (Macdonald et al., 1995). BMPs (Bone Morphogenetic Proteins) also have profound effects on patterning in the nervous system. BMP4, for instance, is involved in the dorsalization of the retina (Koshiba-Takeuchi, 2000). The spatial and temporal expression patterns of Shh and BMP4 suggest a model in which ventral midline-derived Shh and dorsally derived BMP4, either from the dorsal midline of the neural tube or the dorsal optic cup, play roles in establishing the D-V polarity of the optic primordium (Schulte et al., 1999; Zhang and Yang, 2001; Martí and Bovolenta, 2002; Ohkubo et al., 2002).

To elucidate the mechanism involved in dorso-ventral (D-V) patterning in the developing optic vesicle and how it affects ocular development, embryonic transplantation using avian embryos is a powerful approach. A truncated optic vesicle was transplanted inversely without changing the A-P direction in order to study how the D-V polarity is established within the optic vesicle (Araki et al., 2002; Uemonsa et al., 2002); in this transplantation study, the left optic vesicle was cut and transplanted inversely in the right eye cavity of the host chick embryos (Fig. 3). This method ensured that the D-V polarity was reversed while the antero-posterior axis remained normal. The location of the choroid fissure, a ventral structure, was altered from the normal (ventral) to ectopic positions as the embryonic stage of transplantation progressed from 6 to 18 somites. At the same time, expression patterns of Pax2 and Tbx5, marker genes for ventral and dorsal regions of the optic vesicle, respectively, changed concomitantly in a similar way. Ovo explant culturing of the optic vesicle showed that the D-V polarity and choroid fissure formation were already specified by the 10-somite stage. These results indicate that the D-V polarity of the optic vesicle is established gradually between 8- and 14-somite stages under the influence of signals derived from the midline portion of the forebrain. The presumptive signal(s) appeared to be transmitted from proximal to distal regions within the optic vesicle. A severe anomaly was observed in the development of optic vesicles reversely transplanted around the 10-somite stage: optic cup formation was disturbed and subsequently NR and RPE did not develop normally, probably because the once developed D-V polarity in the optic vesicle was forced to be influenced by the other signal from the reverse side of the truncated brain (Fig. 4). It was concluded that the establishment of D-V polarity in the optic vesicle plays an essential role in optic cup formation, the patterning and differentiation of NR and RPE.

Role of Neighboring Tissues in Optic Cup Development

Two different tissues, namely the surface ectoderm and the mesenchyme, surround the optic vesicle and these tissue interactions play substantial roles in defining the developmental fate of the optic vesicle (Fig. 5); one interaction with the surface ectoderm along the proximo-distal direction and the other interaction with the dorsal mesenchymal cells along the dorso-ventral direction.

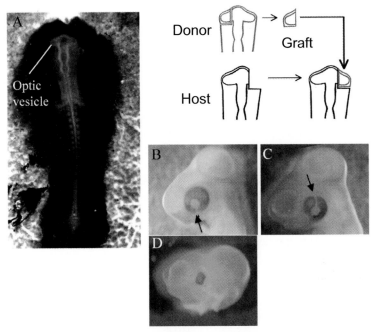

Fig. 3. Embryonic transplantation of early chick embryos. To elucidate how the dorso-ventral polarity is established and what role it plays in eye development, optic vesicles were transplanted reversely into the host cavity. (A) Chick embryo at 10-somite stage. Reverse transplantation was performed in embryos at (B) 6-somite, (C) 15-somite and (D) 10-somite stages, and the embryos were incubated for another two days. When transplantation was done in early embryos, the optic vesicle developed normally, while transplanted later at 15-somite stage, the optic vesicle developed up side down. The arrows indicate the optic fissure, which normally develops at the ventral position. When reversely transplanted at 10-somite stage, the optic cup formation is disrupted and the eye is considerably smaller than the normal eye.

1. Tissue Interaction between the Epidermal Ectoderm and the Optic Vesicle

The ectoderm-neuroepithelium interaction has been studied intensely and the ectoderm is considered to play a crucial role in defining the region with retinal fate through some signaling molecules like FGFs (Fibroblast Growth Factors). FGFs emanate from the overlying surface ectoderm and are considered to induce the adjacent neuroepithelium to assume the neural retinal fate. Using an explant culture of mouse optic vesicles, it was shown that the optic vesicle neuroepithelium is originally bipotential, expressing both NR and RPE marker genes, and that FGF administration leads to a rapid downregulation of Mitf (RPE marker). Conversely, removal of the surface ectoderm results in the maintenance of Mitf expression in the distal optic epithelium and lack of CHX10 expression, suggesting that bipotential optic neuroepithelium is

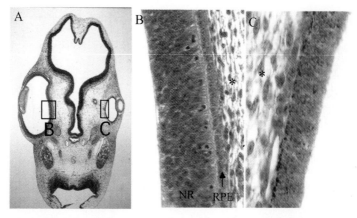

Fig. 4. Transverse view of an operated embryo as shown in Fig. 3D. (A) The transplanted eye is at the right side. Invagination is incomplete and the eye failed to form a bilayered optic cup. The parts indicated by rectangles in A are shown in B and C at a higher magnification. In C, the epithelium consists of a pseudostratified pigmented epithelium. Asterisks in B and C show the mesenchymal tissue that develops to the choroid and sclera.

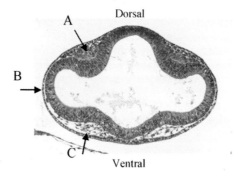

Fig. 5. A transverse section of the proencephalon. The optic vesicle makes contact with different types of tissues at the dorsal (A), distal (B) and ventral (C) regions. These are the dorsal mesenchyme derived from the cephalic neural crest, the surface epidermis and ventral mesenchyme derived from the procaudal mesoderm.

separated into discrete expression domains and, hence, to domain specification (Nguyen and Arnheiter, 2000). Ectopic expression of FGF9 in the proximal region of the optic vesicle, as shown in transgenic mice, results in the extension of neural differentiation into the presumptive RPE domain and in a duplicate neural retina (Zhao et al., 2001). In the developing chick, cells of the optic vesicle display both NR and RPE phenotypes when the surface ectoderm is ablated, indicating that positional cues provided by FGF in the ectoderm organize the bipotential optic vesicle into specific NR and RPE domains (Hyer et al., 1998).

2. Tissue Interaction between the Mesenchyme and the Optic Vesicle

In contrast with ectodermal tissue interaction, interaction between the mesenchyme and neuroepithelium has been sparsely studied (Lopashov, 1963; Johnston et al., 1979), and only recently several studies are elucidating the role of mesenchymal cells in RPE determination using an in vitro organotypic culture method (Fuhrmann et al., 2000; Kagiyama et al., 2005). In the absence of extraocular mesenchymal tissue, RPE development did not occur, as shown by the down regulation of RPE-specific genes such as Mitf. The extraocular mesenchyme also inhibits the expression of NR specific transcription factors like Chx10 and Rax (Bailey et al., 2004). The TGFβ (Transforming Growth Factor β) family member activin can substitute for the extraocular mesenchyme by promoting the expression of RPE-specific genes and, at the same time, downregulating NR-specific genes (Fuhrmann et al., 2000), suggesting that extraocular mesenchyme and possibly an activin-like signal pattern the domains of the optic vesicle into RPE and NR.

Embryonic transplantation of an early eye primordium (optic vesicle) into a wing bud of a more developed embryo results in a well-developed ectopic eye at the wing bud distal tip. The eye was surrounded by numerous mesenchymal cells, many of which were derived from the host tissue, as shown by quail-chick chimeric transplantation (Kagiyama et al., 2005), suggesting that the early optic vesicle has the capability of autonomic development in a foreign milieu and that extraocular mesenchymal cells play an important role in eye development (Fig. 6). This was also confirmed by an in vitro organotypic culture of the optic vesicle, and there is a good correlation between the existence of extraocular mesenchymal cells and RPE development. The extraocular mesenchymal cells under culture conditions

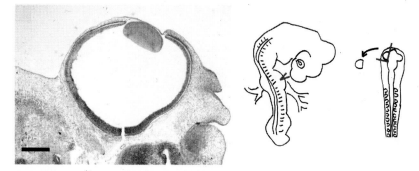

Fig. 6. Embryonic transplantation of the optic vesicle into the distal wing bud. Transplantation was performed as shown in the schematic illustration; optic vesicle at 10-somite stage was inserted into an incision made at the distal part of a wing bud of a host embryo. A well-developed eye is seen embedded in the mesodermal tissue of the wing bud. In this case, quail optic vesicle was transplanted to chick wing bud.

do not proliferate as much as seen in the ocular transplant in the embryonic wing bud, indicating that some additional factors deriving from mesenchymal cells are needed for further development. Using this organotypic culture, it was revealed that the dorsal and ventral halves of optic vesicle are fated to develop into the RPE and NR, respectively, at the early stage of development (Fig. 7). Combined culturing of the optic vesicle either with the dorsal or ventral portion of the head indicates that only the dorsal part-derived factors have an important role in RPE development by upregulating Mitf expression, denoting that the head portion plays an essential role in this fate determination of eye tissues. At the same time, the

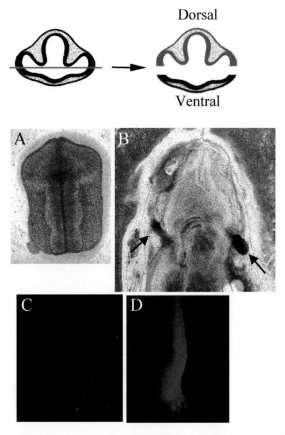

Fig. 7. Organotypic culture of dorsal or ventral half of the optic vesicle. The embryonic brain was cut into the dorsal and ventral halves, which were then laid onto the membrane filter cup and cultured for three days. (A) Dorsal half on the filter cup immediately after transferred on the cup. (B) The same dorsal half cultured for three days. Melanin deposition can be seen at the both sides (arrows), indicating RPE differentiation proceeds. (B, C) are shown at the same magnification. Mitf expression is also found (D), while cRax can be seldom found (C), suggesting that the dorsal region of the optic vesicle is specified to develop into RPE.

dorsal factors suppress Pax2 expression, a gene necessary for the development of the ventral structure of the eye. These studies indicate that factors from the dorsal head portion play important roles in the establishment of dorso-ventral polarity within the optic vesicle, which, in turn, induces, the patterning and differentiation of RPE and NR.

The molecular nature of these substances is still poorly understood, but recent studies have revealed that BMPs, SHH, and FGFs constitute a tripartite signaling center in the developing rostral brain structure. Both SHH and FGF signaling pathways affect retinal development (Russell, 2003). The expression of most BMPs is restricted to the dorsal procencephalon, while SHH expression is largely ventrally restricted (Furuta et al., 1997; Furuta and Hogan, 1998; Crossley, et al., 2001; Trousse et al., 2001). Increased BMP signaling and the loss of SHH expression are correlated with decreased proliferation, increased cell death and hypoplasia of the optic vesicle (Martí and Bovolenta, 2002; Ohkubo et al., 2002). Application of BMP4-containing beads in close proximity to the optic vesicle results in the suppression of NR development, but RPE differentiation proceeds normally, forming a pigmented vesicle (Hyer et al., 2003). These results indicate that BMP4, a dorsally located signal, has an important role in patterning RPE and NR domains in the optic vesicle. It is suggested that cephalic neural crest cells, which migrate towards the proximal region of the optic vesicle affect the development of the optic vesicle (Fuhrmann et al., 2000; Kagiyama et al., 2005). Whether or not migrating neural crest cells influence ocular development by secreting BMP signals should be examined in a further study.

Transdifferentiation of RPE into the Retinal Fate in Developing Chick Embryos: An Avian Model of Retinal Regeneration

The developing chick embryo has been studied as a model for retinal regeneration; by removing neural retina in ovo from early embryos, RPE cells start to proliferate and enter into neural retinal fate (a phenomenon designated as transdifferentiation) (Okada, 1991). All the RPE cells finally transform into a retinal layer with reverse polarity (photoreceptor layer is located at the innermost position facing the vitreous fluid) and they do not transform into RPE again (Coulombre and Coulombre, 1965, 1970). This process of transdifferentiation can be seen only in early embryos (up to about five days of incubation) and requires some neural derivatives like a fragment of retinal tissue to be left in the eye chamber. The molecular nature of substances that activate RPE cells to initiate transdifferentiation were unknown for a long time until Park and Hollenberg (1989, 1991) reported that intraocularly administered FGF-2 actually triggers RPE transdifferentiation. In vitro studies also revealed that FGF-2 causes RPE cells to proliferate and re-differentiate into neurons (Pittack et al., 1997). All the RPE cells differentiated into neural cells in vitro as seen in vivo (Araki et al., 1998), and this contrasts with cultures of newt RPE where some RPE cells remain intact and some transdifferentiate into neural cells. FGF-2-induced RPE transdifferentiation into retinal tissue was also reported in *Xenopus* embryos under culture conditions (Sakaguchi et al., 1997), suggesting that FGF2 is a substance widely functioning in transdifferention. This is not surprising if it is considered that FGFs serve as inductive signals for NR development in the embryonic optic vesicle (Hyer et al., 1998; Zhao et al., 2001; Spence et al., 2004).

RPE transdifferentiation is also reported in mammalian embryos under culture conditions, but this occurs only in the very early stage of development (E13 to E14 at the optic cup stage) and again in the presence of FGFs (Zhao et al., 1995). Whether mammalian RPE retains its transdifferentiation ability in more advanced developmental stages is still not clear, but several in vitro studies have indicated that RPE at later stages quickly becomes unable to fully transdifferentiate (Neill and Barnstable, 1990).

The regulatory mechanism involved in RPE transdifferentiation has been studied intensively at the gene expression level. Mitf, a basic helix-loop-helix-leucine-ziper (bHLHzip) protein, plays a substantial role in the fate determination of RPE cells, as mentioned above. Retrovirus-mediated over expression of Mitf inhibits FGF-2 induced transdifferentiation of cultured chick RPE cells which would otherwise transdifferentiate into both the lens and neural retina, indicating that the downregulation of Mitf expression is essential for RPE transdifferentiation (Mochii et al., 1998a). A link between the intracellular FGF signaling pathway and Mitf has been studied in the developing eye (Nguyen and Arnheiter, 2000; Zhao et al., 2001), and developmental conversion of RPE into neural retina can be mimicked by the activation of downstream intracellular targets of FGF receptors that regulate Mitf expression (Galy et al., 2002).

While tissue interaction between the mesenchyme and optic vesicle plays an important role in newt RPE transdifferentiation, it is unknown whether a similar tissue interaction plays any role in the RPE transdifferentiation of avian embryos. However, it is reported that RPE sheets purely isolated from four-day chick embryos are much less susceptible to FGF-2 under culture conditions than those combined with the underlying periocular connective tissues (Araki et al., 1998), and a study using a unique avian mutant, Silver quail homozygote, shows that such a tissue interaction also has a significant role in RPE transdifferentiation (refer the following section).

An Avian Mutant Model for RPE Transdifferentiation

Another example of a choroid-RPE interaction which plays a role in RPE transdifferentiation comes from an avian mutant, Silver quail mutant (*B/B*) (Homma et al., 1969; Fuji and Wakasugi, 1993).

The Silver quail homozygote mutant has a clear phenotype of plumage color, indicating that melanogenesis is affected in feather melanocytes. Interestingly, the Silver homozygote also has a unique eye defect; during embryonic development, RPE cells autonomously transdifferentiate to neural retina in the homozygous mutant, resulting in a double layer of NR at the posterior region of the eye (one original and the other newly formed retinal layer). This transdifferentiation begins at about four days of incubation and occurs only at the posterior central region of the eye (Fig. 8) (Araki et al., 1998). The newly formed retinal layer has a reverse polarity and the photoreceptor cells are located at the innermost position, as seen in the

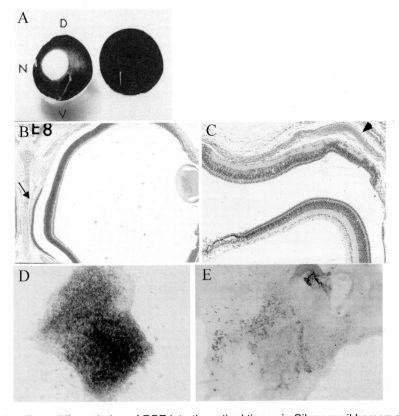

Fig. 8. Transdifferentiation of RPE into the retinal tissue in Silver quail homozygote. (A) Eye cups of day 15 homozygote embryos. At the left of the figure is an eye of a Silver embryo and at the right a wild type embryo. Both are left eyes. A circular unpigmented area is seen in the Silver embryo at the nasal side from the pectene. (B) A micrograph of eight-day embryonic homozygote eye. The region indicated by an arrow shows a thickening of RPE epithelium which has already transdifferentiated to the retinal fate. (C) In 13-day embryo, transdifferentiated neural retina shows well-organized cell layers and a reversed polarity. The original retina begins to degenerate. An arrowhead shows scleral cartilage. (D, E) Culture of RPE sheets from Silver homozygote embryos. Tissues were removed from day 3 embryos and cultured for 9-10 days on filter membranes in the absence (D) or in the presence of FGF2 (E). In (D), the epithelial sheet remains densely pigmented. In the presence of FGF2 (E), RPE cells proliferate, become depigmented, and then transdifferentiate into iodopsin-immunoreactive cone cells.

experimentally transdifferentiated NR. Mitf from Silver homozygote has an amino acid substitution in the basic region and is truncated in the C-terminal region (Mochii et al., 1998b). Since Mitf is expressed in melanocytes, a neural crest-derived cell type, feather color is also affected. Mitf expression is found in several different tissues including RPE, neural crest cells, osteoclasts and mast cells, and the fact that transdifferentiation occurs only in a limited portion of the RPE indicates that some specific tissue interactions between RPE and other neighboring tissues must occur in that area. This is supported by an in vitro experiment where RPE layers from Silver mutant embryos were purely isolated from the choroid, a connective tissue, and cultured alone; RPE cells do not transdifferentiate into neural cells but they actually do so when combined again with the remaining choroids (Araki et al., 1998).

In embryonic transplantation experiments, optic vesicles were truncated from a donor quail mutant and then transplanted into the right place of the chick embryo at the same developmental stage from which the optic vesicle had been removed (Fig. 9) (Araki et al., 2002). This creates a chick embryo

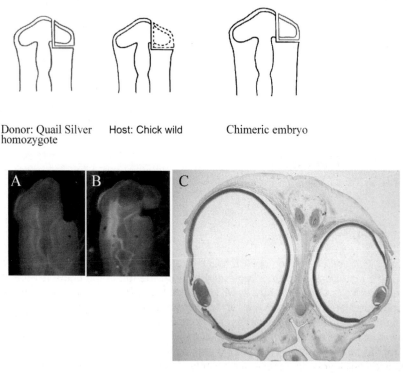

Donor: Quail Silver   Host: Chick wild   Chimeric embryo
homozygote

Fig. 9. A chimeric transplantation of quail and chick embryos. (A, B) Optic vesicles were removed from Silver homozygote embryos (quail) and then transplanted to the cavity of the host chick embryos. (C) A chimeric embryo whose right eye is from quail optic vesicle of Silver homozygote. The size of the eye is smaller than the host chick eye and similar to that of the normal quail eye. Note that there is no indication of transdifferentiation at the posterior region.

with one quail eye whose cornea, lens, neural retina and RPE are all derived from a donor quail optic vesicle. Interestingly, the grafted optic vesicle developed normally without any RPE transdifferentiation. It was found that most of the cells in the choroid (fibroblastic cells and a few melanocytes) were those of the chick, indicating that choroid cells are responsible for the transdifferentiation phenomenon. The choroid develops from neural crest cells migrating towards the optic vesicle and, in the Silver homozygote, neural crest migration was affected (Araki et al., 2002). These results indicate that the choroid cells surrounding the RPE play a principal role in the fate determination of RPE cells and that periocular neural crest cells are affected by mutation and have some defect in conducting the development and maintenance of RPE.

Amphibian Retinal Regeneration

Complete retinal regeneration in adult animals following the surgical removal of whole retinal tissue has been considered to occur only in certain urodele amphibians. This has been best studied in the newt, with over a century since the time when experimental studies of this phenomenon were initiated at the end of the 19th century (Okada, 1991; Mitashov, 1997; Hitchcock et al., 2004; Araki, 2007), although very little information is available on the cellular and molecular mechanisms involved in newt RPE transdifferentiation (Okada, 2004). RPE cells undergo transdifferentiation to produce all cell types constituting the neural retina. The regeneration processes have been fully documented in many studies (Stone, 1950a,b; Hasegawa, 1958; Reyer, 1971, 1977). Recent studies have also suggested that many common features exist between the development and regeneration of the retina at cellular, molecular, and gene expression levels (Kaneko et al., 1999; 2001). A similar retinal regeneration from RPE cells also occurs in some other species including birds and mammals, as already mentioned, but the regenerative capacity is strictly limited to early stages of embryonic development and require exogenously administered factors such as FGFs for the onset of transdifferentiation (Park and Hollenberg, 1989; Zhao et al., 1995; Araki et al., 1998; Del Rio-Tsonis and Tsonis, 2003). No RPE layer develops again, since all the RPE cells transdifferentiate into the retina. Thus, there are many fundamental differences between the newt and other animal species.

Under an organotypic culture condition, newt RPE cells were shown to undergo transdifferentiation into retinal neurons (Ikegami et al., 2002). This work was significant in enabling us to compare RPE cell properties among various animal models under the same in vitro conditions. With this method it was found that anuran (*Xenopus laevis*) RPE cells also transdifferentiate into retinal cells, opening a possibility that the retina can regenerate in the anuran amphibians as it does in the newt (Yoshii et al., 2007). Interestingly, RPE transdifferentiation into neural cells can be seen only when they are cultured with the underlying connective tissue (the choroid) and they never proliferate

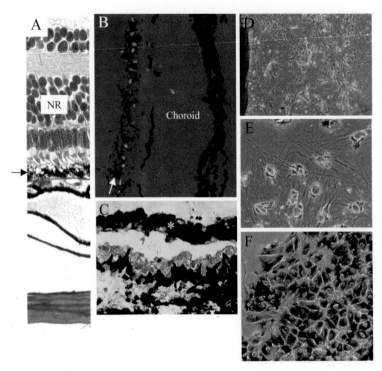

Fig. 10. Newt retinal regeneration in vitro. (A) A micrograph of a normal newt retina and RPE (arrow). Hematoxylin and eosin staining. (B) five days after surgical removal of the retina, RPE cells start proliferation, depigmentation and transdifferentiation. BrdU uptake is first observed at four-five days after retinal removal. Numerous RPE cells are labeled for BrdU (arrow). (C) RPE sheets were isolated from the remaining connective tissues by dispase treatment. Asterisk shows RPE sheet. (D, E, F) Organotypic culture of RPE sheets. In (D), RPE and choroid tissues were co-cultured. RPE cells extend out from the sheet and transdifferentiate into neural cells as shown in (E). When RPE sheets were cultured alone, no neural cells differentiated and the epithelial structure remained intact (F), suggesting that the underlying choroid plays a role in RPE transdifferentiation.

nor differentiate into neural cells when cultured alone (Fig. 10). This indicates that the presence of the choroid is essential for RPE cells to proliferate and differentiate into neural cells, and also that the choroid plays an essential role in newt retinal regeneration (Mitsuda et al., 2005). A pharmacological approach indicates that FGF signaling is most important in RPE cell transdifferentiation (Mitsuda et al., 2005). IGF1 (Insulin-like Growth Factor 1) plays some supplementary role, as has been shown in the case of chick retinal regeneration (Fisher and Reh, 2001); IGF1 alone has no effect on RPE cells, but it has a strong enhancing effect when added with FGF2. RT-PCR of in vivo regenerating newt eyes shows that gene expressions of both FGF2 and IGF1 are upregulated in the choroid after retinal removal (Mitsuda et al., 2005).

Whether or not in vitro cultured RPE cells have the potency to reconstitute the retinal structure has not yet been shown, but the intraocular transplantation of cultured RPE tissues suggest that they have that potency; when RPE cells cultured with the choroid were transplanted into the posterior eye chamber, they regenerated the neural retinal layer in a manner similarly to the normal regeneration process. When cultured RPE sheets alone were transplanted in the same way, they did not change but still remained pigmented. These results indicate that cultured RPE cells retain the capability to develop the retinal structure. Further studies are still necessary to reveal which substances play roles in retinal regeneration which include RPE transdifferentiation and the formation of the multi-layered structure.

Although the most important question has not yet been answered (why the removal of the retina from the newt eye triggers RPE cells to initiate transdifferentiation), the above study revealed for the first time that the tissue interaction between the RPE and the choroid plays an essential role in newt RPE transdifferentiation. There are several possibilities to address the question: one preferable possibility is that persistence of the intact retina-RPE tissue interaction may inhibit RPE cells from responding to the choroid-deriving factors by suppressing the FGF signal pathway within RPE cells; alternatively, the removal of the retina may cause the choroid cells to initiate FGF gene expression. Although there is no information to address these, the latter possibility is not involved in the present case, because the retina is a rich source of FGFs, and there must be some repressive mechanism in RPE cells not to respond to FGFs produced by the retina. It is interesting to note that there is a five-day latent period until RPE cells initiate DNA synthesis (BrdU uptake) during both in vivo regeneration and in vitro trans-differentiation (Ikegami et al., 2002). Cultured RPE cells also appear not to respond to FGF2 by day five of culturing (Mitsuda et al., 2005). These data suggest that RPE cells need some extra time until they become ready to start regeneration after the onset of necessary gene expression. It is of interest to identify which step of the FGF signaling pathway is the crucial one and how signal transduction is suppressed in the normal condition. These questions are crucial in order to understand how retinal regeneration is controlled.

CONCLUSION

We can find many common features between retinal development and regeneration at the cellular, molecular, and gene expression levels. Although only a few studies have been undertaken till date, it has been proved that there are some common features in tissue interactions that control both retinal development and regeneration. Accumulating evidence, including both in vitro cultures and gene expression patterns, now makes it possible to compare various aspects of retinal development with those of regeneration. It is also becoming increasingly important to compare those aspects among a variety of animal models. This will make it possible to address the important question of what mechanisms are involved in those vertebrates, including humans, that does not enable regeneration of retina.

ACKNOWLEDGEMENTS

The author thanks all of his colleagues in the Developmental Neurobiology Laboratory in Nara Women's University for their contributions to this work, particularly those by Ms. Uemonsa, Ikegami-Ueda, Mitsuda, Kagiyama and Hirashima are greatfully acknowledged. A part of this work was funded by a Grant-in-Aid from the Ministry of Education, Culture, Sports, Science and Technology of Japan.

REFERENCES

Adler, R. (2000). A model of retinal cell differentiation in the chick embryo. *Prog. Retina Eye Res.* **19**: 529-557.

Araki, M. (2007). Regeneration of the amphibian retina: Role of tissue interaction and related signaling molecules on RPE transdifferentiation. *Dev. Growth Differ.* **49**: 109-120.

Araki, M., Yamao, M. and Tudzuki, M. (1998). Early embryonic interaction of retinal pigment epithelium and mesenchymal tissue induces conversion of pigment epithelium to neural retinal fate in the *Silver* mutation of the Japanese quail. *Develop. Growth Differ.* **40**: 167-176.

Araki, M., Takano, T., Uemonsa, T., Nakane, Y., Tsudzuki, M. and Kaneko, T. (2002). Epithelia-mesenchyme interaction plays an essential role in transdifferentiation of retinal pigment epithelium of silver mutant quail: Localization of FGF and related molecules and aberrant migration pattern of neural crest cells during eye rudiment formation. *Dev. Biol.* **244**: 358-371.

Bailey, T.J., El-Hodiri, H., Zhang, L., Shah, R., Mathers, P.H. and Jamrich, M. (2004). Regulation of vertebrate eye development by Rx genes. *Int. J. Dev. Biol.* **48**: 761-770.

Baümer, N., Marquardt, T., Stoykova, A., Spieler, D., Treichel, D., Ashery-Padan, R. and Gruss, P. (2003). Retinal pigmented epithelium determination requires the redundant activities of Pax2 and Pax6. *Develop.* **130**: 2903-2915.

Bovolenta, P., Mallamaci, A., Briata, P., Corte, G. and Boncinelli, E. (1997). Implication of OTX2 in pigment epithelium determination and neural retina differentiation. *J. Neurosci.* **17**: 424-425.

Chow, R.L. and Lang, R.A. (2001). Early eye development in vertebrates. *Annu. Rev. Cell Dev. Biol.* **17**: 255-296.

Coulombre, J.L. and Coulombre, A.J. (1965). Regeneration of neural retina from the pigmented epithelium in the chick embryo. *Dev. Biol.* **12**: 79-92.

Coulombre, J.L. and Coulombre, A.J. (1970). Influence of mouse neural retina on regeneration of chick neural retina from chick embryonic pigmented epithelium. *Nature* **228**: 559-560.

Crossley, P.H., Martinez, S., Ohkubo, Y. and Rubenstein, J.L. (2001). Coordinate expression of Fgf8, Otx2, Bmp4, and Shh in the rostral prosencephalon during development of the telencephalic and optic vesicles. *Neurosci.* **108**: 183-206.

Del Rio-Tsonis, K. and Tsonis, P.A. (2003). Eye regeneration at the molecular age. *Dev. Dyn.* **226**: 211-224.

Dragomirov, N.I. (1937). The Influence of the neighboring ectoderm on the organization of the eye rudiment. *Dokl. Akad. Nauk.* **15**: 61-64.

Dütting, D. and Thanos, S. (1995). Early determination of nasal-temporal retinotopic specificity in the eye anlage of the chick embryo. *Dev. Biol.* **167**: 263-281.

Fisher, A. and Reh, T.A. (2001). Transdifferentiation of pigmented epithelial cells: a source of retinal stem cells? *Dev. Neurosci.* **23**: 268-276.

Fuhrmann, S., Levine, E.M. and Reh, T.A. (2000). Extraocular mesenchyme patterns the optic vesicle during early eye development in the embyonic chick. *Develop.* **127**: 4599-4609.

Fuji, J. and Wakasugi, N. (1993). Transdifferentiation from retinal pigment epithelium into neural retina due to *Silver* plumage mutant gene in Japanese quail. *Dev. Growth Differ.* **35**: 487-494.

Furuta, Y., Piston, D.W. and Hogan, B.L.M. (1997). Bone morphogenetic proteins (BMPs) as regulators of dorsal forebrain development. *Develop.* **124**: 2203-2212.

Furuta, Y. and Hogan, B. (1998). BMP4 is essential for lens induction in the mouse embryo. *Genes Dev.* **12**: 3764-3775.

Galy, A., Neron, B., Planque, N., Saule, S. and Eychene, A. (2002). Activated MAPK/ERK kinase (MEK-1) induces transdifferentiation of pigmented epithelium into neural retina. *Dev Biol.* **248**: 251-264.

Hasegawa, M. (1958). Restitution of the eye after removal of the retina and lens in the newt, *Triturus pyrrhogaster*. *Embryologia* **4**: 1-32.

Hatakeyama, J. and Kageyama, R. (2004). Retinal cell fate determination and bHLH factors. *Semin. Cell Dev. Biol.* **15**: 83-89.

Hitchcock, P., Ochocinska, M., Sieh, A. and Otteson, D. (2004). Persistent and injury-induced neurogenesis in the vertebrate retina. *Prog. Retina Eye Res.* **23**: 183-194.

Homma, K., Jinno, M. and Kito, J. (1969). Studies on *Silver*-feathered Japanese quail. *Jpn. J. Zootech. Sci.* **40**: 129-130.

Hyer, J., Mima, T. and Mikawa, T. (1998). FGF1 patterns the optic vesicle by directing the placement of the neural retina domain. *Develop.* **125**: 869-877.

Hyer, J., Kuhlman, J., Afif, E. and Mikawa, T. (2003). Optic cup morphogenesis requires pre-lens ectoderm but not lens differentiation. *Dev. Biol.* **259**: 351-363.

Ikegami, Y., Mitsuda, S. and Araki, M. (2002). Neural cell differentiation from retinal pigment epithelial cells of the newt: An organ culture model for the urodele retinal regeneration. *J. Neurobiol.* **50**: 209-220.

Jean, D., Ewan, K. and Gruss, P. (1998). Molecular regulators involved in vertebrate eye development. *Mech. Dev.* **76**: 3-18.

Johnston, M.C., Noden, D.M., Hazelton, R.D., Coulombre, J.L. and Coulombre, A.J. (1979). Origins of avian ocular and periocular tissues. *Exp. Eye Res.* **29**: 27-43.

Kagiyama, Y., Gotouda, N., Sakagami, K., Yasuda, K. and Araki, M. (2005). Extraocular dorsal signal affects the developmental fate of the optic vesicle and patterns the optic neuroepithelium. *Dev. Growth Differ.* (in press).

Kaneko, Y., Matsumoto, G. and Hanyu, Y. (1999). Pax-6 expression during retinal regeneration in the adult newt. *Dev. Growth Differ.* **41**: 723-729.

Kaneko, Y., Hirota, K., Matsumoto, G. and Hanyu, Y. (2001). Expression pattern of a newt Notch homologue in regenerating newt retina. *Dev. Brain Res.* **128**: 53-62.

Kondoh, H. (2002). Development of the eye. In: *Mouse Development* (Rossant, J. and Tam, P.P.L., eds.), Academic Press, San Diego., 519-538.

Koshiba-Takeuchi, K., Takeuchi, J.K., Matsumoto, K., Momose, T., Uno, K., Hoepker, V., Ogura, K., Takahashi, N., Nakamura, H., Yasuda, K. and Ogura, T. (2000). Tbx5 and the retinotectum projection. *Science* **287**: 134-137.

Lopashov, G.V. (1963). *Developmental Mechanisms of Vertebrate Eye Rudiments*. McMillan (Pergamon), New York.

Martí, E. and Bovolenta, P. (2002). Sonic hedgehog in CNS development: one signal, multiple outputs. *Trends Neurosci.* **25**: 89-96.

Macdonald, R.K., Barth, A., Xu, Q., Holder, N., Mikkola, I. and Wilson, S.W. (1995). Midline signaling is required for Pax gene regulation and patterning of the eyes. *Develop.* **121**: 3267-3278.

Mitashov, V.I. (1997). Retinal regeneration in amphibians. *Int. J. Dev. Biol.* **41**: 893-905.

Mitsuda, S., Yoshii, C., Ikegami, Y. and Araki, M. (2005). Tissue interaction between the retinal pigment epithelium and the choroid triggers retinal regeneration of the newt *Cynopus pyrrhogaster*. *Dev. Biol.* **280**: 122-132.

Mochii, M., Mazaki, Y., Mizuno, N., Hayashi, H. and Eguchi, G. (1998a). Role of Mitf in differentiation and transdifferentiation of chicken pigmented epithelial cell. *Dev. Biol.* **193**: 47-62.

Mochii, M., Ono, T., Matsubara, Y. and Eguchi, G. (1998b). Spontaneous transdifferentiation of quail pigmented epithelial cell is accompanied by a mutation in the *mitf* gene. *Dev. Biol.* **196**: 145-159.

Moshiri, A., Close, J. and Reh, T.A. (2004) Retinal stem cells and regeneration. *Int. J. Dev. Biol.* **48**: 1003-1014.

Neill, J.M. and Barnstable, C.J. (1990). Expression of the cell surface antigens RET-PE2 and N-CAM by rat retinal pigment epithelial cells during development and in tissue culture. *Exp. Eye Res.* **51**: 573-583.

Nguyen, M.-T.T. and Arnheiter, H. (2000). Signaling and transcriptional regulation in early mammalian eye development: a link between FGF and MITF. *Develop.* **127**: 3581-3591.

Ohkubo, Y., Chiangb, C. and Rubenstein, J.L.E. (2002). Coordinate regulation and synergistic actions of BMP4, SHH and FGF8 in the rostral prosencephalon regulate morphogenesis of the telencephalic and optic vesicles. *Neurosci.* **111**: 1-17.

Okada, T.S. (1991). *Transdifferentiation*. Oxford University Press, New York.

Okada, T.S. (2004). From embryonic induction to cell lineages: revisiting old problems from modern study. *Int. J. Dev. Biol.* **48**: 739-742.

Oliver, G. and Gruss, P. (1997). Current views on eye development. *Trends Neurosci.* **20**: 415-421.

Park, C.M. and Hollenberg, M.J. (1989). Basic fibroblast growth factor induces retinal regeneration in vivo. *Dev. Biol.* **134**: 201-205.

Park, C.M. and Hollenberg, M.J. (1991). Induction of retinal regeneration in vivo by growth factors. *Dev. Biol.* **148**: 322-333.

Pittack, C.B., Grunwald, G.B. and Reh, T.A. (1997). Fibroblast growth factors are necessary for neural retina but not pigmented epithelium differentiation in chick embryo. *Develop.* **124**: 805-816.

Reyer, R.W. (1971). DNA synthesis and the incorporation of labeled iris cells into the lens during lens regeneration in adult newts. *Dev. Biol.* **24**: 533-558.

Reyer, R.W. (1977). The amphibian eye: development and regeneration. In: *Hand-book of Sensory Physiology*, (Crescitelli, F., ed.), Springer-Verlag, Berlin, New York, 309-390.

Russell, C. (2003). The roles of hedgehogs and fibroblast growth factors in eye development and retinal cell rescue. *Vis. Res.* **43**: 899-912.

Saha, M.S., Servetnick, M. and Grainger, R.M. (1992). Vertebrate eye development. *Curr. Opin. Genet. Dev.* **2**: 582-588.

Sakaguchi, D.S., Janick, L.M. and Reh, T.A. (1997). Basic fibroblast growth factor (FGF-2) induced transdifferentiation of retinal pigment epithelium: Generation of neurons and glia. *Dev. Dyn.* **209:** 387-398.

Schwarz, M., Cecconi, F., Bernier, G., Andrejewski, N., Kammandel, B., Wagner, M. and Gruss, P. (2000). Spatial specification of mammalian eye territories by reciprocal transcriptional repression of Pax2 and Pax6. *Develop.* **127:** 4325-4334.

Schulte, D., Furukawa, T., Peters, M.A., Kozak, C.A. and Cepko, C. (1999). Misexpression of the Emx-related homeobox genes cVax and mVax2 ventralizes the retina and perturbs the retinotectal map. *Neuron* **24:** 541-553.

Spence, J.R., Madhavan, M., Ewing, J.D., Jones, D.K., Lehman, B.M. and Del Rio-Tsonis, K. (2004). The hedgehog pathway is a modulator of retina regeneration. *Develop.* **131:** 4607-4621.

Stone, L.S. (1950a). Neural retina degeneration followed by regeneration from surviving retinal pigment cells in grafted adult salamander eyes. *Anat. Rec.* **106:** 89-109.

Stone, L.S. (1950b). The role of retinal pigment cells in regenerating neural retinae of adult salamander eyes. *J. Exp. Zool.* **113:** 9-32.

Trousse, F., Esteve, P. and Bovolenta, P. (2001). BMP4 mediates apoptotic cell death in the developing chick eye. *J. Neurosci.* **21:** 1292-1301.

Uemonsa, T., Sakagami, K., Yasuda, K. and Araki, M. (2002). Development of dorsal-ventral polarity in the optic vesicle and its presumptive role in eye morphogenesis as shown by embryonic transplantation and in ovo explant culturing. *Dev. Biol.* **248:** 319-330.

Yoshii, C., Ueda, Y., Okamoto, M. and Araki, M. (2007). Neural retinal regeneration in the anuran amphibian *Xenopus laevis* post-metamorphosis: Transdifferentiation of retinal pigmented epithelium regenerates the neural retina. *Dev. Biol.* **303:** 45-56.

Zhang, X.M. and Yang, X.J. (2001). Temporal and spatial effects of Sonic hedgehog signaling in chick eye morphogenesis. *Dev. Biol.* **233:** 271-290.

Zhao, S., Thornquist, S.C. and Barnstable, C.J. (1995). In vitro transdifferentiation of embryonic rat retinal pigment epithelium to neural retina. *Brain Res.* **677:** 300-310.

Zhao, S., Hung, F.C., Colvin, J.S., White, A., Dai, W., Lovicu, F.J., Ornitz, D.M. and Overbeek, P.A. (2001). Patterning the optic neuroepithelium by FGF signaling and Ras activation. *Develop.* **128:** 5051-5060.

Endocrine Physiology

VI

Endocrine Physiology

CHAPTER 1

Comparative Aspects of the Mammalian Pineal Gland Ultrastructure

Michal karasek and Anna Zielinska

Abstract

The number of papers devoted to pineal morphology, and especially its ultrastructure, is very high. The ultrastructure of the pinealocyte has been examined in various species in many natural and experimental conditions. The aim of this survey is to present briefly the basic data on the ultrastructure of the mammalian pineal, with emphasis on its comparative aspects, based mainly on the authors' long-term experience in investigation of the pineal gland of many mammalian species.

Ultrastructure of pinealocytes shows many common features in majority of the mammalian species, and existing differences are of quantitative nature only. The observed qualitative differences include MBB (membrane-bound bodies) in the pig, inclusion bodies in the cotton rat, abundance of pigment granules in the horse, and calcareous concretions in the monkey and gerbil.

Numerous ultrastructural studies performed in recent years have brought to light much information, which elucidates the correlation of pineal morphology with pineal function. The ultrastructural features of pinealocytes suggest high activity of these cells. Pinealocytes contain numerous organelles, indicating their high metabolic activity (as mitochondria), as well as indicating high secretory activity (as Golgi apparatus).

INTRODUCTION

The discoverer of the pineal gland is believed to be a famous Alexandrian anatomist, Herophilus (325-280 BC) who, localizing the soul in the brain ventricular system, thought that the pineal is a valve regulating the flow of 'pneuma' ('spiritus') from the 3rd to 4th ventricle. Although none of his

Department of Neuroendocrinology, Chair of Endocrinology Medical University of Lodz, 92-216 Lodz, Czechoslowacka 8/10, Poland.
Tel.:/Fax: +48 42 675 7613. E-mail: karasek@csk.umed.lodz.pl

writings survived, they were extensively cited by a Greek doctor Galen (130-200 AD), to whom the authors owe the first written description of the pineal gland. Galen, who named the organ 'konareion' or 'soma konoeides' after its shape, similar to that of a pinecone, thought that pineal is a gland supporting the venous network in the brain. The first pictorial representation of the human pineal organ was provided by Andreas Vesalius Bruxellensis (1514-1564) in his anatomy textbook *De Humani Corpora Fabrica, Libri Septem*, published in Basel in 1543, which constituted a milestone in the history of medicine. Probably the most known ancient theory about the role of the pineal is that of Descartes (1596-1650), the French mathematician and philosopher, who regarded the pineal as the only unpaired structure in the brain, as the "soul's right hand" (and not "the seat of the soul" as is commonly stated). According to Descartes, the pineal functions as a valve controlling flow of 'animal spirits' to and from the brain (Kappers, 1979; Zrener, 1985; Arendt, 1995).

For many forthcoming years, there was a common belief that pineal gland is a functionless, rudimentary organ. However, investigations carried out at the end of the XIX and beginning of XX century brought about some progress in the knowledge of this organ, especially in morphological, ontogenetic and phylogenetic aspects. The year 1958, in which Lerner and coworkers (Lerner et al., 1958) isolated an active compound of this organ – melatonin, unlocks a new epoch of investigations of the pineal gland. Since then, many researchers have become interested in this small, mysterious gland, and in consequence, the pineal gland was studied from many aspects: morphological, biochemical, physiological, pharmacological, and also clinical. Results of these investigations clearly indicated that the pineal gland is not a rudimentary organ, but should be considered as an integral part of the endocrine system.

The number of papers devoted to pineal morphology, and especially its ultrastructure, is very high. The ultrastructure of the pinealocyte has been examined in various species in many natural and experimental conditions (Pevet, 1979, 1981; Vollrath, 1981; Karasek, 1983, 1986, 1992; Bhatnagar, 1992; Karasek and Reiter, 1992, 1996). The objective of this survey is to present briefly the basic data on the ultrastructure of the mammalian pineal, with emphasis on its comparative aspects.

ANATOMY AND HISTOLOGY OF THE PINEAL GLAND

The pineal gland is present in almost all mammalian species. Dissimilarities among species in its shape and location in relation to the third ventricle of the brain help in distinguishing three basic types of the pineal gland (Vollrath, 1979, 1981).

In the first type – A (proximal), the cone-shaped pineal gland lies in the immediate vicinity of the third ventricle.

In the second type – AB (proximo-intermediate) the shape of the pineal gland is elongated, and the length is twice as much or more than its greatest width.

In the third type – ABC (proximo-intermedio-distal), the pineal is very long, more or less rod-like, and substantial amount of pineal tissue lies either superficially or closely related to the cerebellum.

There are also intermediate types of pineal gland in the species in which a certain part of the gland become considerably reduced (in this case capital letters are replaced by small letters of the Greek alphabet) or even lacking (in this case the corresponding letter is omitted). For example, the pineal gland of man is classified as type A, of pig as type AB, of guinea pig as type ABC, of rhesus monkey as type Aβ or Aβγ, of rat as αβC, of gerbil as AβC, of Syrian hamster as αC (Fig. 1).

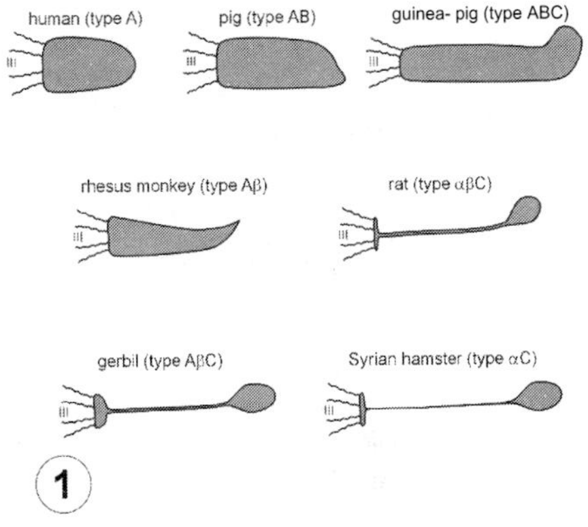

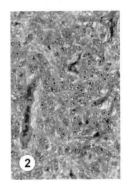

The mammalian pineal gland is a parenchymal organ, surrounded by a connective tissue capsule. In some species (especially in the human) varying amounts of connective tissue penetrate the parenchyma, forming lobules or cell cords. Pinealocyte constitutes the major cellular component of the mammalian pineal gland, amounting to about 90% of the total cells (Fig. 2). In addition to pinealocytes, glial cells are present in varying numbers, and in some species, fibroblasts, mast cells, plasma cells, pigment-containing cells, and nerve cells are occasionally present. The organ is richly vascularized and innervated (Vollrath, 1981; Karasek, 1983).

Fig. 1. Examples of various types (according to Vollrath's classification, 1979) of the pineal gland.

ULTRASTRUCTURE

The description of the pineal gland ultrastructure presented herein is based on the authors' papers (Karasek and Wyrzykowski, 1980; Karasek, 1981, 1983, 1986, 1992, 1997; Karasek and Reiter, 1992, 1996; Karasek and Zielinska, 2000) as well as on publications of other authors (Anderson, 1965; Wolfe, 1965; Welsh and Reiter, 1978; Pevet, 1979, 1981; Vollrath, 1981; Bhatnagar, 1992). All micrographs are the authors', collected over several years of experience investigating the pineal gland of various mammalian species.

Fig. 2. The semi-thin section of the European Bison pineal gland. Toluidine blue staining, × 400.

Pinealocytes

Ultrastructural features of pinealocytes (Figs. 3 and 4) are basically very similar in all mammalian species. These cells are characterized by an irregular shape and the presence of a varying number of cytoplasmic processes emerging from cell bodies. The endings of these processes are generally club-shaped (Fig. 4).

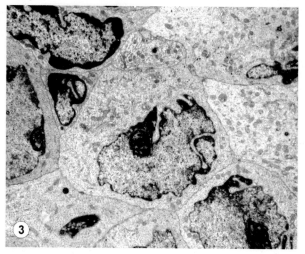

Fig. 3. Low power micrograph of the pinealocyte. Kangaroo rat; × 6000.

The nucleus is usually oval or irregular in shape, with cytoplasmic invagination, which vary in number and depth among pinealocytes of the same species as well as among species. The nucleoplasm, in most species, is characterized by low electron density, chromatin is usually dispersed, although in some species some amount of condensed chromatin is located in an uneven marginal zone of the nucleus, closely related to the nuclear membrane. The nucleolus is usually prominent (Fig. 3).

Mammalian pinealocytes are similar in size (Table 1). In the studies conducted (Karasek and Zielinska, 2000) on eight mammalian species, cross-sectional areas of pinealocytes varied from 44.55 μm^2 in the European Bison to 66.68 μm^2 in the mouse, and those of nucleus were between 12.10 μm^2 (in the Syrian hamster) and 29.83 μm^2 (in the horse).

Cytoplasm of pinealocytes contains typical cellular organelles. There are some quantitative differences in the organelles in various mammalian species (Table 2).

The pinealocytes are characterized by an abundance of mitochondria, which are the most numerous organelles. Mitochondria are usually round, oval or elongated in shape and the size 0.5-1.5 μm^2. A matrix of majority of mitochondria is of moderate electron density with a relatively large number of cristae (Fig. 5).

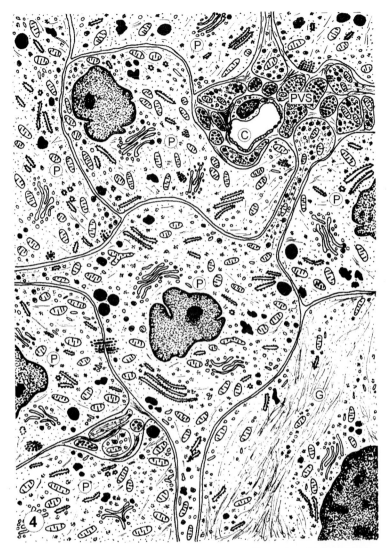

Fig. 4. Schematic representation of the mammalian pineal ultrastructure. P – pinealocyte; G – glial cell; C – capillary; PVS – perivascular space containing numerous adrenergic nerve terminals and endings of pinealocyte processes.

The Golgi apparatus is usually well developed and consists of a system of sacs which are associated with numerous vesicles of varying diameter. Most of the vesicles are clear, although dense-core vesicles, are also present (Fig. 6). Dense-core vesicles produced within the Golgi apparatus (Fig. 7), are observed throughout the perikaryon (Fig. 8), being especially abundant in the endings of pinealocyte processes (Figs. 9 and 10). The number of dense-core vesicles varies among species (Table 3). They are especially abundant in the

Table 1. Cross-sectional areas of pinealocytes and their nuclei in various mammalian species

Species	Pinealocyte [μm^2]	Nucleus [μm^2]	Nucleus [%]
Mouse	67.9[2]; 100.0[4]	15.9[2]; 31.6[4]	23.4[2]; 31.6[4]
Rat	41.4[1]; 42.4[2]; 105.0[6]; 40.5[7]	24.3[1]; 12.5[2]; 18.2[6]; 11.2[7]	58.7[1]; 29.5[2]; 17.3[6]; 27.5[7]; 22.3[9]
Syrian hamster	56.6[2]; 54.5[10]	12.1[2]; 11.9[10]	21.4[2]; 21.8[10]
Gerbil	59.4[2]	24.4[2]	41.1[2]; 28.2[11]
Pig	53.8[2]	22.7[2]	42.2[2]
Sheep	43.9[2]	19.0[2]	43.3[2]
Horse	47.0[2]	29.8[2]	63.4[2]
European bison	40.6[2]	21.7[2]	53.6[2]
Cotton rat	87.2[5]	25.9[5]	29.7[5]
Djungarian hamster	-	-	23.9[3]
White-footed mouse	65.4[8]	25.0[8]	38.2[8]

According to: [1]Trouillas et al., 1984; [2]Karasek and Zielinska, 2000; [3]Karasek et al., 1982a; [4]Karasek et al., 1983a; [5]Karasek et al., 1983c; [6]Karasek et al., 1988b; [7]Karasek et al., 1990; [8]King et al., 1982; [9]Krstic, 1977; [10]Swietoslawski and Karasek, 1993; [11]Welsh et al., 1979.

Table 2. Comparison of relative volumes of various organelles (mitochondria – MIT, Golgi apparatus – GA, granular endoplasmic reticulum – GER, lysosomes – LYS, and lipid droplets – LD) in pinealocytes of various species

Species	MIT [%]	GA [%]	GER [%]	LYS [%]	LD [%]
Mouse	7.7[1]; 7.9[3]	1.2[1]; 1.6[3]; 0.3[13]	1.2[1]; 0.6[3]	0.4[1]; 0.5[3]; 0.3[13]	0.4[1]; 0.6[3]; 1.3[13]
Rat	9.8[1]; 9.0[5]; 11.0[6]; 6.1[8]	1.9[1]; 0.6[5]; 0.7[6]; 1.5[8]	1.5[1]; 3.0[5]; 0.9[6]; 5.4[8]	0.8[1]; 0.3[5]; 1.3[6]	1.6[1]; 1.5[5]; 1.2[8]
Syrian hamster	11.0[1]	3.8[1]	1.6[1]	1.4[1]	0.8[1]
Gerbil	18.3[1]; 9.1[14]	11.9[1]; 0.6[14]	1.2[1]; 4.3[14]	0.9[1]; 0.5[14]	0[1]
Pig	9.4[1]; 9.5[10]; 12.0[11]; 10.0[12]	5.7[1]; 1.0[10]; 2.0[11]; 3.0[12]	1.5[1]; 2.0[10]; 3.2[11]; 2.2[12]	0.5[1]; 1.5[10]; 0.4[11]; 0.5[12]	0[1]
Sheep	12.4[1]; 11.5[9]	10.5[1]; 2.0[9]	0.9[1]; 4.0[9]	0.7[1]; 0.6[9]	2.9[1]; 0.9[9]
Horse	17.6[1]	8.8[1]	1.4[1]	0.6[1]	2.4[1]
European bison	23.7[1]	7.0[1]	1.5[1]	0.5[1]	1.3[1]
Cotton rat	12.0[4]	3.4[4]	0.9[4]	0.2[4]	-
Chipmunk	-	1.7[2]	1.7[2]	0.2[2]	-
White-footed mouse	6.1[7]	1.7[7]	1.2[7]	0.5[7]	-

According to: [1]Karasek and Zielinska, 2000; [2]Karasek et al., 1982b; [3]Karasek et al., 1983a; [4]Karasek et al., 1983c; [5]Karasek et al., 1984; [6]Karasek et al., 1990; [7]King et al., 1982; [8]Krstic, 1977; [9]Lewczuk et al., 1993; [10]Przybylska et al., 1990; [11]Przybylska et al., 1994a; [12]Przybylska et al., 1994b; [13]Redins et al., 2001; [14]Welsh et al., 1979.

pinealocytes of the mouse, the Syrian hamster, and the white-footed mouse (above 5/50 µm²), and less numerous in other mammalian species (from 0.5/50 µm² to 3/50 µm²).

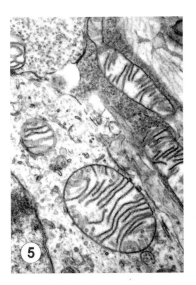

Fig. 5. Mitochondria in the pinealocyte cytoplasm. Cat; × 16000.

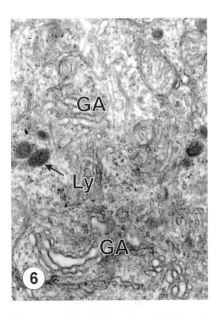

Fig. 6. Golgi apparatus (GA) and lysosomes (Ly) in the pinealocyte cytoplasm. Rat; × 25000.

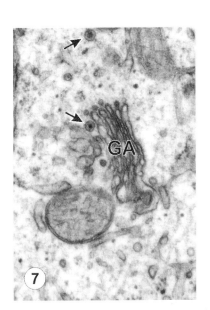

Fig. 7. Golgi apparatus (GA) associated with dense-core vesicles (arrows) in the pinealocyte cytoplasm. Mouse; × 42000.

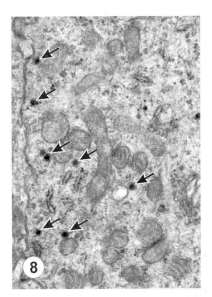

Fig. 8. Numerous dense-core vesicles (arrows) in the pinealocyte cytoplasm. Kangaroo rat; × 23000.

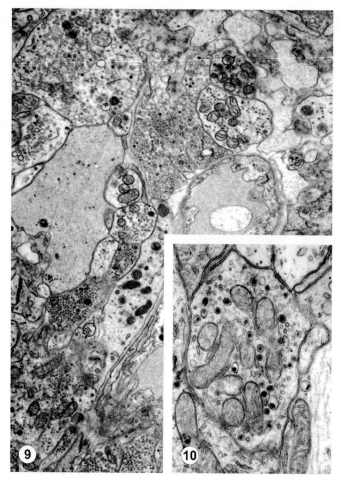

Fig. 9. Numerous endings of pinealocyte processes in the perivascular space. Mouse; × 15000.

Fig. 10. Numerous dense-core vesicles in the ending of pinealocyte process. Mouse; × 34000.

A consistent component of the mammalian pinealocyte is a moderately developed granular endoplasmic reticulum. In the majority of the pinealocytes, isolated cisternae of granular endoplasmic reticulum are typically present, although in some pinealocytes complexes of parallel-oriented cisternae can be found (Fig. 11). Smooth endoplasmic reticulum is present in mammalian pinealocytes in varying amounts.

Lysosomes, round or oval in shape and measuring 0.2-1.5 µm in diameter are observed in many pinealocytes (Fig. 6), although their number shows great variation among species.

Characteristic structures for pinealocytes are "synaptic" ribbons. These structures consist of a few electron-dense rods, measuring 20-60 nm in width

Table 3. "Synaptic" ribbons (SR) in pinealocytes of various species

Species	SR (n/20.000 μm²)
Mouse	0^4; 0^9
Rat	12.3^2; 22.1^4; 15.8^5; 11.2-30.8^6; 17.0^7; 15.8-42.4^8; 10.4^{11}
Syrian hamster	44.0^1; 33.1^4
Gerbil	11.5^4
Pig	1.2^4; 0.7^{10}
Sheep	2.9^4; 0^{10}
Horse	5.4^3; 6.5^4
European bison	6.1^4
Cow	0^{10}
Chipmunk	251.0^5
Cotton rat	9.8^5
Fox	10.0^5
Djungarian hamster	46.0^5
Ground squirrel	136.0^5

According to: [1]Hewing, 1980; [2]Karasek, 1976; [3]Karasek and Cozzi, 1990; [4]Karasek and Zielinska, 2000; [5]Karasek et al., 1983b; [6]Karasek et al., 1988a,b; [7]Kurumado and Mori, 1977; [8]Saidapur et al., 1991; [9]Satoh and Vollrath, 1988; [10]Struwe and Vollrath, 1990; [11]Vollrath and Welker, 1984.

and 0.2-2.0 μm in length. The rod is surrounded by a single layer of electron-lucent vesicles measuring 30-50 nm in diameter (Fig. 12). The number of these structures considerably varies among mammalian species. In some mammals (e.g., mouse and cow) "synaptic" ribbons are absent, whereas in others (e.g., ground squirrel and chipmunk) they are especially abundant, reaching up to 251/ 20000 μm² (Table 4). Apart from "synaptic" ribbons in the pinealocytes, "synaptic" spherules are also present, consisting of a dark core (the diameter of which ranges is from 90 nm to 180 nm) surrounded by a layer of electron-lucent vesicles measuring about 30 nm in diameter. The number of "synaptic" spherules is significantly lower than that of "synaptic" ribbons.

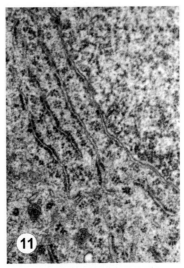

Fig. 11. Parallel-oriented cisternae of granular endoplasmic reticulum in the pinealocyte cytoplasm. Rat; × 28000.

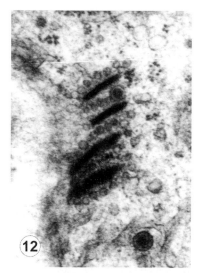

Fig. 12. "Synaptic" ribbons in the pinealocyte cytoplasm. Ground squirrel; × 65000.

Table 4. Dense-core vesicles (DCV) in various mammalian species

Species	DCV (n/50 µm^2)
Mouse	6.4[2]; 9.9[3]; 4.5[4]
Rat	3.0[2]; 1.0[3]; 3.0[6]; 2.5[7], 2.8[8]
Syrian hamster	6.3[1]; 5.0[2]; 5.2[13]
Gerbil	1.2[2]
Pig	1.0[2]; 2.5[10]; 1.2[11]; 1.5[12]
Sheep	1.1[2]; 1.5[9]
Horse	1.0[2]
European bison	1.1[2]
Cat	0.8[3]
Fox	0.5[3]
Cotton rat	1.7[3]; 6.4[5]
Ground squirrel	0.9[3]
Djungarian hamster	2.2[3]
White-footed mouse	5.1[3]
Chipmunk	1.1[3]

According to: [1]Dombrowski and McNulty, 1984; [2]Karasek and Zielinska, 2000; [3]Karasek et al., 1982a; [4]Karasek et al., 1983a; [5]Karasek et al., 1983c; [6]Karasek et al., 1984; [7]Karasek et al., 1990; [8]Krstic, 1977; [9]Lewczuk et al., 1993; [10]Przybylska et al., 1990; [11]Przybylska et al., 1994a; [12]Przybylska et al., 1994b; [13]Swietoslawski and Karasek, 1993.

Lipid droplets are one of the consistent components of the pinealocytes (Fig. 13), although their number shows great variations among species. Glycogen particles, which are abundant in some species (Fig. 14), are absent in other species. Pigment granules, which are observed in some species (particularly in old animals), are especially numerous in the horse pinealocytes (Fig. 15) (Cozzi and Ferrandi, 1984; Cozzi, 1986; Karasek and Zielinska, 2000). Moreover, cilia, centrioles, multivesicular bodies, microtubules and microfilaments are present, in various numbers, in the pinealocytes.

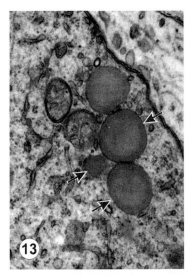

Fig. 13. Lipid droplets (arrows) in the pinealocyte cytoplasm. Rat; × 33000.

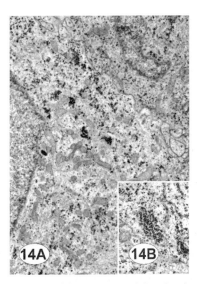

Fig. 14. Glycogen particles in the pinealocyte cytoplasm. Brush mouse; A − × 8000, B − × 10000.

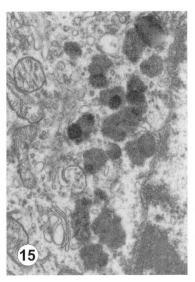

Fig. 15. Lipofuscin granules in the pinealocyte cytoplasm. Horse; × 21000.

In pinealocytes of some mammalian species are characteristic for these species were described below.

Consistent components of the cotton rat pinealocytes are not membrane-bound inclusion bodies, 2-6 μm in diameter, usually round or oval in shape, and showing large variation in internal structure. Some of them are composed of finely granular material interspersed with irregularly sized less electron-dense areas, whereas others have finely granular material located peripherally, and a clear central zone in which electron-dense particles or accumulation of highly electron-dense material is present (Fig. 16) (Matsushima et al., 1979; Karasek et al., 1983d).

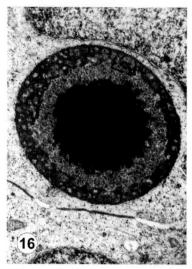

Fig. 16. Inclusion body in the pinealocyte cytoplasm. Cotton rat; × 16000.

Membrane-bound bodies (MBB) are a characteristic feature of the pig pinealocytes (Karasek and Wyrzykowski, 1980; Wyrzykowski et al., 1985). Two types of MBB are distinguished. MBB -1 (Fig. 17) characterizes the complicated internal structure (fine or coarse granular material, amorphous dense aggregates, homogenous dense material or "flocculent" material), whereas MBB-2 (Fig. 18) contains regular or

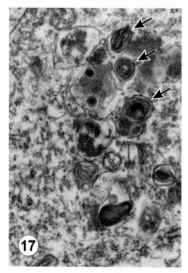

Fig. 17. Membrane-bound bodies (MBB-1; arrows) in the pinealocyte cytoplasm. Pig; × 30000.

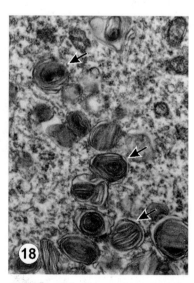

Fig. 18. Membrane-bound bodies (MBB-2, arrows) in the pinealocyte cytoplasm. Pig; × 20000.

irregular lamellar structures. The involvement of MBB in the secretory processes of pig pinealocytes was suggested (Karasek and Wyrzykowski, 1980; Wyrzykowski et al., 1985).

Glial Cells

Glial cells are the second largest group of cells in the mammalian pineal gland. However, the number of these cells greatly varies among mammalian species (from 0% up to 12%). Mostly fibrous or protoplasmic astrocytes are present. The number of fibrous astrocytes is greater than protoplasmic.

The glial cells (Fig. 19) are characterized by the presence of numerous filaments occurring throughout the cytoplasm and processes (glial fibres). The nucleus is regular in shape, oval or spherical and is characterized by high electron density. Nucleolus is usually small. The cytoplasm contains cell organelles localized mainly in the perinuclear region. The granular endoplasmic reticulum and Golgi apparatus are well developed. Mitochondria are usually smaller in size than mitochondria in pinealocytes, and the matrix shows higher electron density. The glial cells contain a great number of lysosomes, which are irregular in shape and heterogenic in content.

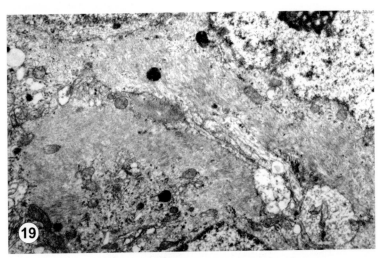

Fig. 19. Glial cell. Fox; × 15000.

Nerve Cells

Occasionally, nerve cells are present in the pineal gland of some mammalian species. An ultrastructure of these cells is similar among all of species where they have been found (ferret, rabbit, ground squirrel and chipmunk [Karasek, 1983]). Nerve cells are characterized by the large, electron-lucent nucleus, highly developed granular endoplasmic reticulum, Golgi apparatus, presence of numerous lysosomes, granular vesicles ca. 100 nm in diameter, and numerous microtubules and microfilaments. Numerous unmyelinated nerve fibres and endings often surround the perikarya of nerve cells.

Nerve Fibres

The mammalian pineal gland contains mainly postganglionic sympathetic nerve fibres originating in the superior cervical ganglia. The number of the sympathetic nerve fibres varies greatly among species (Karasek et al., 1983b). These fibres are situated mainly in perivascular areas (Figs. 20, 21 and 23), but some of them are also located within the pineal parenchyma, adjacent to pinealocytes. The endings of sympathetic fibres contain numerous clear vesicles (40-80 nm in diameter) and granular vesicles (small, 40-80 nm in diameter, possessing a dense core of ca. 25 nm, and large, 90-150 nm in diameter with a dense core of 50-70 nm) (Fig. 21). Apart from sympathetic nerves, myelinated nerve fibres coming from the central nervous system reach the pineal gland (Fig. 22).

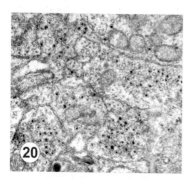

Fig. 20. Numerous adrenergic nerve endings in the perivascular space. Fox; × 23000.

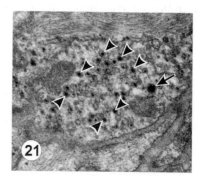

Fig. 21. Adrenergic nerve ending containing large (arrow) and small (arrowheads) granulated vesicles and clear vesicles in the perivascular space. Fox; × 31000.

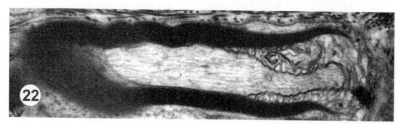

Fig. 22. Myelinated nerve in the perivascular space. Chipmunk; × 14000.

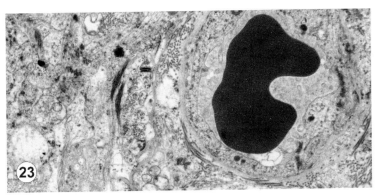

Fig. 23.   Non-fenestrated capillary surrounded by wild perivascular space. European bison; × 10000.

Blood Supply

The blood supply of the pineal gland is very copious. The capillary network is very well developed. There are differences among species in the ultrastructure of the pineal capillaries. Some of the species have capillaries with nonfenestrated endothelium (Fig. 23), whereas the endothelium of capillaries of the other mammalian species is fenestrated (Figs. 24 and 25). Capillaries of all species are surrounded by perivascular spaces enclosed by two thin laminas. The basal lamina forms the inner lamina of perivascular space and outer one is formed by the basal lamina covering the pineal parenchyma. The outer lamina often is discontinuous. The perivascular space contains primarily nerve fibres and their endings, terminals of pinealocyte processes, collagen, and glial fibres (Figs. 23 and 24). There are apparent differences in the width of perivascular spaces among species (Karasek, 1983; Karasek and Reiter, 1996).

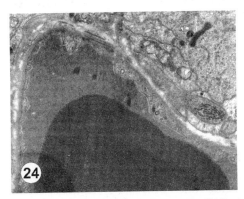

Fig. 24.   Fenestrated capillary and perivascular space limited by inner and outer basement membrane. Rat; × 10000.

Calcareous Deposits

In the pineal gland of a limited number of mammalian species (especially humans, ungulates, and some rodents) calcerous deposits (acervuli, brain sand, and corpora arenacea) are found. They vary in size and are mainly located within the stroma of the gland, however intracellular localization is also observed in some species (e.g., in gerbil). Under an electron microscope, the images of calcareous deposits appear to be either amorphous or composed of concentric laminae (Fig. 26). Most of them consist of needle-shaped crystals which are randomly or radially embedded in an amorphous matrix of moderate electron density (Fig. 27). Calcareous deposits are chemically composed of hydroxyapatite and calcium carbonate (Krstic and Golaz, 1977; Karasek, 1983; Krstic, 1986).

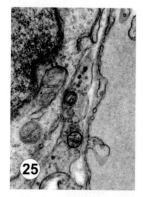

Fig. 25. Fenestrated capillary. Chipmunk; × 19000.

Fig. 26. Fragment of the calcareous concretion. Rhesus monkey; × 40000.

Fig. 27. Fragment of the calcareous concretion in the pinealocyte cytoplasm. Gerbil; × 32000.

CONCLUSION

Ultrastructure of pinealocytes shows many common features in majority of the mammalian species, and existing differences are of quantitative nature only. Volumes or the numbers of some cell structures (mitochondria, Golgi apparatus, lipid droplets, dense-core vesicles, and "synaptic" ribbons) show distinct interspecies dissimilarities, whereas those of other cell structures (granular endoplasmic reticulum, and lysosomes) are similar in all species. The observed qualitative differences include MBB in the pig, inclusion bodies in the cotton rat, abundance of pigment granules in the horse, and calcareous concretions in the monkey and gerbil.

Numerous ultrastructural studies performed in recent years have brought to light much information, which elucidates the correlation of pineal morphology with pineal function. The ultrastructural features of pinealocytes suggest high activity of these cells. Pinealocytes contain numerous organelles indicating their high metabolic activity (as mitochondria), as well indicating high secretory activity (as Golgi apparatus).

ACKNOWLEDGEMENTS

The work by the authors and this review was supported by Medical University of Lodz, Project No. 503-138-2. The help of Mr. Jacek Swietoslawski, M.Sc., is greatly appreciated.

REFERENCES

Anderson, E. (1965). The anatomy of bovine and ovine pineals. Light and electron microscopic studies. *J. Ultrastr. Res.* **suppl. 8**: 1-80.

Arendt, J. (1995). *Melatonin and the Mammalian Pineal Gland.* Chapman and Hall, London.

Bhatnagar, K.D. (1992). The ultrastructure of mammalian pinealocytes: A systemic investigation. *Micr. Res. Techn.* **21**: 85-115.

Cozzi, B. and Ferrandi, B. (1984). Fine structure and histochemistry of the equine pineal gland, with special reference to the possible functional role of the electron-dense intrapinealocyte bodies. *Clin. Veter.* **107**: 337-346.

Cozzi, B. (1986). Cell types in the pineal gland of the horse: an ultrastructural and immunocytochemical study. *Anat. Rec.* **216**: 165-174.

Dombrowski, T.A. and McNulty, J.A. (1984). Morphometric analysis of the pineal complex of the golden hamster over a 24-hour light:dark cycle. *Am. J. Anat.* **171**: 359-368.

Hewing, M. (1980). Synaptic ribbons in the pineal system of normal and light deprived golden hamsters. *Anat. Embryol.* **159**: 71-80.

Kappers, J.A. (1979). Short history of pineal discovery and research. *Progr. Brain Res.* **52**: 3-22.

Karasek, M. (1976). Quantitative changes in number of "synaptic" ribbons in rat pinealocytes after orchidectomy and in organ culture. *J. Neural Transm.* **38**: 149-157.

Karasek, M. and Wyrzykowski, Z. (1980). The ultrastructure of pinealocytes in the pig. *Cell Tissue Res.* **211**: 151-161.

Karasek, M. (1981). Some functional aspects of the ultrastructure of rat pinealocytes. *Endocrinol. Exp.* **15**: 17-34.

Karasek, M., Hurlbut, E.C., Hansen, J.T. and Reiter, R.J. (1982a). Ultrastructure of pinealocytes of the kangaroo rat (*Dipodomys ordi*). *Cell Tissue Res.* **226**: 167-175.

Karasek, M., King, T.S., Hansen, J.T. and Reiter, R.J. (1982b). A quantitative ultrastructural study of the pinealocyte of the chipmunk (*Tamias striatus*) during the daytime and at night. *J. Neurosci. Res.* **7**: 397-401.

Karasek, M. (1983). Ultrastructure of the mammalian pineal gland: its comparative and functional aspects. *Pineal Res. Rev.* **1**: 1-48.

Karasek, M., Bartke, A. and Hansen, J.T. (1983a). Influence of prolactin on pinealocytes of the mouse with hereditary hypopituitarism: a quantitative ultrastructural studies. *Mol. Cell. Endocrinol.* **29**: 101-108.

Karasek, M., King, T.S., Brokaw, J., Hansen, J.T., Petterborg, L.J. and Reiter, R.J. (1983b). Inverse correlation between "synaptic" ribbon number and the density of adrenergic nerve endings in the pineal gland of various mammals. *The Anatomical Record* **205**: 93-99.

Karasek, M., Petterborg, L.J., King, T.S., Hansen, J.T. and Reiter, R.J. (1983c). Effect of superior cervical ganglionectomy on the ultrastructure of the pinealocyte in the cotton rat (*Sigmodon hispidus*). *Gen. Comp. Endocrinol.* **51**: 131-137.

Karasek, M., Smith, N.K.R., King, T.S., Petterborg, L.J., Hansen, J.T. and Reiter, R.J. (1983d). Inclusion bodies in pinealocytes of the cotton rat (*Sigmodon hispidus*). An ultrastructural study and X-ray microanalysis. *Cell Tissue Res.* **232**: 413-420.

Karasek, M., Bartke, A. and Doherty, P.C. (1984). Effects of experimentally induced chronic hyperprolactinemia on the ultrastructure of pinealocytes in male rats. *J. Pineal Res.* **1**: 237- 244.

Karasek, M. (1986). Quantitative aspects of the ultrastructure of the mammalian pinealocyte. *Adv. Pineal Res.* **1**: 9-18.

Karasek, M. (1987). Functional ultrastructure of the mammalian pinealocyte. *Adv. Pineal Res.* **2**: 19-33.

Karasek, M., Lewinski, A. and Vollrath, L. (1988a). Precise annual changes in the numbers of "synaptic" ribbons and spherules in the rat pineal gland. *J. Biol. Rhythms* **1**: 41-48.

Karasek, M., Marek, K. and Pevet, P. (1988b). Influence of a short light pulse at night on the ultrastructure of the rat pinealocyte: a quantitative study. *Cell Tissue Res.* **254**: 247-249.

Karasek, M. and Cozzi, B. (1990). Synaptic ribbons in the pineal gland of the horse. *J. Pineal Res.* **8**: 355-358.

Karasek, M., Stankov, B., Lucini, F., Scaglione, F., Esposti, G., Mariani, M. and Fraschini, F. (1990). Comparison of the rat pinealocyte ultrastructure with melatonin concentrations during daytime and night. *J. Pineal Res.* **9**: 252-257.

Karasek, M. (1992). Ultrastructure of the mammalian pinealocyte under natural and experimental conditions: quantitative aspects. *Micr. Res. Techn.* **21**: 116-123.

Karasek, M. and Reiter, R.J. (1992). Morphofunctional aspects of the mammalian pineal gland. *Micr. Res. Techn.* **21**: 136-157.

Karasek, M. and Reiter, R.J. (1996). Functional morphology of the mammalian pineal glang. In: *Monographs on Pathology of Laboratory Animals. Endocrine System* (2nd edition) (Jones, T.C., Capen, C.C. Mohr, U., eds.), Springer, Berlin, 193-204.

Karasek, M. (1997). *Szyszynka.* Wyd. Naukowe PWN, Warszawa- Łódź.

Karasek, M. and Zielinska, A. (2000). Comparative quantitative ultrastructural study of pinealocytes in eight mammalian species. *Neuroendocrinol. Lett.* **21**: 195-202.

King, T.S., Karasek, M., Petterborg, L.J., Hansen, J.T. and Reiter, R.J. (1982). Effects of advancing age on the ultrastructure of pinealocytes in the male white-footed mouse (*Peromyscus leucopus*). *J. Exp. Zool.* **224**: 127-134.

Krstic, R. (1977). Glande pineale de rat. Analyse morphometrique aux microscopes photonique et electronique. *Labirint, Dept. Anat. Sarajevo* 121-124.

Krstic, R. and Golaz, J. (1977). Ultrastructural and X-ray microprobe comparison of gerbil and human pineal acervuli. *Experientia* **33**: 507-508.

Krstic, R. (1986). Pineal calcification: its mechanism and significance. *J. Neural Transm.* Suppl. **21**: 415-432.

Kurumado, K. and Mori, W. (1977). A morphological study of the circadian cycle of the pineal gland of the rat. *Cell Tissue* **182**: 565-568.

Lerner, A.B., Case, J.D., Takahashi, Y., Lee, T.H. and Mori, N. (1958). Isolation of melatonin, pineal factor that lightens melanocytes. *J. Am. Chem. Soc.* **80**: 2587.

Lewczuk, B., Przybylska, B., Udala, J., Koncicka, A. and Wyrzykowski, Z. (1993). Ultrastructure of the ram pinealocytes during breeding season in natural short and in artificial long photoperiods. *Folia Morphol. (Warsz.)* **52**: 133-142.

Matsushima, S., Morisawa, Y., Petterborg, L.J., Zeagler, J.W. and Reiter, R.J. (1979). Ultrastructure of pinealocytes of the cotton rat, *Sigmodon hispidus. Cell Tissue Res.* **204**: 407-416.

Pevet, P. (1979). Secretory processes in the mammalian pinealocyte under natural and experimental conditions. *Progr. Brain Res.* **52**: 149-192.

Pevet, P. (1981). Ultrastructure of the mammalian pinealocyte. In: *The Pineal Gland, Vol. 1, Anatomy and Biochemistry* (Reiter, R.J., ed.), Boca Raton, CRC Press, 121-154.

Przybylska, B., Wyrzykowski, Z., Wyrzykowska, K. and Karasek, M. (1990). Ultrastructure of pig pinealocytes in various stages of the sexual cycle: a quantitative study. *Cytobios* **64**: 7-14.

Przybylska, B., Wyrzykowski, Z. and Kaleczyc, J. (1993). Effect of ovariectomy followed by administration of ovarian steroid hormones on the pig pinealocyte ultrastructure. *Steroids* 58: 466-471.

Przybylska, B., Lewczuk, B., Dusza, L. and Wyrzykowski, Z. (1994a). Effect of long-term administration of melatonin on ultrastructure of pinealocytes in gilts. *Folia Morphol. (Warsz.)* **53**: 129-136.

Przybylska, B., Lewczuk, B., Wyrzykowski, Z. and Karasek, M. (1994b). Effects of p-chlorophenylalanine, amiflamine and melatonin treatment on the ultrastructure of pinealocytes in *Sus scrofa. Cytobios* **77**: 233-246.

Redins, G.M., Redins, C.A. and Novaes, J.C. (2001). The effect of treatment with melatonin upon the ultrastructure of the mouse pineal gland: a quantitative study. *Braz. J. Biol.* **61**: 679-684.

Saidapur, S.K., Seidel, A. and Vollrath, L. (1991). No differences in pineal synaptic ribbon and spherule numbers in different stocks and strains of laboratory rats. *J. Exp. Anim. Sci.* **34**: 7-11.

Satoh, Y. and Vollrath, L. (1988). Lack of "synaptic" ribbons in the pineal gland of BALB/C mice. *J. Pineal Res.* **5**: 13-17.

Struwe, M.C.W. and Vollrath, L. (1990). "Synaptic" bodies in the pineal glands of the cow, sheep and pig. *Acat Anat.* **139**: 335-340.

Swietoslawski, J. and Karasek, M. (1993). Day-night changes. In the ultrastructure of pinealocytes in the Syrian hamster: a quantitative study. *Endokr. Pol.* **44**: 81-87.

Trouillas, J., Morel, Y., Pharabos, M., Cordier, G., Girod, C. and Andre, J. (1984). Morphofunctional modifications associated with the inhibition by estradiol of MtTF4 rat pituitary tumor growth. *Cancer Research* **44**: 4046-4052.

Vollrath, L. (1979). Comparative morphology of the vertebrate pineal complex. *Progr. Brain Res.* **52**: 25-38.

Vollrath, L. (1981). *The Pineal Organ.* Springer, Berlin.

Vollrath, L. and Welker, A. (1984). No correlation of pineal "synaptic" ribbon numbers and melatonin formation in individual rat pineal glands. *J. Pineal Res.* **1**: 187-195.

Welsch, M.G. and Reiter, R.J. (1978). The pineal gland of the gerbil, *Meriones unguiculatus.* I. An ultrastructural study. *Cell Tissue Res.* **193**: 323-336.

Welsh, M.G., Cameron, I.L. and Reiter, RJ. (1979). The pineal gland of the gerbil, *Meriones unguiculatus.* II. Morphometric analysis over a 24-hour period. *Cell Tissue Res.* **204**: 95-109.

Wolfe, D.E. (1965). The epiphyseal cell: an electron-microscopic study of its intercellular relationships and intercellular morphology in the pineal body of the albino rat. *Progr. Brain Res.* **10**: 332-386.

Wyrzykowski, Z., przybylska, B. and Wyrzykowska, K. (1985). Ultrastructure and topography of dense bodies in pinealocytes of castrated males of domestic pig. *Folia Morphol. (Warsz.)* **44**: 175-185.

Zrenner, C. (1985). Theories of pineal function from classical antiquity to 1900: a history. *Pineal Res. Rev.* **3**: 1-40.

The Ultimobranchial Gland in Poikilotherms: Morphological and Functional Aspects

Ajai Kumar Srivastav[1,*], *Seema Yadav*[1], *Sunil Kumar Srivastav*[1] *and Nobuo Suzuki*[2]

Abstract

The intention of this chapter is to provide the morphological and functional aspects of ultimobranchial gland (which secretes calcitonin) in poikilotherms. During development these glands are derived from the pharyngeal pouch epithelium. In protochordatres calcitonin-like cells (in small groups of two or three) have been observed in the endostylar region of the pharynx. These cells contain large number of basically situated electron dense granules. A molecule closely resembling human calcitonin (CT) have been extracted from the nervous system of several protochordates. In Amphioxus calcitonin-immunoreactive cells are distributed in the latter half of the mid gut caecum. The presence of calcitonin-like molecules in the nervous system of protochordates suggests that it has some function related with the nervous system of these vertebrates. In cyclostomes a discrete ultimobranchial gland (UBG) is not present although a molecule resembling human CT was extracted from the nervous system of Myxine. Moreover, in Lampreys and Hagfishes CT-immunoreactive cells were encountered in the brain. In cartilaginous and bony fishes UBG are small discrete organ commonly found near the anterior end of the heart. In amphibians, the galnds are paired in apodans and anurans whereas in urodeles it is present only on left side. Adult reptiles possess one (on the left side) or two glands situated anterior to heart.

Histologically, the UBG in poikilotherms is usually composed of a follicular structure and may contain single cell type or there may be two or three cell types. In some species mucous cells, goblet cells and ciliated cells are also present in addition to simple cuboidal or pseudostratified columnar cells. The

[1]Department of Zoology, D.D.U. Gorakhpur University, Gorakhpur 273 009, India.
[2]Noto Marine Laboratory, Faculty of Science, Kanazawa University, Ogi-Uchiura, Ishikawa 927-0553, Japan.
Corresponding author: Tel.: + 91 551 2501173. E-mail: ajaysrivastav@hotmail.com

lumen of the follicle may contain a colloid-like substance with desquamated cell or debris.

The functional aspect of CT is not well clear in poikilotherms. Contradictory reports have been obtained regarding the effect of exogenous CT – no effect, hypocalcemic or hypercalcemic effect. Several reports regarding fish CT have been published which indicates that it is more effective in mammals as compared to teleosts. In various assays native form of mammalian CT are many times less potent than those obtained from lower vertebrates.

INTRODUCTION

In vertebrates, calcium particularly its ionic form, plays a vital role in a variety of biological processes – permeability of cell membranes to ions, neuronal excitability, muscle contraction, cell adhesion, responses of organisms to fluctuations in acid-base balance, and is also essential for ultimate initiation of many endocrine events, such as hormone release and responses of their definitive effectors. Minor changes in calcium level, particularly the ionized fraction, severely affects these physiological processes. Thus, the calcium concentration in the extracellular fluids is maintained within narrow limits, usually between 2.25 and 2.75 m mol/L (Simkiss, 1967). Several hormones have been implicated in the control of calcium in vertebrates, which are mainly parathyroid hormone (PTH), calcitonin (CT), vitamin D-related steroids, stanniocalcin and prolactin. The endocrine systems for the regulation of calcium differ between the aquatic and land vertebrates. Fish possess a unique system, which is more complex than that of land vertebrates as they (fish) are in constant contact with surrounding water, which provides an inexhaustible supply of calcium that can be taken up through the body surfaces particularly the gills. In land vertebrates, exchange of calcium through the body surface is not possible and food is the only source of calcium for them.

The ultimobranchial gland (UBG) has been ascribed various names by the earlier investigators. Van Bemmelen (1886) described it as a 'suprapericardial body' in elasmobranchs and chimaeroids with regard to its location on the pericardium. Maurer (1888) found the gland as an independent structure posterior to the gills and named it as 'postbranchialer korper'. Nusbaum-Hilarowicz (1916) termed this gland as 'suboesophageal gland' because in deep-sea fishes it lies under the digestive tract. Greil (1905) for the first time employed the term 'ultimobranchial body', which is most widely used because it describes the embryological origin (from the epithelium of the last branchial arch) and the anatomical position of the gland in most vertebrates.

Copp et al. (1962) were the first to discover the hypocalcemic hormone supposed to be antagonistic to the parathyroid hormone. This hormone was later named as calcitonin, which is secreted by the calcitonin (C) cells in mammals (Nunez and Gershon, 1978; Dacke, 1979; Srivastav and Rani, 1988). During development, C cells originate in the neuroectodermal cells of neural crest and later migrate to various organs where they finally settle (Le Douarin and Le Lievre, 1970; Le Douarin et al., 1974).

In mammals, these cells migrate and become incorporated into the thyroid, which is the principal site of C cell settlement, however, in addition they may also settle in the parathyroid and thymus (Nunez and Gershon, 1978; Kameda, 1981; Srivastav and Swarup, 1982; Srivastav and Rani, 1988). In non-mammalian vertebrates, the C cells do not merge with the thyroid but form a discrete organ, the ultimobranchial gland (Dacke, 1979; Wendelaar Bonga and Pang, 1991).

Kameda (1995) studied the development of ultimobranchial anlage in chick embryos by immunohistochemistry with the antibodies to class III beta-tubulin, TuJ1, human leukemic cell-line (HNK-1), and protein gene product (PGP) 9.5. The results indicate that the ultimobranchial anlage was formed by the extension of the ventral portion of the fourth pharyngeal pouch and the neuronal cells from the distal vagal ganglion enter into the ultimobranchial anlage and give rise to C cells i.e., C cells differentiate along the neuronal anlage.

The objective of this chapter is to provide the existing information on the morphological and functional aspects of the ultimobranchial gland in poikilotherms.

Protochordates

Thorndyke and Probert (1979) have observed calcitonin-like cells in the endostylar region of the pharynx of *Styela clava* occupying a position immediately adjacent to the iodine-binding thyroidal cells. These cells generally occur in small groups of two or three and contain a large number of basically situated electron dense granules. Girgis et al. (1980) have immunologically localized and chromatographically extracted a molecule from the nervous system of several protochordates, which very closely resembles human calcitonin. In *Ciona intestinalis*, calcitonin-like immunoreactivity has been found with Peroxidase-anti-peroxidase (PAP) method in the cells of alimentary tract epithelium, as well as in the nerve cells and nerve fibres in the connective tissue underlying the alimentary tract epithelium (Fritsch et al., 1980). In amphioxus, *Branchiostoma belcheri*, CT-immunoreactive cells were distributed concentrically in about 1 mm region of the latter half of the mid-gut (Urata et al., 2001). These cells were also present in the mid-gut caecum, but dispersed throughout the length and their number was less as compared to the cells in the mid-gut (Urata et al., 2001).

The calcitonin-like cells in the *Styela clava* degranulated when the animals were exposed to elevated calcium levels (sea water to which 0.5% $CaCl_2$ had been added) (Thorndyke and Probert, 1979). Girgis et al. (1980) have suggested that the presence of human CT-like molecules in the nervous systems of primitive chordates indicates that they have some function in the nervous system of these vertebrates and the bone-regulating function of CT may have arisen much later in the vertebrates. Urata et al. (2001) have suggested that considering the morphology and the characteristics of

CT-immunoreactive cells in amphioxus, these cells could be related to some phenomenon of the digestive processes.

Agnatha

In cyclostomes, a discrete UBG is not present. However, a molecule closely resembling human CT (immunologically and chromatographically) was extracted from the nervous systems of *Myxine* (Girgis et al., 1980). Sasayama et al. (1991) have located immunoreactive calcitonin-producing cells in the brain of lampreys and hagfishes by the peroxidase-antiperoxidase method using antisalmon CT antiserum. Studies regarding the functional aspects of these immunoreactive cells are lacking. Suzuki (1995) reported the presence of a calcitonin-like substance in the plasma of hagfish (*Eptatratus burgeri*), which is very similar to salmon CT.

In lamprey, *Lampetra japonica* (which undergoes spawning migration from seawater to freshwater), a CT-like substance was not detected in the plasma of seawater-adapted specimens (both sexes), whereas it was detected in the plasma of freshwater-adapted specimens (both sexes) by enzyme-linked immunosorbent assay using antisalmon CT serum (Suzuki, 2001). This substance increased with the increased gonado-somatic index, which led Suzuki (2001) to suggest that in lamprey the CT-like substance may have some relationship with gonadal maturation.

Elasmobranchs (Chondrichthyes)

In Atlantic stingray *Dasyatis sabina* and bonnethead shark *Sphyrna tiburo*, the UBG is located between the dorsal musculature of the pericardial cavity and the ventral pharyngeal epithelium (Gelsleichter, 2001, 2004; Nichols et al., 2003). It is a paired structure, which consists of small round follicles with cuboidal or columnar epithelial walls and central lumina. In the shark *Squalus acanthias*, granulated and nongranulated cell types are present in the glandular parenchyma (Kito, 1970). Sasayama et al. (1984) have reported that in UBG of *Dasyatis akajei*, most of the pharyngeal cells reacted positively to the anti-human antiserum. The tendency for immunoreactivity was greater in the basal portion of the cytoplasm of follicle cells.

CT from the UBG of shark (Copp et al., 1967) and ray (Sasayama et al., 1992) provoked hypocalcemia when administered into mammals. In elasmobranchs, conflicting results have been reported regarding the effect of CT – hypocalcemia (stingray – Srivastav et al., 1998a), hypercalcemia (leopard shark – Glowacki et al., 1985) and no effect (shark – Hayslett et al., 1971). The observed effects of CT (either hypocalcemia or hypercalcemia) demonstrate that CT plays a role in calcium homeostasis of elasmobranchs (possessing a cartilaginous skeleton) by a mechanism different from that on bones (Srivastav et al., 1998a).

Nichols et al. (2003) have found increased serum CT concentrations in the mature female bonnethead shark, *Sphyrna tiburo* during early stages of gestation. During the same period, immunoreactive CT was also detected in the duodenum and pancreas of embryonic *Sphyrna tiburo*. These results suggest that the CT obtained from endogenous and/or maternal sources might function in regulating yolk digestion in embryonic shark and may reflect an important role for this hormone in embryonic development.

Osteichthyes

The UBG in teleosts is a single or paired structure, usually situated either in the interseptum between the pericardial and abdominal cavities or between the oesophagus and sinus venosus (Krawarik, 1936; Sehe, 1960; Oguri, 1973; Lopez et al., 1976; Takagi and Yamada, 1977; Yamane, 1978; Hooker et al., 1979; Swarup and Ahmad, 1979; Zaccone, 1980; Robertson, 1986; Srivastav et al., 1999, 2002; Sasayama et al., 2001). McMillan et al. (1976) have found small venules following a more or less straight course from the gland and opening directly into the sinus venosus. They have suggested that this relationship provides the rapid release of hormone into the general circulation. The gland may contain either clusters of cells (Fig. 1), follicles (only single or several small follicles), cell strands or both follicles and clusters of cells (Fig. 2) (Robertson, 1986; Sasayama et al., 1995, 1999, 2001; Srivastav et al., 1999, 2002). The follicular lumina of UBG are filled with cell debris and colloid-like material (Camp, 1917; Watzka, 1933; Sehe, 1960, McMillan et al., 1976; Takagi and Yamada, 1977; Hooker et al., 1979; Srivastav et al., 1999, 2002; Sasayama et al., 2001).

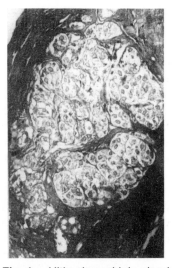

Fig. 1. Ultimobranchial gland of *Amphipnous cuchia* exhibiting parenchyma, which is composed of cords and clumps of cells. Hematoxylin-eosin × 100.

UBG in teleosts have been reported to contain a single cell type (Srivastav et al., 1999, 2002), two cell types (Peignoux-Deville et al., 1975; McMillan et al., 1976; Hooker et al., 1979; Swarup and Ahmad, 1979; Zaccone, 1980) or even three cell types (Takagi and Yamada, 1977). Ultrastructurally the UBG has been reported to contain granular (secretory) and nongranular (supporting) cells (Hooker et al., 1979; Robertson, 1986). The granular cells are columnar and filled with numerous electron-opaque granules, which are concentrated near the basement membrane and the lumen. These cells possess less electron dense ovoid nuclei, rough endoplasmic reticulum and Golgi profiles always near the nuclei. Microtubules are usually found in the apex of those cells which make contact with the lumen. Mitochondria are distributed throughout the granular cells and microfilaments are occasionally seen (Hooker et al., 1979).

The nongranular cells are irregularly shaped and contain no secretory granules. Their cytoplasmic organelles consist of mitochondria, numerous microfilaments, many free ribosomes, short segments of smooth and rough endoplasmic reticulum and occasional Golgi profiles (Hooker et al., 1979). The apical surface has long microvilli and generally a single cilium is associated with each cell. Nongranular cells predominantly line the luminal surface, while the basal surface demonstrates a more intermittent occurrence of these cells. The nongranular cells have extensive interdigitations with adjacent similar cells and are frequently joined together by junctional complexes and desmosomes. McMillan et al. (1976) have shown immunohistochemically that CT is localized primarily in the epithelium of UBG of trout and the cells at the luminal surface are devoid of CT. Hooker et al. (1979) suggested that immunopositive cells are granular cells and the CT free cells are the supporting cells. Moreover, there is no evidence of CT reactivity in the luminal contents of the follicles, which suggests that release into the lumen is not a mechanism for secretion of CT.

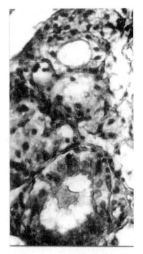

Fig. 2. Ultimobranchial gland of *Heteropneustes fossilis* showing follicles. Hematoxylin-eosin X 100.

In addition to the UBG, CT-immunoreactive cells have also been found in the intestinal epithelium of goldfish, which are mainly distributed in the anterior part (Okuda et al., 1999). The nucleus of these cells is located in the basal portion of the serosal side and the cytoplasm is elongated to the luminal side. From the anterior part of goldfish intestine, cDNA fragments with the same nucleotide sequence as that of goldfish ultimobranchial calcitonin gene were amplified by reverse transcriptase-polymerase chain reaction (RT-PCR). Okuda et al. (1999) have suggested that the function of intestinal CT-immunoreactive cells might be to restrain the acute absorption of nutrients and not to control blood calcium levels.

For establishing a correlation between the UBG and calcium homeostasis in fish, experiments have been performed which involve either the extirpation of UBG or administration of CT. In Japanese eels (*Anguilla anguilla*) electrocautery of the UBG produced contradictory results – (i) significant elevation of plasma calcium level (Chan et al., 1969) and (ii) a decrease in plasma calcium level (Chan, 1972). Removal of UBG in silver eels maintained in calcium-enriched tap water resulted in a 25% increase in plasma calcium within two weeks, which returned to normal level after four weeks (Lopez et al., 1976). In partially ultimobranchialectomized goldfish (*Carassius auratus*) transferred from freshwater to 30% seawater, the plasma calcium levels increased more than those of sham-operated controls (Fenwick, 1975). Dacke

(1979) has questioned surgical removal of UBG in eels (as the calcitonin secreting cells are diffused within the oesophagus and pericardium – Orimo et al., 1972) and stated that until the efficacy of the surgical technique has been demonstrated, it will be difficult to interpret the results of studies using this method.

Several investigators have tried to demonstrate a hypocalcemic effect of calcitonin in fish, however, the results were variable (Dacke, 1979; Wendelaar Bonga and Pang, 1991). Administration of mammalian CT or salmon CT in many freshwater or marine fish caused either no effect on plasma calcium levels (Pang, 1971; Srivastav and Swarup, 1980; Wendelaar Bonga, 1980; Hirano et al., 1981; Srivastav et al., 1989, 1998b; Singh and Srivastav, 1993) or hypercalcemia (Fouchereau-Peron et al., 1987; Oughterson et al., 1995). Conversely, hypocalcemia following CT administration was reported for freshwater European eel (Chan et al., 1968; Lopez et al., 1976), low-calcium freshwater adapted sticklebacks (Wendelaar Bonga, 1981 – only the ultrafiltrable calcium was reduced) and mudskippers, which have been made hypercalcemic by taking them out of the water (Fenwick and Lam, 1988). The hypocalcemic response of CT in fish maintained in acalcic freshwater (trout – Lopez et al., 1971; catfish – Singh and Srivastav, 1993; freshwater mud eel – Srivastav et al., 1998c) indicates that the availability of environmental calcium may be an important factor in response of CT, possibly by lowering the endogenous level of this hormone. Wendelaar Bonga and Pang (1991) have suggested that the absence of an effect in most experiments may not only have been caused by the absence of any hypocalcemic potency of CT preparations used, but also by the effective counteraction of hypercalcemic control mechanisms. In this regard it is interesting to note that CT provoked hypocalcemia in hypophysectomized freshwater catfish (Srivastav et al., 1998b). Keeping in view the inconsistency regarding the effects of CT in fish, Wendelaar Bonga and Pang (1991) have concluded that although CT may have hypocalcemic effects under some conditions, it is doubtful whether it has a hypocalcemic function in fish.

In fishes a definite role for CT has yet to be established, but it is of interest to note that the UBG exhibits inactivity after CT treatment, though the blood calcium contents are not affected (Peignoux-Deville et al., 1975; Wendelaar Bonga, 1980; Srivastav et al., 1989). Moreover, the gland depicts hyperactivity in response to experimentally induced hypercalcemia (Lopez and Peignoux-Deville, 1973; Peignoux-Deville et al., 1975; Swarup and Srivastav, 1984; Srivastav et al., 1997, 2002).

Keeping in view contradictory reports of CT, some investigators have attributed other functions to this hormone in fishes. Chan et al. (1968), Yamauchi et al. (1978) and Wendelaar Bonga and Pang (1991) have suggested that in fishes CT has a role in skeletal protection during the times of excessive calcium demand. Activation of osteoblast and increased mineral deposition in bone have been reported following administration of salmon CT in tilapia (Wendelaar Bonga and Lammers, 1982) and starved goldfish (Shirozaki and

Mugiya, 2000, 2002). CT has also been reported to suppress the activities of osteoclasts in the scale (Suzuki et al., 2000).

It has also been suggested that a correlation exists between plasma CT levels and gonadal activity. During natural or experimentally induced sexual maturation (particularly in the female fish), the UBG exhibits increased secretory activity (Lopez et al., 1968; Peignoux-Deville and Lopez, 1970; Oguri, 1973; Peignoux-Deville et al., 1975; Yamane and Yamada, 1977; Yamane, 1977, 1981). During the reproductive period, the plasma CT levels in female rainbow trout were found to be elevated (Bjornsson et al., 1986; Fouchereau-Peron et al., 1990). The level increased until ovulation and decreased sharply thereafter. In salmonids, injections of 17-β estradiol increased plasma CT levels (Bjornsson et al., 1989). These authors have suggested that in fish a reproductive role for CT is more likely than a calcium regulatory role. However, direct evidence for a reproductive role in fish is still missing.

Suzuki et al. (1999) have compared calcium and calcitonin levels in freshwater eels fed a high calcium-consomme solution into the stomach, and in freshwater eels transferred from freshwater to seawater. They have found a correlation between plasma CT and plasma calcium raised by dietary calcium in the consomme form, but it does not participate in the initial processes of seawater adaptation.

Amphibia

In apodans, the UBG arises from the seventh pharyngeal pouch (Marcus, 1908, 1922) and attains maximal size in the late larval stage (Klumpp and Eggert, 1935). The glands are paired and composed of follicles and clusters of cells, which are separated by connective tissue strands. The cells forming clusters and in the basal portion of follicles contain numerous electron-dense, polymorphic secretory granules, well-developed rough endoplasmic reticulum and Golgi membranes, microtubules and small bundles of microfilaments. In larger follicles at the luminal border, relatively flat cells are present which lack secretory granules (Welsch and Schubert, 1975).

In urodeles, the UBG is present only on the left side (Maurer, 1888; Baldwin, 1918; Uhlenhuth and McGowan, 1924; Wilder, 1929; Uchiyama, 1980; Oguro et al., 1983; Takagi and Yamada, 1986). The right gland degenerates during the development. However, in *Amphiuma* and *Necturus* the gland is paired (Wilder, 1929). The gland is located dorso-lateral to the pericardial cavity (Oguro et al., 1983) and has a tubular structure consisting of follicles lined with pseudostratified epithelium. The epithelium is formed by granulated cells (some of which may contain mucoid-like materials, arranged at the periphery [base] of the follicles) and non-granulated cells bordering the central lumen (Coleman and Phillips, 1972; Takagi and Yamada, 1986). Presence of immunoreactive CT in urodelan UBG has also been demonstrated (Sasayama et al., 1984; Matsuda et al., 1989).

In anurans, the UBG arise from the sixth pharyngeal pouch at the time when the operculum is partly closed (Robertson and Swartz, 1964b; Coleman and Phillips, 1974; Coleman, 1975; Sasayama et al., 1976). It is usually located beneath the pharyngeal epithelium on both sides of the glottis (Robertson, 1971a, 1986; Oguro et al., 1984a). The gland consists of follicle(s) lined by pseudostratified epithelium (Robertson, 1968a, b, 1988; Coleman, 1975; Boschwitz, 1976; Swarup and Krishna, 1980; Sasayama et al., 1990; Srivastav and Rani, 1991a, b). The follicular lumen contains cellular debris and the coagulum, which consists of three components – a granular heterogeneous component which is acid mucopolysaccharide with a carbohydrate-protein complex, a sudanophilic component, and a fine colloidal material which is mucoprotein (Robertson and Swartz, 1964a; Robertson, 1971a).

Three cell types (Fig. 3) have been described in the follicular epithelium (Robertson, 1968a, b; Das and Swarup, 1975a; Swarup and Krishna, 1980; Srivastav and Rani, 1991a, b). These cell types are – (i) basal stem cells (comprising less than 4% of UBG parenchyma) containing a large round or slightly ovoid nucleus with a prominent nucleolus and has no free surface bordering the central follicular cavity, (ii) the secretory or storage cells (comprising 60% of UBG parenchyma) possessing an elongated nucleus with a fine reticular chromatin pattern, and (iii) the degenerating cells which possess an irregular, dense staining nucleolus with a coarse clumping of chromatin material. Ultrastructurally, the basal cells possess a well-developed dilated, rough endoplasmic reticulum, inconspicuous Golgi apparatus and large mitochondria with round or ovoid profiles. The secretory cells possess a scant, rough endoplasmic reticulum, numerous dense-cored membrane-bound granules in the apical and basal cytoplasm and Golgi apparatus with polarity of granule formation towards the centre. The degenerating cells lacked cytoplasmic granules or had a few within the basal cytoplasm and contain lysosomes, lipid bodies and associated myelin figures in the apical cytoplasm (Robertson and Bell, 1965; Robertson, 1971a).

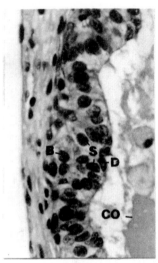

Fig. 3. Ultimobranchial gland of *Rana tigrina* showing pseudostratified epithelium, colloid-like material (CO) in the lumen and cell types – basal (B), secretory (S), and degenerating (D). Hematoxylin-eosin × 400.

On the basis of autoradiographic study after the injection of tritiated thymidine, Robertson (1967) suggested that the UBG parenchyma undergoes a continuous renewal of exhausted secretory cells by the division and sequential differentiation of basal stem cells. CT immunoreactivity is more confined towards the peripheral area, whereas the degenerating cells and colloid lack such response (Sasayama et al., 1984, 1990; Treilhou-Lahille et al., 1984; Robertson, 1988). This indicated that CT is secreted directly into the blood capillaries, which are present in the connective tissue capsule.

It will be of interest for future studies on the functional significance of UBG to be carried out in apodans, as presently there is no available data regarding this. A little more is known of the physiological significance of UBG in urodeles than apodans, but nevertheless there is relative paucity of information.

No studies have been have been performed on the effects of ultimobranchialectomy (UBX) on calcium regulation of urodelans. Administration of salmon CT in *Cynops pyrrhogaster* failed to affect blood calcium concentrations. However, in parathyroidectomized newts maintained in a calcium-rich environment, a rapid increase in serum calcium levels was partially suppressed by administration of salmon CT (Oguro et al., 1978). In *Ambystoma mexicanum*, a single injection of synthetic eel CT was effective in reducing calcium influx rates, however, plasma calcium levels remained unchanged (Kingsbury and Fenwick, 1989). It is important to note that in both the above mentioned studies using fish CT (fish CT is more potent than other CTs), the blood calcium content remained unaltered. Although urodelans seem unresponsive to exogenous CT with respect to their blood calcium content, the UBG extract of *Cynops pyrrhogaster*, *Onychodactylus japonicus* and *Hynobius nigrescens* showed hypocalcemic effect in rats (Uchiyama, 1980; Oguro et al., 1983). Information regarding the effect of CT on calcium exchange occurring through gut, bone and kidney of urodeles is lacking.

In anuran tadpoles, CT suppresses the rapid increase in calcium level in body fluids (Sasayama and Oguro, 1976; Sasayama, 1978; Srivastav et al., 2000). UBX bullfrog tadpoles maintained in a calcium-rich environment exhibited an increased calcium content in serum and coelomic fluids (Sasayama and Oguro, 1976), however, the serum calcium level remained unaffected in intact tadpoles (Sasayama and Oguro, 1976). In UBX tadpoles, the development of hypercalcemia could be prevented by either implantation of UBG or treatment with salmon CT (Sasayama and Oguro, 1976; Sasayama, 1978). In non-feeding bullfrog tadpoles, administration of salmon CT reduced calcium uptake through the gut and skin (Baldwin and Bentley, 1980). CT had no influence on the renal handling of calcium in bullfrog tadpoles (Sasayama and Oguro, 1982). In premetamorphic tadpoles of *Rana catesbeiana*, UBX caused depletion of calcium reserves in the endolymphatic sacs and impairment of bone mineralization (Robertson, 1971b). The UBG enhances deposition of calcium into endolymphatic sacs during larval development, which is later utilized for the mineralization of bones when the animal ceases to feed during the metamorphic climax (Guardabassi, 1960; Pilkington and Simkiss, 1966; Pilkington, 1967; Robertson, 1971b; Sasayama et al., 1976; Sasayama and Oguro, 1985).

In adult *Rana pipiens*, UBX resulted in a slight hypercalcemia and depletion of calcium deposits in the endolymphatic sacs over a period of about six weeks (Robertson, 1969a). This was followed by a significant hypocalcemia and demineralization of skeletal elements (Robertson, 1969a). UBX in *Rana nigromaculata* had no effect on the calcium content of femur, vertebra, cartilage and soft tissues, however, in lime sacs it decreased (calcium content of lime sac was 56% at 15 days and 29% at 35 days) as compared to controls (Oguro et al., 1984a). These studies clearly indicate that the UBG plays a role in calcium deposition in lime sacs.

In *Rana pipiens*, UBX provoked a two-fold increase in urinary calcium, which returned to normal within 24 hours after transplantation of UBG (Robertson, 1969b). This suggests that CT inhibits renal calcium excretion. In UBX *Rana pipiens*, Robertson (1969c) studied the healing of closed fractures of tibiofibulae by using X-ray and histological techniques and noticed that the absence of UBG (CT) did not interfere with the primary cartilaginous callus formation during the first six weeks. In normal frogs, the original callus was replaced by well-calcified endochondral trabeculae, whereas UBX frogs displayed a retardation of secondary callus formation, which manifested as poorly formed trabeculae.

The mRNA of bullfrog otoconin-22 (a major source of protein of aragonitic otoconia considered to be involved in the formation of $CaCO_3$ crystals) has been expressed in the endolymphatic sac. UBX induced a prominent decrease in otoconin-22 mRNA levels of bullfrog lime sacs, the levels exhibited a significant increase after supplementation of synthetic salmon CT (Yaoi et al., 2003). These findings suggest that CT regulates expression of bullfrog otoconin-22 mRNA via the calcitonin receptor-like protein on the endolymphatic sacs, thus stimulating the formation of calcium carbonate crystals in the lumen of endolymphatic sacs.

Contradictory results have been obtained regarding the effects of exogenous CT in anurans (Bentley, 1984; Srivastav et al., 2000). Salmon CT caused either no effect (*Xenopus laevis* – McWhinnie and Scopelliti, 1978; *Bufo marinus* – Bentley, 1983) or provoked hypocalcemia (*Bufo marinus* and *Bufo boreas* – Boschwitz and Bern, 1971; *Rana cyanophlyctis* – Krishna and Swarup, 1985a, b; *Rana tigrina* – Srivastav and Rani, 1989a, b; *Bufo andersoni* – Hasan et al., 1993). The effect of CT has been related with dose (hypocalcemia at 1 µU/g body wt and hypercalcemia at 1 mU/g body wt – McWhinnie and Scopelliti, 1978) and season (hypocalcemia in April and hypercalcemia in June-July, Guardabassi et al., 1968; Dore et al., 1969).

CT induced hypocalcemia in anurans cannot be explained by urinary calcium loss, since Bentley (1983) found no effect of exogenous CT on urinary calcium excretion of *Bufo marinus*. Homeoplastic transplantation of UBG into UBX *Rana pipiens* prevented hypercalciuric response, which led Robertson (1969b) to suggest that CT promotes calcium conservation instead of causing its excretion. Renal calcitonin binding sites were completely absent in amphibians (Bouizer et al., 1989), thus indicating that the regulation of kidney function by CT appeared late in vertebrate evolution.

CT enhances deposition of calcium in endolymphatic sacs (Fig. 4) of anurans (Oguro et al., 1984a; Krishna and Swarup, 1985b, Srivastav and Rani, 1989a; Hasan et al., 1993). The uptake of ^{45}Ca by lime sacs in CT treated UBX *Rana nigromaculata* was higher than in the controls, while bones, cartilage and skin remained unresponsive to such treatment (Oguro et al., 1984a). In *Rana pipiens*, calcium loss occurred across the skin which remained unchanged when frogs were treated with CT (Baldwin and Bentley, 1981). However, in flux studies, CT stimulated Ca^{++} influx in both intact frogs and isolated skin (Stiffler et al., 1998).

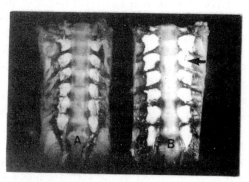

Fig. 4.   Photograph of the vertebral columns of *Rana tigrina* after 15 days of vehicle (A) or calcitonin (B) treatment. Note increased densities of calcium deposits in the paravertebral lime sacs (arrow) in calcitonin-treated frog.

UBG extracts of anurans provoked hypocalcemia in rats (Copp and Parkes, 1968; Oguro and Uchiyama, 1980; Oguro et al., 1981; Oguro and Sasayama, 1985; Yoshida et al., 1997). Conspecific UBG extracts also induced hypocalcemia in *Rana nigromaculata* (Oguro and Uchiyama, 1980; Oguro and Sasayama, 1985). Oguro et al. (1984b) have noticed hypocalcemia in rats after administration of brain extract of bullfrog and suggested that the immunologically detected CT in the bullfrog brain (Yui et al., 1981) was physiologically active. Sasayama et al. (1990) observed that CT immunoreactive neurons in the brain of the tree frog (*Hyla arborea*) did not participate in the regulation of serum calcium even after administration of calcium-rich solution. They suggested that UBG is responsible for systemic control of calcium.

Reports on the response of anuran UBG to hypercalcemia and hypocalcemia are also available, which indicate the activation of the gland in the former and inactivation in the latter condition. Hyperactivity of the gland, expressed by cellular hypertrophy, increased epithelial height and formation of secondary follicles, have been noticed in response to hypercalcemia induced by either vitamin D and/or $CaCl_2$ treatment (Robertson, 1968a, b, 1971b, 1988; Boschwitz, 1973; Das and Swarup, 1975a; Srivastav, 1980; Swarup and Krishna, 1980; Pathan and Nadkarni, 1986; Srivastav and Rani, 1991b). Inactivity/hypoactivity of the gland, indicated by decreased nuclear volume, which diminished the number of ultimobranchial cells and reduced the height of follicular epithelium, has been demonstrated after glucagon (Srivastav and Swarup, 1984) and CT (Boschwitz, 1973; Krishna and Swarup, 1985a; Srivastav and Rani, 1989b).

Reptiles

In reptiles, the UBG is usually unilateral (Crocodilia and Lacertilia) or bilateral (Chelonia and Ophidia). Different locations have been reported for UBG in this group – anterior to the heart (Crocodilia – Hammar, 1937; and Lacertilia – Maurer, 1899; Francescon, 1929; Watzka, 1933; Peters, 1941; Sehe, 1965; Moseley et al., 1968; Das and Swarup, 1975b; Sasayama and Oguro, 1992; Yamada et al., 2001) and in close proximity to the caudal parathyroid (parathyroid IV) gland (Chelonia – Francescon, 1929; Watzka, 1933; Clark, 1971, 1972; Khairallah and Clark, 1971; Takagi and Yamada, 1982; Boudbid et al., 1987 and Ophidia – Clark, 1971, Yoshihara et al., 1979; Singh and Kar, 1985; Kagwade and Padgaonkar, 1992; Srivastav and Rani, 1992a, b; Srivastav et al., 1994, 1998d).

Histologically the gland consists of two epithelial components – follicles of various sizes and cell clumps scattered in the loose connective tissue between the follicles (Fig. 5). The lining of the follicles may contain squamous, cuboidal, ciliated, goblet or pseudostratified columnar cells (Fig. 6) (Clark, 1971; Das and Swarup, 1975b; Anderson and Capen, 1976; Yoshihara et al., 1979; Boudbid et al., 1987; Srivastav and Rani, 1992a, b; Srivastav et al., 1994, 1998d; Yoshidaterasawa et al., 1998). The lumen may be either empty or may contain homogeneous or heterogeneous colloid-like material, desquamated cells or cell debris (Fig. 6) (Sehe, 1965; Das and Swarup, 1975b; Singh and Kar, 1982; Suzuki et al., 1984; Boudbid et al., 1987; Srivastav and Rani, 1992a, b; Yoshidaterasawa et al., 1998; Srivastav et al., 1998d; Yamada et al., 2001).

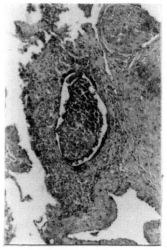

Fig. 5. Ultimobranchial gland of *Natrix piscator* showing compact clumps and follicle. Hematoxylin-eosin × 50 (after Srivastav et al., 1998d).

Ultrastructurally the UBG of *Pseudemys scripta* and *Chrysemys picta* contains a single cell type in the follicular epithelium, which is well equipped for protein and polysaccharide synthesis (Khairallah and Clark, 1971). These cells have well developed Golgi regions, enlarged endoplasmic reticulum cisternae, a large number of free ribosomes and numerous small electro-dense granules (150-250 μ) bound by a smooth surfaced membrane. Numerous large cytoplasmic bodies or granules

(800-1,000 mµ in diameter) found in the luminal region of cells in close association with apical plasmalemma are also present. In soft-shelled turtles, Takagi and Yamada (1982) have found three kinds of cells forming the follicles – light, dark and semi-dark cells. Light cells possess few mitochondria and granules, whereas dark and semi-dark cells are extremely rich in glycogen and mitochondria. Some semi-dark cells are rich in granules having a diameter of 250-400 nm. Most of the cells facing the lumen are light cells with microvilli projecting into the lumen.

Electron microscopy reveals that UBG follicles of iguana are composed of pseudostratified columnar to simple cuboidal cells (Anderson and Capen, 1976). Morphological similarity has been noticed between the C cells and interfollicular cords. The Golgi complex with prosecretory granules was located near the

Fig. 6. A higher magnification of Fig. 5 depicting cell clumps, epithelium of the follicle and heterogeneous colloid-like material in the lumen. Hematoxylin-eosin × 400 (after Srivastav et al., 1998d).

nucleus in an apical position, whereas in C cells of follicles the granules aggregated in the basal position of the cells. The cells lining the epithelium had less electron dense cytoplasm and less developed cytoplasmic organelles than that of C cells. In the lining of UBG follicle sometimes apically situated lamellar bodies were found. The luminal content was granular, variable in electron density and possessed lamellar bodies, membrane-bound large vesicles and degenerating cell organelles. The granules of C cells were extruded into the lumen (Anderson and Capen, 1976).

Ultrastructural studies on UBG of terrestrial snake *Elaphe quadrivirgata* reveals three types (Fig. 7) of parenchymal cells – small granular cells, mucous cells and ciliated cells (Suzuki et al., 1984; Srivastav et al., 1998d). The small granular cells (calcitonin secreting cells) possess a high electron density with spherical granules (100-200 nm) in the basal portion (Figs. 8 and 9). The mucous cells are present on the surface of the follicular lumen and have mucous granules towards the luminal side (Fig. 9). Ciliated cells possess no granules (Suzuki et al., 1984). In the sea snake *Laticauda semifasciata*, the parenchymal cells have been differentiated as small granular cells (granules of 100-200 nm) and large granular cells (granules of 300-500 nm). Both cells possess granules in the basal portion. The small granular cells have been considered as calcitonin secreting cells, whereas the function of large granular cells is not clear and they possibly secrete a substance other than CT (Suzuki et al., 1984). In sea snake both cell types possess large aggregates of tonofilaments (100 Å thick).

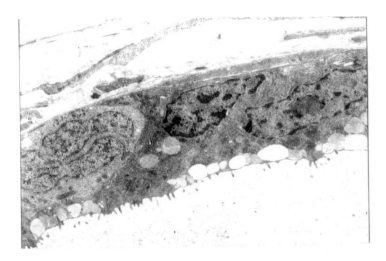

Fig. 7. Ultrastructure of ultimobranchial gland of the terrestrial snake, *Elaphe quadrivirgata* showing different cell types. × 7600 (after Srivastav et al., 1998d).

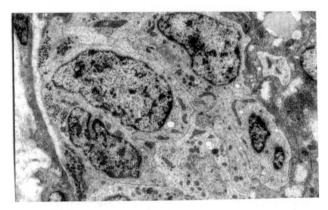

Fig. 8. Ultrastructure of ultimobranchial gland of *Elaphe quadrivirgata* showing small granular cells having almost spherical granules in basal portion. × 6800 (after Srivastav et al., 1998d).

The UBG of skink (*Eumeces okadae*), sea snake (*Laticauda semifasciata*) and rat snake (*Elaphe quadrivirgata*) did not show any immunoreactivity to anti-salmon or anti-human CT antiserum (Sasayama et al., 1984). It indicates that there may be immunological differences between the CT molecules of reptiles and other vertebrates (Sasayama et al., 1984). In turtle, Boudbid et al. (1987) reported that only cell clumps exhibit immunoreactivity to anti-salmon CT. The UBG of Japanese lizard and snakes were positive to antiserum against pig calcitonin, but negative to antiserum against synthesized human CT, which suggests that the configuration of amino acid in the Japanese lizard

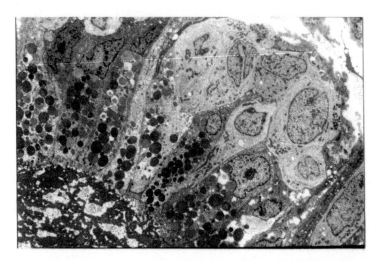

Fig. 9. Ultrastructure of ultimobranchial gland of the snake *Elaphe quadrivirgata* exhibiting small granular cells and mucous cells. Mucous cells are in the surface of follicular lumen and possess mucous granules on luminal side. × 3800 (after Srivastav et al., 1998d).

and snake CT is similar to that of pig CT (Yamada et al., 2001). CT immunoreactivity has also been demonstrated in the brain, lung and ultimobranchial cells of the lizard (*Lacerta muralis*). CT has also been localized in the hypothalamus and median eminence of the lizard brain (McInnes et al., 1982). These studies point towards a possible role of CT as a neurotransmitter in reptiles.

Injections of mammalian CT had no effect on plasma calcium in turtles (Clark, 1968), snakes (Clark, 1971) and lizards (Dix et al., 1970; Kiebzak and Minnick, 1982). Contrary to this, injections of synthetic salmon CT provoked marked hypocalcemia in young iguanas (Kline, 1981) and snakes (Srivastav et al., 1986; Srivastav and Rani, 1989a; Hasan et al., 1992). This response diminished rapidly with age in chuckwalla (Kline, 1982). Administration of CT prevented hypercalcemia caused by UBX (Hasan et al., 1992). Long-term injections of CT induced bone formation and growth in hatchling turtles (Dimond et al., 1972; Belanger et al., 1973; Copp et al., 1973). CT completely prevented parathyroid hormone-induced changes (increased bone resorption) in hatchling turtles (Belanger et al., 1973). CT also affected body growth when hatchlings were reared on low-calcium diet i.e., they grow at a slower rate as compared to turtles kept on a high-calcium diet. Administration of CT increased growth rates of low-calcium hatchlings to a similar level of a high calcium group (Dimond et al., 1972; Copp et al., 1973). These studies indicate a role of CT in reptiles.

Extracts of UBG from turtles (Clark, 1968; Copp and Parkes, 1968), lizards (Moseley et al., 1968) and snakes (Uchiyama et al., 1978, 1981; Yoshihara et al.,

1985) provoked hypocalcemia into rats. By using bioassay, Uchiyam et al. (1981) have shown that UBG of hatching and young snake possess an enormous (2,700 mU/kg body wt) hypocalcemic activity in rats as compared to the adults (460 mU/kg body wt) in terms of relative activity per unit body weight.

Light microscopic and ultrastructural studies have also been performed to determine the functional aspect of UBG in reptiles. Experimentally induced hypercalcemia resulted in increased activity of the gland in young lizards (Dubewar and Suryawanshi, 1977; Dubewar et al., 1978) and snakes (Srivastav and Rani, 1992a; Srivastav et al., 1994), however, UBG of monitor lizard (Das and Swarup, 1975b) and aged *Uromastix hardwickii* (Dubewar et al., 1978) were unresponsive to such treatment. Ultrastructural studies revealed that snake UBG responds to high calcium treatment by releasing CT granules (Suzuki et al., 1982). Further, the increased activity in UBG cells of sea snake has been attributed to the fact that living in seawater, which is rich in calcium (Suzuki et al., 1984). Ultrastructural studies revealed a decreased release and increased intracellular storage of CT in lizard, *Iguana iguana* in response to dietary induced hypocalcemia. CT administration provoked hypocalcemia leading to inactivity of UBG of lizard (Pandey and Swarup, 1989) and snake (Srivastav and Rani, 1992b; Srivastav et al., 1998d).

After reviewing the existing literature on the ultimobranchial gland in vertebrates, it could be concluded that the gland is derived during development from the pharyngeal pouch epithelium and secretes calcitonin regardless of the shape of its structural unit (follicular or compact) and being separately located (non-mammalian vertebrates) or embedded in the thyroid (mammals). It is interesting to note that CT-like substances have also been found in invertebrates, as well as in the brain, liver, gut, thymus and other tissues of vertebrates. The presence of CT-like cells in the pharynx of crustaceans and protochordates seems to be of importance as the C cells originate from the ultimobranchial gland, which is a pharyngeal wall derivative in vertebrates. This indicates that CT-like cells in protochordates are precursors of C cells. The demonstration of human CT-like molecules in the unicellular organisms *Eschericha coli*, *Candida albicans* and *Aspergillus fumigatus* clearly indicates that CT is an ancient hormone which appeared early in evolution.

CT is known to participate in a large and diverse set of effects, but in most cases these were seen in response to supraphysiological doses of the hormone. Also, their physiological relevance is in suspect. CT plays a role in calcium homeostasis, at least in part by effects on bone, by suppressing its resorption, and the kidney, by inhibiting tubular calcium reabsorption. There seems to be species differences in the effects of CT. In some vertebrates, particularly fish, CT appears to play a significant role in calcium homeostasis, however in mammals, particularly humans, it has at best a minor role in regulating blood calcium levels. Mammalian CT at physiological doses is not essential as there is no readily apparent pathology due to chronically increased (medullary thyroid cancer) or decreased (surgical extirpation of thyroid gland) levels of CT. Probably, changes in amino acid sequence during the evolution are responsible for the loss of activity in mammals, as fish CT is 40 times more potent than human CT.

REFERENCES

Anderson, M.P. and Capen, C.C. (1976). Ultrastructural evaluation of parathyroid and ultimobranchial glands in iguanas with experimental nutritional osteodystrophy. *Gen. Comp. Endocrinol.* **30**: 209-222.

Baldwin, F.M. (1918). Pharyngeal derivatives of ambystoma. *J. Morph.* **30**: 605-680.

Baldwin, G.F. and Bentley, P.J. (1980). Calcium metabolism in bullfrog tadpoles. *J. Exp. Biol.* **88**: 357-365.

Baldwin, G.F. and Bentley, P.J. (1981). A role of skin in the Ca metabolism of frogs? *Comp. Biochem. Physiol.* **68**(A): 181-182.

Belanger, L.F., Dimond, M.T. and Copp, D.H. (1973). Histological observations on bone and cartilage of growing turtles treated with calcitonin. *Gen. Comp. Endocrinol.* **20**: 297-304.

Bentley, P.J. (1983). Urinary loss of calcium in an anuran amphibian (*Bufo marinus*) with a note on the effects of calcemic hormones. *Comp. Biochem. Physiol.* **76**(B): 717-719.

Bentley, P.J. (1984). Calcium metabolism in the amphibia. *Comp. Biochem. Physiol.* **79**(A): 1-5.

Bjornsson, B.Th., Haux, C., Forlin, L. and Deftos, L.J. (1986). The involvement of calcitonin in the reproductive physiology of the rainbow trout. *J. Endocrinol.* **108**: 17-23.

Bjornsson, B.Th., Haux, C., Bern, H.A. and Deftos, L.J. (1989). 17β-estradiol increases plasma calcitonin levels in salmonid fish. *Endocrinol.* **125**: 1754-1760.

Boschwitz, D. and Bern, H.A. (1971). Prolactin, calcitonin, and blood calcium in the toads, *Bufo boreas* and *Bufo marinus*. *Gen. Comp. Endocrinol.* **17**: 586-588.

Boschwitz, D. (1973). The antagonistic effects of exogenous calcitonin and calcium on the ultimobranchial body of *Bufo viridis* (Amphibia: Anura). *J. Herpetol.* **7**: 195-200.

Boschwitz, D. (1976). Seasonal changes in the histological structure of the ultimobranchial body of the toad *Bufo viridis*. *Br. J. Herpetology* **5**: 719-726.

Boudbid, H., Leger, A.F., Pidoux, E., Volle, G.E., Tahoulet, J., Moukhtar, M.S. and Treilhou-Lahille, F. (1987). Localization of a salmon calcitonin-like molecule in one type of ultimobranchial cells in the freshwater turtle, *Pseudemys scripta. Gen. Comp. Endocrinol.* **65**: 415-422.

Bouizar, Z., Khattab, M., Taboulet, J., Rostene, W., Milhaud, G., Treilhou-Lahille, F. and Moukhtar, M.S. (1989). Distribution of renal calcitonin binding sites in mammalian and nonmammalian vertebrates. *Gen. Comp. Endocrinol.* **76**: 364-370.

Camp, W.E. (1917). The development of the suprapericardial (postbranchial, ultimobranchial) body in *Squalus acanthias*. *J. Morphol.* **28**: 369-411.

Chan, D.K.O., Chester Jones, I. and Smith, R.N. (1968). The effect of mammalian calcitonin on the plasma levels of calcium and inorganic phosphate in the European eel (*Anguilla anguilla*). *Gen. Comp. Endocrinol.* **11**: 243-245.

Chan, D.K.O. (1972). Hormonal regulation of calcium balance in teleost fish. *Gen. Comp. Endocrinol.* **3**: 411-420.

Clark, N.B. (1968). Calcitonin studies in turtles. *Endocrinol.* **83**: 1145-1148.

Clark, N.B. (1971). The ultimobranchial body of reptiles. *J. Exp. Zool.* **178**: 115-124.

Clark, N.B. (1972). Calcium regulation in reptiles. *Gen. Comp. Endocrinol.* (Suppl.) 3: 430-440.

Coleman, R. and Phillips, A.D. (1972). Ultimobranchial gland ultrastructure of larval axolotls, *Ambystoma mexicanum* Shaw, with some observations on the newt, *Pleurodeles waltlii* Micahelles. *Z. Zellforsch. Mikrosk. Anat.* **134**: 183-192.

Coleman, R. and Phillips, A.D. (1974). The development and fine structure of the ultimobranchial glands in larval *Rana temporaria* L. *Cell Tissue Res.* **148**: 69-82.

Coleman, R. (1975). The development of fine structure of ultimobranchial glands in larval anurans. II. *Bufo viridis, Hyla arborea,* and *Rana ridibunda. Cell Tissue Res.* **164**: 215-232.

Copp, D.H., Cameron, E.C., Cheney, B.A., Davidson, A.G.F. and Henze, K.G. (1962). Evidence for calcitonin – a new hormone from the parathyroid that lowers blood calcium. *Endocrinol.* **70**: 638-649.

Copp, D.H., Cockroft, D.W. and Kueh, Y. (1967). Calcitonin from ultimobranchial glands of dogfish and chicken. *Science* **158**: 924-925.

Copp, D.H. and Parkes, C.O. (1968). Extraction of calcitonin from ultimobranchial tissue. In: *Parathyroid Hormone and Thyrocalcitonin* (Calcitonin) (Talmage, R.V. and Belanger, L.F., eds.), Excerpta Med. Found., Amsterdam. 74-84.

Copp, D.H., Belanger, L.F., Dimond, M.T. and Watts, E.G. (1973). Possible significance of calcitonin in nonmammals. In: *Clinical Aspects of Metabolic Bone Disease* (Frame, B., Parfitt, A.M., Duncan, H., eds.), Amsterdam: Excerpta Medica, 171-177.

Dacke, C.G. (1979). *Calcium Regulation in Sub-mammalian Vertebrates.* Academic Press, New York.

Das, V.K. and Swarup, K. (1975a). Effect of calcium-rich environment on the ultimobranchial glands of *Rana tigrina. Arch. Anat. Mikrosk.* **64**: 261-272.

Das, V.K. and Swarup, K. (1975b). The ultimobranchial body of *Varanus monitor. Arch. Biol.* (Bruxelles). **86**: 163-176.

Dimond, M.T., Belanger, L.F., Rogers, J. and Copp, D.H. (1972). Effect of calcitonin administration on growing turtles. *Physiologist* **15**: 120.

Dix, M.W., Pang, P.K.T. and Clark, N.B. (1970). Lizard calcium metabolism: lack of effect of mammalian calcitonin on serum calcium or phosphate levels in normal and parathyroidectomized lizards *Anolis carolinensis. Gen. Comp. Endocrionol.* **14**: 243-247.

Dubewar, D.M. and Suryawanshi, S.A. (1977). Effect of induced hypercalcemia on ultimobranchial gland of the lizard, *Calotes versicolor. Z. Mikrosk. Anat. Forsch.* **91**: 704-708.

Dubewar, D.M., Suryawanshi, S.A. and Rege, U.G. (1978). Effect of experimental hypercalcemia on the ultimobranchial gland of lizard with special reference to age. *Z. Mikrosk. Anat. Forsch.* **92**: 547-552.

Fenwick, J.C. (1975). Effect of partial ultimobranchialectomy on plasma calcium concentration and on some related parameters in goldfish (*Carassius auratus* L.) during acute transfer from freshwater to 30% seawater. *Gen. Comp. Endocrinol.* **25**: 60-63.

Fenwick, J.C. and Lam, T.J. (1988). Effects of calcitonin on plasma calcium and phosphate in the mud skipper, *Periophthalmodon schlosseri* (Teleostei), in water and during exposure to air. *Gen. Comp. Endocrinol.* **70**: 224-230.

Fouchereau-Peron, M., Arlot-Bonnemains, Y., Moukhtar, M.S. and Milhaud, G. (1987). Calcitonin induces hypercalcemia in grey mullet and immature freshwater and seawater adapted rainbow trout. *Comp. Biochem. Physiol.* **87**(A): 1051-1053.

Fouchereau-Peron, M., Arlot-Bonnemains, Y., Maubras, L., Milhaud, G. and Moukhtar, M.S. (1990). Calcitonin variations in male and female trout, *Salmo gairdneri* during the annual cycle. *Gen. Comp. Endocrinol.* **78**: 159-163.

Francescon, A. (1929). Corpo ultimobranchials nei rettile. *Arch. Ital. Anat. Embriol.* **26**: 387-400.

Gelsleichter, J. (2001). The ultimobranchial gland and calcitonin production in the Atlantic stingray, *Dasyatis sabina*. American Elasmobranch Society 2001 Annual Meeting, Pennsylvania.

Gelsleichter, J. (2004). Hormonal regulation of elasmobranch physiology. In: *Biology and Ecology of Sharks and Their Relatives* (Carrier, J., Musick, J. and Heithaus, M., eds.), CRC Press, Boca Raton, F.L.

Girgis, S.I., Galan Galan, F., Arnett, T.R., Rogers, R.M., Bone, Q., Ravazzole, M. and MacIntyre, I. (1980). Immunoreactive human calcitonin-like molecule in the nervous systems of protochordates and cyclostomes, Myxine. *J. Endocrinol.* **87**: 375-382.

Glowacki, J., O'Sullivan, J., Miller, M., Wilkje, D.W. and Deftos, L.J. (1985). Calcitonin produces hypercalcemia in leopard sharks. *Endocrinol.* **166**: 827-829.

Greil, A. (1905). Uber die anlage Lungen sowie der ultimobranchialen (postbranchialen, supraperikardialen) Korper bei anuren Amphibien. *Anat. Hefts Arb. Anat.* Inst. Wiesbaden **29**: 453-506.

Guardabassi, A. (1960). The utilization of the calcareous deposits in the endolymphatic sacs of *Bufo bufo bufo* in the mineralization of the skeleton. Investigations by means of ^{45}Ca. *Z. Zellforsch. Mikrosk. Anat.* **51**: 278-282.

Hammar, J.A. (1937). Zur Bildungsgeschichte der Krokodile. *Z. Mikrosk. Anat. Forsch.* **41**: 75-87.

Hasan, N., Das, V.K. and Swarup, K. (1992). Effect of salmon calcitonin administration on serum calcium and inorganic phosphate levels of intact and ultimobranchialectomized freshwater snake, *Natrix piscator* (Schneider). *Nat. Acad. Sci. Letters* **15**: 97-101.

Hasan, N., Pandey, A.K. and Swarup, K. (1993). Response of serum calcium and inorganic phosphate levels of *Bufo andersoni* to salmon calcitonin administration. *J. Adv. Zool.* **14**: 70-75.

Hayslett, J.P., Epstein, M., Spector, D., Myers, J.D., Murdaugh, H.V. and Epstein, F.H. (1971). Effect of calcitonin on sodium metabolism in *Squalus acanthias* and *Anguilla rostrata*. *Bull. Mt. Desert Is. Biol. Lab.* **11**: 33-35.

Hirano, T., Hasegawa, S., Yamauchi, H. and Orimo, H. (1981). Further studies on the absence of hypocalcemic effects of eel calcitonin in the eel, *Anguilla japonica*. *Gen. Comp. Endocrinol.* **43**: 42-50.

Hooker, W.H., McMillan, P.J. and Thaete, L.G. (1979). Ultimobranchial gland of the trout (*Salmo gairdneri*). II. Fine structure. *Gen. Comp. Endocrinol.* **38**: 275-284.

Kagwade, M.P. and Padgaonkar, A.S. (1992). Parathyroid and ultimobranchial glands of the smooth water snake, *Enhydris enhydris* (Schneider). *Biological Struct. Morphogen.* **41**: 171-173.

Kameda, Y. (1981). Distribution of C cells in parathyroid gland IV and thymus IV of different mammals studied by immunoperoxidase method using anti-calcitonin and anti-C-thyroglobulin antisera. *Kawasaki Med. J.* **7**: 97-111.

Kameda, Y. (1995). Evidence to support the distal vagal ganglion as the origin of C cells of the ultimobranchial gland in the chick. *J. Comp. Neurol.* **359**: 1-14.

Khairallah, L.H. and Clark, N.B. (1971). Ultrastructure and histochemistry of the ultimobranchial body of freshwater turtles. *Z. Zellforsch.* **13**: 311-321.

Kiebzak, G.M. and Minnick, J.E. (1982). Effects of calcitonin on electrolyte excretion in the lizard *Dipsosaurus dorsalis. Gen. Comp. Endocrinol.* **48**: 232-238.

Kingsbury, D.L. and Fenwick, J.C. (1989). The effect of eel calcitonin on calcium influx and plasma ion levels in axolotls, *Ambystoma mexicanum. Gen. Comp. Endocrinol.* **75**: 135-138.

Kitoh, J. (1970). Electron microscopic studies of the ultimobranchial body of the elasmobranchs. *Arch. Histol. Jap.* **31**: 269-281.

Kline, L.W. (1981). A hypocalcemic response to synthetic salmon calcitonin in the green iguana, *Iguana iguana. Gen. Comp. Endocrinol.* **44**: 476-479.

Kline, L.W. (1982). An age-dependent response to synthetic salmon calcitonin in the chuckwalla *Sauromalus obesus. Canad. J. Zool.* **60**: 1359-1361.

Klumpp, W. and Eggert, B. (1935). Die Schilddruse und die branchiogenen organe von *Ichthyophis glutinosus* L. *Zeitschr. Wiss. Zol.* **146**: 329-381.

Koyama, T., Makita, T. and Enomoto, M. (1984). Parathyroid morphology in rats after administration of active vitamin D_3. *Acta. Pathol. Jpn.* **34**: 313-324.

Krawarik, F. (1936). Uber eine bisher unbekannte Druse ohno Ausfuhrungsgong bei den heimischen Knochenfischen. *Z. Mikrosk. Anat. Forsch.* **39**: 555-609.

Krishna, L. and Swarup, K. (1985a). Response of parathyroid gland, ultimobranchial body, paravertebral lime-sacs and serum calcium level to salmon calcitonin administration in *Rana cyanophlyctis* (Anura: Amphibia). *Herpetologica* **41**: 65-70.

Krishna, L. and Swarup, K. (1985b). Effect of different doses of salmon calcitonin on serum calcium level of *Rana cyanophlyctis. Bolm. Fisiol. Anim. Sao Paulo* **9**: 67-70.

Le Douarin, N. and Le Lievre, C. (1970). Demonstration de l'origin neural des cellules a calcitonine du corps ultimobranchial chez l'embryon de poulet. *C.R. Hebd. Seances Acad. Sci. Ser.* **270**: 2857-2860.

Le Douarin, N., Fontaine, J. and Le Lievre, C. (1974). New studies on the neural crest origin of the avian ultimobranchial glandular cells: Interspecific combinations and cytochemical characterization of C cells based on the uptake of biogenic amine precursors. *Histochemistry* **38**: 297-305.

Lopez, E., Peignoux-Deville, J. and Bagot, E. (1968). Etude histophysiologique du corps ultimobranchial dun teleosteen, *Anguilla anguilla* L. au cours d'hypercalcemia experimentales. *C.R. Acad. Sci. Paris* **267**(D): 1531-1534.

Lopez, E., Chartier-Baraduc, M.M. and Deville, J. (1971). Mise en evidence de l'action de la calcitonine porcine sur l'os de la truite, *Salmo gairdneri* soumise a un traitement demineralisant. *C.R. Acad. Sci. Paris* **272**: 2600-2603.

Lopez, E. and Peignoux-Deville, J. (1973). Effect of prolonged administration of synthetic salmon calcitonin (SCT) on vertebral bone morphology and on the ultimobranchial body (UBB) activity of the mature female eel *Anguilla anguilla* L. In: *Proc. 9th European Symposium on Calcified Tissue* (Czitober, A. and Eschberger, J., eds.), H. Egerman, Eacta Publication, Vienna, 169-174.

Lopez, E., Peignoux-Deville, J., Lallier, F., Martelly, E. and Milet, C. (1976). Effects of calcitonin and ultimobranchialectomy (UBX) on calcium and bone metabolism in the eel, *Anguilla anguilla* L. *Calcif. Tissue Res.* **20**: 173-186.

MacInnes, D.G., Laszlo, I., MacIntyre, I. and Fink, G. (1982). Salmon calcitonin in lizard brain: A possible neuroendocrine transmitter. *Brain Res.* **251**: 371-373.

Marcus, H. (1908). Beitrage zur kenntnis der gymnophionen. I. Uber das Schlundspaltengebiet. *Arch. Mikr. Anat.* **21**: 695-774.

Marcus, H. (1922). Der kehlkopf bei Hypogeophis. *Anat. Anz.* **55**: 188-202.

Matsuda, K., Sasayama, Y., Oguro, C. and Kambegawa, A. (1989). Calcitonin immunoreactive cells found in the extra-ultimobranchial areas of the salamander, *Hynobius nigrescens*, during larval development. *Zool. Sci.* **6**: 611-614.

Maurer, F. (1888). Schilddruse Thymus and Kimenreste der Amphibien. *Morph. Yahrb.* **13**: 296-382.

Maurer, F. (1899). Die schilddruse thymus und andere schlundspalten derivate bei der eidechse. *Morph. Jahrb.* **27**: 119-172.

McMillan, P.J., Hooker, W.M., Roos, B.A. and Deftos, L.J. (1976). Ultimobranchial gland of the trout (*Salmo gairdneri*). I. Immuno-histology and radioimmuno-assay of calcitonin. *Gen. Comp. Endocrinol.* **28**: 313-319.

McWhinnie, D.J. and Scopelliti, T.G. (1978). A comparative study of the influence of calcitonin on plasma mineral and volume dynamics in the amphibians, *Rana pipiens* and *Xenopus laevis*, and the rat. *Comp. Biochem. Physiol.* **61**(A): 487-495.

Moseley, J.M., Matthews, E.W., Breed, R.H., Galante, L., Tse, A. and MacIntyre, I. (1968). The ultimobranchial origin of calcitonin. *Lancet* **1**: 108-110.

Nichols, S., Gelsleichter, J., Manire, C.A. and Cailliet, G.M. (2003). Calcitonin-like immunoreactivity in serum and tissues of the bonnethead shark, *Sphyrna tiburo*. *J. Exp. Zool.* **298**: 150-161.

Nunez, E.A. and Gershon, M.D. (1978). Cytophysiology of thyroid parafollicular cells. *Int. Rev. Cytol.* **52**: 1-80.

Nusbaum-Hilarowicz, J. (1916). Uber den Bau des Darmkanals beieingen Tiefseeknoch enfischen. *Anat. Anz.* **48**: 497-506.

Oguri, M. (1973). Seasonal histologic changes in the ultimobranchial gland of goldfish. *Bull. Jap. Soc. Sci. Fish.* **39**: 851-858.

Oguro, C., Uchiyama, M., Pang, P.K.T. and Sasayama, Y. (1978). Calcium homeostasis in urodele amphibians. In: *Comparative Endocrinology* (Gaillard, P.J. and Boer, H.H., eds.), Elsevier/North-Holland, Amsterdam, 269-272.

Oguro, C. and Uchiyama, M. (1980). Comparative endocrinology of hypocalcemic regulation in lower vertebrates. In: *Hormones, Adaptation and Evolution* (Ishii, S., Hirano, T. and Wada, M., eds.), Jap. Sci. Soc. Press, Tokyo, 113-121.

Oguro, C., Nagai, K.I., Tarui, H. and Sasayama, Y. (1981). Hypocalcemic factor in the ultimobranchial gland of the frog, *Rana rugosa*. *Comp. Biochem. Physiol.* **65**(A): 95-97.

Oguro, C., Tarui, H. and Sasayama, Y. (1983). Hypocalcemic potency of the ultimobranchial gland in some urodele amphibians. *Gen. Comp. Endocrinol.* **51**: 272-277.

Oguro, C., Fujimori, M. and Sasayama, Y. (1984a). Changes in the distribution of calcium in the frog, *Rana nigromaculata*, following ultimobranchialectomy and calcitonin administration. *Zool. Sci.* **1**: 82-88.

Oguro, C., Nogawa, H. and Sasayama, Y. (1984b). Occurrence of a physiologically active calcitonin in the brain of the bullfrog, *Rana catesbeiana*. *Zool. Sci.* **1**: 841-843.

Oguro, C. and Sasayama, Y. (1985). Endocrinology of hypocalcemic regulation in anuran amphibians. In: *Current Trends in Comparative Endocrinology* (Lofts, B. and Holmes, N., eds.), Hong Kong Univ. Press, Hong Kong, 839-841.

Okuda, R., Sasayama, Y., Suzuki, N., Kambegawa, A. and Srivastav, Ajai K. (1999). Calcitonin cells in the intestine of goldfish and comparison of the number of cells among saline-fed, soup-fed, or high Ca soup-fed fishes. *Gen. Comp. Endocrinol.* **113**: 267-273.

Orimo, H., Fujita, T., Yoshikawa, M., Watanabe, S., Otani, M. and Jinnosuke, J. (1972). Ultimobranchial calcitonin of the eel, *Anguilla japonica. Endocrinol. Jap.* **19**: 299-302.

Oughterson, S.M., Munoz-Chapuli, R., Andres, V.D., Lawson, R., Heath, S. and Davies, D.H. (1995). The effects of calcionin on serum calcium levels in immature brown trout *Salmo trutta. Gen. Comp. Endocrinol.* **97**: 42-48.

Pandey, A.K. and Swarup, K. (1989). Response of the ultimobranchial body, serum calcium and inorganic phosphate levels to porcine calcitonin administration in *Varanus flavescens* (Gray). *Folia Biol.* (Krakow) **37**: 97-100.

Pang, P.K.T. (1971). Calcitonin and ultimobranchial gland in fishes. *J. Exp. Zool.* **178**: 89-100.

Pathan, K.M. and Nadkarni, V.D. (1986). Effect of calcium-rich environment on the ultimobranchial bodies of *Rana cyanophlyctis* treated with calciferol. *Proc. Indian Natl. Sci. Acad.* **52**(B): 575-579.

Peignoux-Deville, J. and Lopez, E. (1970). Le corps ultimobranchil du salmon *Salmo salar.* Elude histophysiologique a diverse etapes de son cycle vital en eau douce. *Arch. Anat. Microsc. Exp.* **59**: 393-402.

Peignoux-Deville, J., Lopez, E., Lallier, F., Bagot, E.M. and Milet, C. (1975). Responses of ultimobranchial body in eels (*Anguilla anguilla* L.) maintained in seawater and experimentally matured to injections of synthetic salmon calcitonin. *Cell Tissue Res.* **164**: 73-83.

Peters, H. (1941). Morphologische und experimentelle untersuchungen uber die epithelkorper bei eidechsen. *Z. Mikrosk. Anat. Forsch.* **49**: 1-40.

Pilkington, J.B. and Simkiss, K. (1966). The mobilization of the calcium carbonate deposits in the endolymphatic sacs of metamorphosing frogs. *J. Exp. Biol.* **45**: 329-341.

Pilkington, J.B. (1967). Changes in composition of bone during the metamorphosis of *Rana temporaria. Calcif. Tissue Res.* **1**: 246-248.

Robertson, D.R. and Swartz, G.E. (1964a). The development of the ultimobranchial body in the frog *Pseudacris nigrita triseriata. Trans. Am. Microsc. Soc.* **83**: 330-337.

Robertson, D.R. and Swartz, G.E. (1964b). Observations on the ultimobranchial body in *Rana pipiens. Anat. Rec.* **148**: 219-230.

Robertson, D.R. and Bell, A.L. (1965). The ultimobranchial body in *Rana pipiens.* I. The fine structure. *Z. Zellforsch. Mikrosk. Anat.* **66**: 118-129.

Robertson, D.R. (1967). The morphology of innervation of the ultimobranchial body in the goldfish. *Anat. Rec.* **157**: 510.

Robertson, D.R. (1968a). The ultimobranchial body in *Rana pipiens.* IV. Hypercalcemia and glandular hypertrophy. *Z. Zellforsch.* **85**: 441-452.

Robertson, D.R. (1968b). The ultimobranchial body in *Rana pipiens.* VI. Hypercalcemia and secretory activity. Evidence for origin of calcitonin. *Z. Zellforsch.* **85**: 453-465.

Robertson, D.R. (1969a). The ultimobranchial body in *Rana pipiens.* VIII. Effects of extirpation upon calcium distribution and bone cell types. *Gen. Comp. Endocrinol.* **12**: 479-490.

Robertson, D.R. (1969b). The ultimobranchial body in *Rana pipiens.* IX. Effect of extirpation and transplantation on urinary calcium excretion. *Endocrinol.* **84**: 1174-1178.

Robertson, D.R. (1969c). The ultimobranchial body in *Rana pipiens.* X. Effect of glandular extirpation on fracture healing. *J. Exp. Zool.* **172**: 425-442.

Robertson, D.R. (1971a). The endocrinology of amphibian ultimobranchial glands. *J. Exp. Zool.* **178**: 104-114.

Robertson, D.R. (1971b). Cytological and physiological activity of the ultimobranchial glands in the premetamorphic anuran, *Rana catesbeiana. Gen. Comp. Endocrinol.* **16**: 329-341.

Robertson, D.R. (1986). The ultimobranchial body. In: *Vertebrate Endocrinology: Fundamentals and Biomedical Implications* (Pang, P.K.T. and Schreibman, M.P., eds.), Vol. I, Academic Press, 235-259.

Robertson, D.R. (1988). Immunohistochemical and morphometric analysis of calcitonin distribution in anuran ultimobranchial glands. *Gen. Comp. Endocrinol.* **71**: 349-358.

Sasayama, Y. and Oguro, C. (1976). Effects of ultimobranchialectomy on calcium and sodium concentrations of serum and coelomic fluid in bullfrog tadpoles under high calcium and high sodium environment. *Comp. Biochem. Physiol.* **54**(A): 35-37.

Sasayama, Y., Noda, H. and Oguro, C. (1976). On the development of the ultimobranchial gland in an anuran *Rana japonica japonica. Develop. Growth Differ.* **18**: 467-471.

Sasayama, Y. (1978). Effects of implantation of the ultimobranchial glands and the administration of synthetic salmon calcitonin on serum Ca concentration in ultimobranchialectomized bullfrog tadpoles. *Gen. Comp. Endocrinol.* **34**: 229-233.

Sasayama, Y. and Oguro, C. (1982). Urine calcium concentrations in bullfrog tadpoles kept in tap water or high calcium environment with special reference to ultimobranchialectomy. *Comp. Biochem. Physiol.* **77**(A): 309-311.

Sasayama, Y., Oguro, C., Yui, R. and Kambegawa, A. (1984). Immuno-histochemical demonstration of calcitonin in ultimobranchial glands of some lower vertebrates. *Zool. Sci.* **1**: 755-758.

Sasayama, Y. and Oguro, C. (1985). The role of the ultimobranchial glands on calcium balance in bullfrog tadpoles. In: *Current Trends in Comparastive Endocrinology* (Lofts, B. and Holmes, W.N., eds.), Hong Kong Univ. Press, Hong Kong, 837-838.

Sasayama, Y., Yoshihara, M., Fujimori, M. and Oguro, C. (1990). The role of the ultimobranchial gland in calcium metabolism of amphibians and reptiles, (Epple, A., Scanes, C.G. and Stetson, M.H., eds.), *Prog. Clin. Biol. Res.* **342**: 592-597.

Sasayama, Y., Koizumi, T., Oguro, C., Kambegawa, A. and Yoshizawa, H. (1991). Calcitonin immunoreactive cells are present in the brains of some cyclostomes. *Gen. Comp. Endocrinol.* **84**: 284-290.

Sasayama, Y. and Oguro, C. (1992). Ultimobranchial gland. In: *Atlas of Endocrine Organs*, (Matsumoto, A. and Ishii, S., eds.), Springer-Verlag, Berlin, 83-92.

Sasayama, Y., Suzuki, N., Oguro, C., Takei, Y., Takahashi, A., Watanabe, T.X., Nakajima, K. and Sakakibara, S. (1992). Calcitonin of the stingray: comparison of the hypocalcemic activity with other calcitonins. *Gen. Comp. Endocrinol.* **86**: 269-274.

Sasayama, Y., Suzuki, N. and Magtoon, W. (1995). The location and morphology of the ultimobranchial gland in medaka, *Oryzias latipes. Fish Biol. J. Medaka* **7**: 43-46.

Sasayama, Y., Matsubara, T. and Takano, K. (1999). Topography and immunohistochemistry of the ultimobranchial glands in some subtropical fishes. *Ichthyol. Res.* **46**: 219-222.

Sasayama, Y., Takemura, M. and Takano, K. (2001). Ultimobranchial glands in the teleost (*Plecoglossus altivelis ryukyuensis*): special references to changes of gland volume with maturation. *Okajimas Folia Anat. Jpn.* **78**: 101-106.

Sehe, C.T. (1960). Studies on the ultimobranchial bodies and thyroid gland in vertebrates – fishes and amphibians. *Endocrinol.* **67**: 671-676.

Sehe, C.T. (1965). Comparative studies on the ultimobranchial body in reptiles and birds. *Gen. Comp. Endocrinol.* **5**: 45-59.

Shirozaki, F. and Mugiya, Y. (2000). Effects of salmon calcitonin on calcium deposition on and release from calcified tissues in fed and starved goldfish *Carassius auratus*. *Fisheries Science* **66**: 695.

Shirozaki, F. and Mugiya, Y. (2002). Histomorphometric effects of calcitonin on pharyngeal bone in fed and starved goldfish *Carassius auratus*. *Fisheries Science* **68**: 269.

Simkiss, K. (1967). *Calcium in Reproductive Physiology*. Chapman and Hall, London.

Singh, R. and Kar, I. (1982). Ultimobranchial gland of the freshwater snake, *Natrix piscator* Schneider. *Gen. Comp. Endocrinol.* **48**: 1-6.

Singh, R. and Kar, I. (1985). Calcium and phosphate metabolism in Snake (*Natrix piscator*): role of parathyroid, ultimobranchial and thyroid gland. *Arch. Biol.* **96**: 73-80.

Singh, S. and Srivastav, A.K. (1993). Effects of calcitonin administration on the serum calcium and inorganic phosphate levels of the fish *Heteropneustes fossilis* maintained either in artificial freshwater, calcium-rich freshwater or calcium-deficient freshwater. *J. Exp. Zool.* **265**: 35-39.

Srivastav, A.K. (1980). Cytological activity of ultimobranchial body of froglets of *Rana cyanophlyctis* in response to calcium-rich environment. *Arch. Biol.* **91**: 439-444.

Srivastav, A.K. and Swarup, K. (1980). Serum calcium of *Heteropneustes fossilis* (Teleost) in response to calcitonin treatment. *Nat. Acad. Sci. Letters* **3**: 373-375.

Srivastav, A.K. and Swarup, K. (1982). Effect of calcitonin on calcitonin cells, parathyroid glands and serum electrolytes in the house shrew, *Suncus murinus*. *Acta Anatomica* **114**: 81-87.

Srivastav, A.K. and Swarup, K. (1984). Ultimobranchial body and parathyroid gland of *Rana tigrina* in response to glucagon administration. *Arch. Anat. Microsk. Morphol. Expt.* **73**: 69-74.

Srivastav, A.K., Srivastav, S.P., Srivastav, S.K. and Swarup, K. (1986). Calcitonin effects on serum calcium levels of the freshwater snake, *Natrix piscator*. *Gen. Comp. Endocrinol.* **61**: 436-437.

Srivastav, A.K. and Rani, L. (1988). Mammalian calcitonin cells: Retrospect and prospect. *Biological Struct. Morphogen.* **1**: 117-123.

Srivastav, S.P., Swarup, K., Singh, S. and Srivastav, A.K. (1989). Effects of calcitonin administration on ultimobranchial gland, Stannius corpuscles and prolactin cells in male catfish, *Clarias batrachus*. *Arch. Biol.* **100**: 385-392.

Srivastav, A.K. and Rani, L. (1989a). Serum calcium and inorganic phosphorus level of the freshwater snake, *Sinonatrix piscator* (formerly *Natrix piscator*) in response to calcitonin treatment. *Can. J. Zool.* **67**: 1335-1337.

Srivastav, A.K. and Rani, L. (1989b). Influence of calcitonin administration on serum calcium and inorganic phosphate of the frog, *Rana tigrina*. *Gen. Comp. Endocrinol.* **74**: 14-17.

Srivastav, A.K. and Rani, L. (1991a). Response of ultimobranchial body, parathyroid gland, serum calcium and serum phosphate in the frog, *Rana tigrina* after prolactin administration. *Europ. Arch. Biol.* **102**: 159-163.

Srivastav, A.K. and Rani, L. (1991b). Response of ultimobranchial body and parathyroid gland of the frog, *Rana tigrina* to administration of various vitamin D analogs. *Europ. Arch. Biol.* **102**: 153-158.

Srivastav, A.K. and Rani, L. (1992a). Ultimobranchial body and parathyroid glands of the freshwater snake, *Natrix piscator* in response to vitamin D_3 administration. *J. Exp. Zool.* **262**: 255-262.

Srivastav, A.K. and Rani, L. (1992b). Effects of calcitonin administration on the ultimobranchial body and parathyroid glands of the freshwater snake, *Natrix piscator*. *European Arch. Biol.* **103**: 181-185.

Srivastav, A.K., Rani, L. and Sasayama, Y. (1994). Influence of prolactin administration on the ultimobranchial body and parathyroid glands of the freshwater snake, *Natrix piscator. Okajimas Folia Anat. Jpn.* **71**: 59-66.

Srivastav, A.K., Singh, S., Srivastav, S.K. and Suzuki, N. (1997). Ultimobranchial gland of the catfish, *Heteropneustes fossilis* treated with vitamin D_3 and maintained either in artificial freshwater, calcium-rich freshwater or calcium-deficient freshwater. *Asian J. Exp. Sciences* **11**: 43-53.

Srivastav, A.K., Srivastav, S.K., Sasayama, Y. and Suzuki, N. (1998a). Salmon calcitonin induced hypocalcemia and hyperphosphatemia in an elasmobranch, *Dasyatis akajei. Gen. Comp. Endocrinol.* **109**: 8-12.

Srivastav, S.K., Srivastav, A.K. and Suzuki, N. (1998b). Influence of calcitonin on serum calcium levels of intact or hypophysectomized freshwater catfish, *Heteropneustes fossilis. General Comp. Endocrinol.* **112**: 141-145.

Srivastav, A.K., Tiwari, P.R., Srivastav, S.K., Sasayama, Y. and Suzuki, N. (1998c). Serum calcium and phosphate levels of calcitonin treated freshwater mud eel, *Amphipnous cuchia* kept in different calcium environments. *Netherlands J. Zool.* **48**: 189-198.

Srivastav, A.K., Das, V.K., Srivastav, S.K., Sasayama, Y. and Suzuki, N. (1998d). Morphological and functional aspects of reptilian ultimobranchial gland. *Anat. Histol. Embryol.* **27**: 359-364.

Srivastav, A.K., Tiwari, P.R. and Suzuki, N. (1999). Morphology and histology of ultimobranchial gland of the freshwater mud eel, *Amphipnous cuchia. J. Adv. Zool.* **20**: 17-20.

Srivastav, A.K., Das, V.K., Srivastav, S.K. and Suzuki, N. (2000). Amphibian calcium regulation: Physiological aspects. *Zoologica Poloniae* **45**: 9-36.

Srivastav, A.K., Tiwari, P.R., Srivastav, S.K. and Suzuki, N. (2002). Responses of the ultimobranchial gland to vitamin D3 treatment in freshwater mud eel, *Amphipnous cuchia*, kept in different calcium environments. *Anat. Histol. Embryol.* **31**: 257-261.

Stiffler, D.F., Yee, J.C. and Tefft, J.D. (1998). Responses of frog skin, *Rana pipiens*, calcium ion transport to parathyroid hormone, calcitonin, and vitamin D_3. *Gen. Comp. Endocrinol.* **112**: 191-199.

Suzuki, K., Yoshizawa, H., Yoshihara, M., Sasayama, Y. and Oguro, C. (1982). Ultrastructural studies on the ultimobranchial glands in some lower tetrapods. Comp. Endocrinology of Calcium Regulation. Japan Sci. Soc. Press, Tokyo, 115-119.

Suzuki, K., Yoshizawa, H., Yoshihara, M., Sasayama, Y. and Oguro, C. (1984). Ultrastructure of snake ultimobranchial glands. In: *Endocrine Control of Bone and Calcium Metabolism* (Cohn, D.V., Potts, J.T. Jr. and Fujita, T., eds.), Elsevier Science Publishers, B.V. 203-205.

Suzuki, N. (1995). Calcitonin-like substance in plasma of the hagfish, *Eptatretus burgeri* (Cyclostomata). *Zool. Sci.* **12**: 607-610.

Suzuki, N., Suzuki, D., Sasayama, Y., Srivastav, Ajai K., Kambegawa, A. and Asahina, K. (1999). Plasma calcium and calcitonin levels in eels fed a high calcium solution or transferred to seawater. *Gen. Comp. Endocrinol.* **114**: 324-329.

Suzuki, N., Suzuki, T. and Kurokawa, T. (2000). Suppression of osteoclastic activities by calcitonin in the scales of goldfish (freshwater teleost) and nibbler fish (seawater teleost). *Peptides* **21**: 115-124.

Suzuki, N. (2001). Calcitonin-like substance in the plasma of cyclostomata and its putative role. *Comp. Biochem. Physiol. B Biochem. Mol. Biol.* **129**: 319-326.

Swarup, K. and Ahmad, N. (1979). Ultimobranchial body of *Notopterus notopterus* in relation to calcium and sodium-rich environments. *Z. Mikrosk. Anat. Forsch.* **93**: 662-672.

Swarup, K. and Krishna, L. (1980). Studies of ultimobranchial body of *Rana cyanophlyctis* in response to experimental hypercalcemia. *Z. Mikrosk. Anat. Forsch.* **94**: 818-824.

Swarup, K. and Srivastav, S.P. (1984). Structure and behaviour of ultimobranchial gland in response to vitamin D_3 induced hypercalcemia in male *Clarias batrachus*. *Arch. Anat. Microsc. Exp.* **73**: 223-229.

Takagi, I. and Yamada, K. (1977). An electron microscopic study of the ultimobranchial body of the crucian carp (*Carassius carassius*). *Okajimas Folia Anat. Jan.* **54**: 205-228.

Takagi, I. and Yamada, K. (1982). Light and electron microscopic studies on ultimobranchial body of softshelled turtle. *Okajimas Folia Anat. Jpn.* **59**: 231-246.

Takagi, I. and Yamada, K. (1986). Histological structure and calcitonin secretion in the Japanese newt ultimobranchial body in the urodele amphibians. *Okajimas Folia Anat. Jpn.* **63**: 23-26.

Thorndyke, M.C. and Probert, L. (1979). Calcitonin-like cells in the pharynx of the ascidian *Styela clava*. *Cell Tissue Res.* **203**: 301-309.

Treilhou-Lahille, F., Jullienne, A., Aziz, M., Beaumont, A. and Moukhtar, M.S. (1984). Ultrastructural localization of immunoreactive calcitonin in the two cell types of the ultimobranchial gland of the common toad (*Bufo bufo* L.). *Gen. Comp. Endocrinol.* **53**: 241-251.

Uchiyama, M., Yoshihara, M., Murakami, T. and Oguro, C. (1978). Presence of a hypocalcemic factor in the ultimobranchial gland of the snake. *Gen. Comp. Endocrinol.* **36**: 59-62.

Uchiyama, M. (1980). Hypocalcemic factor in the ultimobranchial glands of the newt, *Cynops pyrrhogaster*. *Comp. Biochem. Physiol.* **66**(A): 330-334.

Uchiyama, M., Yoshihara, M., Murakami, T. and Oguro, C. (1981). Calcitonin content in the ultimobranchial gland of the snake *Elaphe climacophora*: Comparison of adults, young and hatchlings. *Gen. Comp. Endocrinol.* **43**: 259-262.

Uhlenhuth, E. and McGowan, F. (1924). The growth of the thyroid and postbranchial body of the salamander, *Ambystoma opacum*. *J. Gen. Phys.* **6**: 597-602.

Van Bemmelen, J.F. (1886). Uber vermuthliches redimenters Kiemenspalten bei Elasmobranchier. *Mitt. Zool. Sta. Neapel* **6**: 165-184.

Watzka, M. (1933). Vergleichende untersuchungen uber den ultimobranchialen korper. *Z. Mikrosk. Anat. Forsch.* **34**: 485-533.

Welsch, U. and Schubert, C. (1975). Observations on the fine structure, enzyme histochemistry, and innervation of parathyroid gland and ultimobranchial body of *Chthonerpeton indistinctum* (Gymnophiona, Amphibia). *Cell Tissue Res.* **164**: 105-119.

Wendelaar Bonga, S.E. (1980). Effect of synthetic salmon calcitonin and low ambient calcium on plasma calcium, ultimobranchial cells, Stannius bodies and prolactin cells in the teleost *Gasterosteus aculeatus*. *Gen. Comp. Endocrinol.* **40**: 99-108.

Wendelaar Bonga, S.E. (1981). Effect of synthetic salmon calcitonin on protein-bound and free plasma calcium in the teleost *Gasterosteus aculeatus*. *Gen. Comp. Endocrinol.* **43**: 123-126.

Wendelaar Bonga, S.E. and Lammers, P.I. (1982). Effects of calcitonin on ultrastructure and mineral content of bone and scales of the ichlid teleost, *Sarotherodon mossambicus*. *Gen. Comp. Endocrinol.* **48**: 60-70.

Wendelaar Bonga, S.E. and Pang, P.K.T. (1991). Control of calcium regulating hormones in the vertebrates: Parathyroid hormone, calcitonin, prolactin and stanniocalcin. *Int. Rev. Cytol.* **128**: 139-213.

Wilder, M.C. (1929). The significance of the ultimobranchial body: A comparative study of its occurrence in urodeles. *J. Morphol.* **47**: 283-333.

Yamada, K., Sakai, K. and Yamada, S. (2001). A morphological and immunohistochemical study of the ultimobranchial body in the Japanese lizard and the snake. *Okajimas Folia Anat. Jpn.* **78**: 161-168.

Yamane, S. (1977). Sexual differences in histology of the ultimobranchial gland of mature Japanese eels (*Anguilla japonica*). *Zool. Mag.* **26**: 261-263.

Yamane, S. and Yamada, J. (1977). Histological changes of the ultimobranchial gland through the life history of the masu salmon. *Bull. Jpn. Soc. Sci. Fish.* **43**: 375-386.

Yamane, S. (1978). Histology and fine structure of the ultimobranchial gland in the zebrafish, *Brachydanio rerio*. *Bull. Fac. Fish. Hokkaido Univ.* **29**: 212-221.

Yamane, S. (1981). Sexual differences in ultrastructure of the ultimobranchial gland of mature eels (*Anguilla japonica*). *Bull. Fac. Fish. Hokkaido Univ.* **32**: 1-5.

Yamauchi, H., Matsuo, M., Yoshida, A. and Orimo, H. (1978). Effect of eel calcitonin on serum electrolytes in the eel *Anguilla japonica*. *Gen. Comp. Endocrinol.* **34**: 343-346.

Yaoi, Y., Suzuki, M., Tomura, H., Sasayama, Y., Kikuyama, S. and Tanaka, S. (2003). Molecular cloning of otoconin-22 complementary deoxyribonucleic acid in the bullfrog endolymphatic sac: effect of calcitonin on otoconin-22 messenger ribonucleic acid levels. *Endocrinol.* **144**: 3287-3296.

Yoshida, A., Kaiya, H., Takei, Y., Watanabe, T.X., Nakajima, K., Suzuki, N. and Sasayama, Y. (1997). Primary structure and bioactivity of bullfrog calcitonin. *Gen. Comp. Endocrinol.* **106**: 1-6.

Yoshidaterasawa, K., Hayakawa, D., Chen, H., Emura, S., Tamada, A., Okumura, T. and Shoumura, S. (1998). Fine structure of the parathyroid and ultimobranchial glands of the snake, *Elaphe quadrivirgata*. *Okajimas Folia Anat. Jpn.* **75**: 141-153.

Yoshihara, M., Uchiyama, M. and Murakami, T. (1979). Notes on the ultimobranchial glands of some Japanese snakes. *Zool. Mag.* **88**: 180-184.

Yoshihara, M., Uchiyama, M., Murakami, T., Yoshizawa, H. and Oguro, C. (1985). Calcitonin content in the ultimobranchial gland of snake: Comparison of pre-laying and post-laying females. In: *Current Trends in Comparative Endocrinology* (Lofts, B. and Holmes, W.N., eds.), Hong Kong University Press, Hong Kong, 849-850.

Yui, R., Yamada, Y., Kayamori, R. and Fujita, T. (1981). Calcitonin immunoreactive neurons in the brain of the bullfrog, *Rana catesbeiana* with special reference to their liquor-contacting and neurosecretory nature. An immunochemical and immunohistochemical study. *Biomed. Res.* **2**: 208-216.

Zaccone, G. (1980). Histochemical identification of fish ultimobranchial cells with polypeptide hormone producing APUD cells. *Acta Histochem.* **67**: 13-16.

CHAPTER 3

Melatonin and Immunomodulation: Involvement of the Neuro-endocrine Network

C. Haldar[1],[*], S.S. Singh[1], S. Rai[1], K. Skwarlo-Sonta[2], J. Pawlak[2] and M. Singaravel[1]

Abstract

Many neurotransmitters, neuroendocrine factors and hormones can drastically change immune function. Environmental stimulus to the nervous system affects the immune system and vice versa, essentially via the endocrine system. The present chapter deals with seasonal fluctuation in immunity, endocrine functions and the incidence of opportunistic diseases documented in a variety of species. Changes in immune functions appear to be mediated by day length (dark phase) and secretion of melatonin (MEL) from pineal gland. On the basis of findings, we may suggest that MEL secretion induced by short photoperiod acts as blaster to the immune function in winter to help the individuals to cope with seasonal stresses that would otherwise compromise immune function to critical levels.

The mechanisms underlying action of MEL involve a complex neuroendocrine network. Opioid peptides, particularly β-endorphin and Met-enkephalin have been implicated as immunomodulators. MEL stimulates the production of Interleukin-2 and γ-interferon. It has been demonstrated that MEL could act directly on the target cells through high-affinity G-protein coupled membrane-bound receptors (MT1 and MT2) described in primary and secondary lymphoid organs of various mammalian species. There is a reciprocal communication between the HPA (Hypothalamo-Pituitary-Adrenals) axis and the immune system. The precise mechanism by which immune system affects HPA axis is unknown, but it probably involves the release of diverse cytokines from activated

[1]Pineal Research Laboratory, Department of Zoology, Banaras Hindu University, Varanasi 221 005, India.
[2]Department of Animal Physiology, Faculty of Biology, Warsaw University, Miecznikowa 1, 02-096 Warsaw, Poland.
[*]*Corresponding author*: Tel.: + 91 542 2307149 ext. 125(0), Fax: + 91 542 2368174.
E-mail: chaldar@bhu.ac.in, chaldar 2001@yahoo.com

immune cells. MEL completely counteracts thymus involution and immunological depression induced by stress or glucocorticoid treatment.

Generally, thyroid hormone enhances immune function by promoting thymocyte maturation and differentiation. Receptors for thyrotropin (TSH) with many similarities to TSH-binding sites on thyroid cells have been found on lymphocytes. We have noted an interesting feature of the functional relationships between thyroxin and MEL supporting our idea of "Trade off" hypothesis between those two hormones in the control of the immune status. Testosterone generally suppresses the immune function. Castration resulted in increased immunity. Siberian hamsters and squirrels kept in short day, decreased estradiol level in females, which can be responsible for a winter decline in the immune activity suggesting again the major role of MEL as an immuno-enhancer.

INTRODUCTION

Maintenance of health depends to a significant extent on the ability of the exposed host to respond appropriately and, eventually, to adapt to environmental stressors. It is now well established that inappropriate or maladaptive response to such stressors weakens the body's resistance to other stimuli from the environment, such as pathogenic organisms, which have an impact on the body via redundant and reciprocal interactions between the body and the brain. Anatomically and functionally interconnected nervous, endocrine and immune systems utilize a large array of chemical messengers including hormones, cytokines and neurotransmitters, and express specific receptors able to recognize chemical messages sent by a particular component of this homeostasis keeping systems (Besedovsky and del Rey, 1996). It is, therefore, obvious that many neurotransmitters, neuroendocrine factors and hormones can drastically change the immune function, and on the contrary cytokines derived from immunocompetent cells can profoundly affect the central nervous system. As a consequence, any environmental stimulus to the nervous system will affect the immune system and *vice versa*, essentially via the endocrine system. In this conceptual basis it should not be surprising to note that the day/night photoperiod, which constitutes a basic environmental cue for any organism, can also influence the immune hematopoietic system. As for many other adaptive responses, a major mediator of this influence seems to be the pineal gland, which: (1) transduces the light/dark cycle into the circadian synthesis, and (2) releases melatonin (Yellon et al., 1999; Guerrero and Reiter, 2002).

The study of the immune system is quite common among clinicians and physicians, but the basic biologists usually examined the interactions between endocrine organs and immunity. Furthermore, complete understanding of the mechanisms underlying the immunological and hematopoietic action of many neurotransmitters and neurohormones, including melatonin is still a far cry. Even between the basic biologists, the interactions of the endocrine and immune systems, which protect the seasonal breeders from natural challenges, has received very little attention.

Seasonal Challenges to Immune System

Seasonal fluctuation in immune functions and the incidence of opportunistic diseases have been well documented in a variety of species including humans (Nelson and Drazen, 1999). The number of circulating leukocytes, spleen and thymus masses in mice (*Mus*), rats (*Rattus*), rabbits (*Lepus*), dogs (*Canis*), ground squirrels (*Citellus*), voles (*Microtus*), deer mice (*Peromyscus maniculatus*), cotton rats (*Sigmodon hispidus)* and humans were reported to be elevated during autumn and winter (Lochmiller and Ditchkoff, 1999). Environmental pressures might have led to the increased sophistication of vertebrate anticipatory (adaptive) immune system, perhaps due to the enhanced threat of infections in these complexes. In long living animals, the development of a finely tuned immune system with circulating effector cells was favored. The challenges of the environment also come from the variations in ecofactors (photoperiod, temperature, rainfall/humidity), which is responsible for providing food and shelter to improve the status of the animal for survival.

Tropical mammals and birds expressing seasonal diseases such as conjunctivitis and dermal infection during monsoon (high temperature and humidity) present a relationship between annual rhythm in melatonin and immune status (Haldar et al., 2001; Haldar and Singh, 2001). Moreover, in diurnal tropical rodent *Arvicanthis ansorgei*, living in environmental conditions where the annual day length does not change dramatically, the changes in the pineal arylalkylamine N-acetyl transferase (AA-NAT) activity and melatonin content were found recently (Garidou-Boof et al., 2005). The immune function in this species was not examined to date, however, it is well accepted that their seasonal reproductive cycle is controlled by the annual variations in day length and modulated by ambient temperature, rainfall and food availability. It can be supposed that due to seasonality in their resistance against pathogens, animals indicated the difference in their immune function efficiency.

Photoperiodic Modulation of Immune Function

Seasonal changes in the immune functions appear to be mediated by day length, since in deer mice and Syrian hamsters, short-day exposure increases splenic mass and elevates total lymphocyte and macrophage count (Nelson and Drazen, 1999), as well as attenuates the development of infections (Bilbo et al., 2002). Also, in laboratory strains of rats (*Rattus norvegicus*), which are traditionally considered to be reproductively non responsive to photoperiodic information (Nelson and Blom, 1994), seasonal changes in immunity could be observed. Maintaining rats in constant darkness (DD) for four weeks increased thymic mass and the number of thymocytes, in comparison to the control animals maintained in normal day length (Mahmoud et al., 1994). Contrarily, constant light (LL) exposure for four weeks decreased the thymic mass in Wistar rats (Mahmoud et al., 1994). Wurtman and Weisel (1969) reported that photoperiod also influences the splenic weight in rats. However, photoperiod appears to be more effective in influencing the immune function in seasonal breeders like hamsters or deer mice, i.e., species in which reproduction strongly depends on changing lighting conditions (Demas et al., 1996; Nelson et al., 1998). For example, splenic mass, total splenic lymphocyte number and macrophage count as well as white blood cell (WBC) number, were significantly higher in hamsters exposed to short-day, as compared to animals exposed to long photoperiod (Brainard et al., 1987). However, in this species, photoperiod neither affected thymic weight nor antibody production (Brainard et al., 1987). Deer mice maintained in short photoperiod displayed faster healing rates than long-day animals (Nelson and Blom, 1994). The pineal gland and its principal hormone, melatonin can also affect lymphatic tissue sizes. Exposure of male and female hamsters to short-days or daily afternoon melatonin injections elevated splenic mass, which could be prevented in short-day hamsters by pinealectomy (Guerrero and Reiter, 2002).

The Pineal Gland-Immune Network

The study of the two-way relationship between the pineal gland, melatonin and immune system has raised considerable importance in recent years (Guerrero and Reiter, 2002; Skwarlo-Sonta, 2002; Skwarlo-Sonta et al., 2003; Carillo-Vico et al., 2005). Experiments have shown that there is a functional relationship between the pineal gland and the immune system. Circadian rhythms are present in most, if not all, immune functions e.g., the peak level (acrophase) of circulating lymphocytes is observed close to that of melatonin in normal (24 hr) environment (Haldar et al., 2006). It has been reported that the circadian rhythm of mononuclear cells, T and B lymphocytes and the serum levels of melatonin, particularly affect the functionality of the natural killer cells in rats (McNaulty et al., 1990). Both surgical pinealectomy (Csaba and Barath, 1975; Del Gobbo et al., 1989), and functional (permanent lighting) and pharmacological (evening administration of the β-adrenergic blocker, propranolol) inhibition of the pineal gland function in mice (Maestroni et al., 1987) resulted in a depressed cellular and humoral immune response and IL-2 production. Exogenous melatonin administration in the evening reversed this effect, suggesting that circadian melatonin synthesis is mandatory for optimal immune functions (Rai and Haldar, 2003).

Moreover, a parallel pattern of the diurnal rhythms of melatonin and thymic hormones (thymosin α_1 and thymulin) was demonstrated in rats and humans. Pinealectomy caused a decrease in both hormone content in rat thymus and serum, reversed by melatonin injections (Molinero et al., 2000). The pineal gland exerted a stimulatory action on thymic growth, since the administration of pineal extracts might cause in the peripubertal mice, around 60 days of age, a thymic hyperplasia, suggesting an involvement of sex steroid hormones (Vermeulen et al., 1993). Disruption of the nighttime peak of melatonin following pinealectomy completely abolished the proliferation of bone marrow progenitors for granulocytes and macrophages (colony-forming units granulocyte-macrophage (CFU-GM)); Haldar et al., 1991, 1992a, b, c). As predicted, melatonin receptors have been identified on circulating lymphocytes (Calvo et al., 1995), thymocytes and splenocytes of rats (Rafii-El-Idrissi et al., 1995) and Indian palm squirrel (Rai, 2004), suggesting a direct effect of melatonin on the immune system function.

On the basis of findings, it could be suggested that melatonin secretion induced by short photoperiod acts as blaster to the immune function of these animals in winter to help the individuals to cope with seasonal stresses (low ambient temperature) that would otherwise compromise the immune function to critical levels. On the other hand, in seasonally breeding and hibernating Siberian hamsters *Phodopus sungorus,* field studies showed decreased immune activity during short-day period of the year (autumn and winter), in contradiction to laboratory studies, in which many immune parameters are elevated in short-day (Nelson et al., 1995). In the authors' model of experimental peritonitis, hamsters kept in short-day laboratory conditions presented a decreased inflammatory reaction measured as the reactive oxygen species (ROS) production in peritoneal leukocytes compared to long-day animals. Also, splenocyte proliferation in the latter was significantly higher (Pawlak et al., 2005). Melatonin addition *in vitro* further inhibited both ROS production and splenocyte mitogenic response. These data suggest that the influence of melatonin on the immune system depends on the species and experimental model used. It shows also that mechanisms underlying its action in the immune system involve a complex neuro-endocrine network.

In one of the authors' experimental models, the tropical Indian palm squirrel *F. pennanti,* short-day related increase in immunity was noted, suggesting a strong influence of endogenous melatonin in maintenance of immune status in this seasonal breeder (Haldar et al., 2001; Rai, 2004). The authors further interested in immune status of seasonally breeding laboratory animals, when observed that a certain part of the year they are susceptible to more infections and diseases and these are particularly the transitional phases between two reproductive periods i.e., progressive and regressive phases (Singh, 2003; Rai, 2004). Indian palm squirrels, during the reproductively active phase in summer months, are healthy even though the peripheral level of the so-called immunoenhancing hormone-melatonin, is low. The favorable environmental conditions with enough food, shelter for young ones and photoperiod along with internal high gonadal steroids could also be responsible for their good health. On the other hand, during reproductively inactive phase when the environmental conditions are not that favorable for squirrels, the internal high melatonin enhances the immune function to keep them healthy and to surpass winter (Haldar et al., 2004). Transitional periods, however, are crucial for this small rodent i.e., gonadal regressive and progressive phases when neither melatonin is having a threshold to enhance immunity, nor are the decreasing gonadal steroids. Work is in progress to examine the sensitivity of melatonin receptors on the lymphoid tissues during those two transitional phases and to find means to improve the immune status of the squirrels. The findings clearly revealed that endogenous melatonin had a positive effect on rhythmic function of some immune parameters such as CFU-GM of bone marrow (Haldar et al., 1992a, b, c). The authors assessed the effect of melatonin on immune parameters in a seasonally breeding animal, the Indian palm squirrel *F. pennanti* and bird *P. asiatica* during reproductive phases *in vivo* as well as *in vitro*. Melatonin treated squirrels and birds showed an increase in percent lymphocyte count (% LC) and peripheral blood total leukocyte count (TLC) and percent stimulation ratio (%SR) of thymocytes and splenocytes, which was low following pinealectomy in squirrels. Melatonin administration recovered all the decreased immune parameters to the control level (Rai and Haldar, 2003). MEL supplementation also enhanced all immune parameters in both *in vivo* and *in vitro*, suggesting that the immune system was sensitive during the reproductively active phase to MEL when peripheral melatonin level was low and the role of melatonin could be important for the maintenance of the immune status of seasonal breeders.

MECHANISM OF IMMUNOMODULATION BY MELATONIN

The mechanism by which melatonin modulates the immune function is still not resolved. However, the following hypotheses have been suggested:

1. The Pineal Gland–Immune System – Opioid Network

Opioids have been suggested to be the mediators of melatonin action on the immune system. Naltrexone – the opioid antagonist, abolished the immunoenhancing and antistress effects of melatonin (Maestroni et al., 1999). Dynorphin B and β-endorphin mimicked the immunological effects of melatonin. Melatonin also stimulated the release of opioid peptides by activated T-helper thymocytes (Maestroni and Conti, 1990). These melatonin-induced immuno-opioids (MIIO) mediated the immunoenhancing and antistress effects of melatonin and cross-reacted immunologically with anti β-endorphin and anti-met-enkephalin antisera in laboratory rodents (Maestroni and Conti, 1990). Recently, proopiomelanocortin (POMC) and enkephalin genes expression in immune cells (peritoneal leukocytes and splenocytes) in exogenous melatonin treated chicken with experimental peritonitis have been demonstrated (Majewski et al., 2005).

Opioid peptides, particularly β-endorphin and Met-enkephalin, have also been implicated as immunomodulators, since they can affect several immune mechanisms. However, it is difficult to conclude whether, as a whole, these peptides are immunosuppressive or immunostimulant since the effects reported differ depending on the type of immune process studied, the cell source and type, the species, the concentrations of the opioid peptide used, and the experimental conditions i.e., *in vivo* or *in vitro* (Homo Delarche and Durant, 1994). Actually, melatonin-induced endogenous opioid demonstrated in chickens seem to be responsible for a pro-inflammatory effect observed in this experimental model (Majewski et al., 2005).

2. Lymphokines as Mediators of Melatonin Action on the Immune System

The production of Interleukin-2 and γ-interferon is stimulated by melatonin (Caroleo et al., 1992). As interleukins and interferons are known stimulators of natural killer cells' activity and other immune cells, it is possible that these lymphokines mediated the observed effects of melatonin on the mammalian immune responses. Recent data presented a synergistic effect of melatonin and lipopolysaccharide conditioned medium (containing induced IL-1), on the T-cell genesis (thymus) and also on functional B and T lymphocytes (spleen). An explanation could be that initially melatonin binds to specific melatonin receptors in helper T cells and/or monocytes stimulating the production of either IFN- γ, IL-2, melatonin induced immuno opioid (MIIO), IL-1, IL-6 and IL-12, which in turn might upregulate the immune response. However, these explanations deserve to be studied further to note their balancing effect on immune status under adverse environmental conditions (Haldar et al., 2004).

3. Direct Melatonin Action

a. Melatonin Receptors within Immune System

Different aspects of melatonin effects within immune system was extensively reviewed and discussed by Carillo-Vico and co-workers (2005). It has been demonstrated that melatonin acts directly on the target cells through high-affinity G-protein coupled membrane-bound receptors (Dubocovich and Markowska, 2005). This type of receptor has been described in: primary and secondary lymphoid organs of various mammalian species (Lopez-Gonzalez et al., 1993; Poon et al., 1994), rat splenocytes (Rafii-El-Idrissi et al., 1995), human peripheral lymphocytes, T-helper lymphocytes in bone marrow (Maestroni, 1995), splenocytes, thymocytes and bone marrow lymphocytes of Indian palm squirrel (Rai, 2004).

Moreover, being a highly lipophilic molecule, melatonin easily penetrates all biological barriers, including cell membrane, and therefore is able to exert its action also within immune cells, using RZR/ROR (retinoid Z receptor/ retinoid orphan receptor) nuclear receptors, described also in immune cells (Rafii-El-Idrissi et al., 1998). These receptors seem to mediate some effects of melatonin on cytokine production, cell proliferation and oncostasis (Garcia-Maurino et al., 2000; Winczyk et al., 2001; Treeck et al., 2006).

b. Melatonin as an Antioxidant

Melatonin is also well-known as a potent antioxidant due to its particular property of free radical scavenging (Reiter et al., 2000a), demonstrated also within the immune system (Reiter et al., 2000b). Thus, melatonin exhibits a strong anti-inflammatory activity, counteracting liposachharide (LPS)-induced nitric oxide synthase (NO production) and abolishes a rise in lipid peroxidation in both *in vivo* and *in vitro* inflammation models. It has been also demonstrated that melatonin protects against oxidative damages through the stimulation of anti-oxidative enzymes, i.e., reduced glutathione level (Gitto et al., 2001). Melatonin also inhibits adhesion molecules and pro-inflammatory cytokines (like TNF-α) (Reiter et al., 2000b).

4. Melatonin Action through Neuro-endocrine Network

It is now considered that both neural and endocrine factors work together to maintain the immune system within a safe operating limit, to prevent the over activation of the immune system to avoid the destruction of self tissue and cells (Fig. 1).

(a) Hormones of the Hypothalamo-Pituitary-Adrenal Axis

When the hormonal regulation of the immune system is considered, the main focus should be on glucocorticoids, which have a well-documented

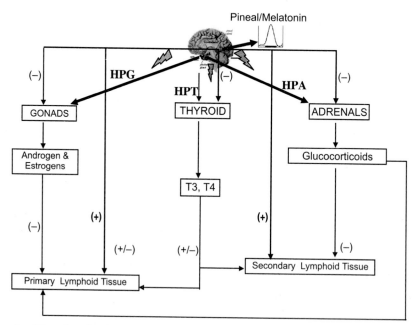

Fig. 1. Pineal-endocrine control of immune system (HPG – hypothalamo-pituitary-gonadal, HPT – hypothalamo-pituitary thyroid and HPA – hypothalamo-pituitary-adrenal axis).

immunoinhibitory activity (Black, 1994). Adrenalectomy increases lymphatic organ masses and B-cell activity (del Rey et al., 1984). Corticoids can act directly on immune cells through their receptors (Werb et al., 1978) reducing, for example, production of cytokines (Munk and Guyre, 1986). However, the upstream components of the hypothalamo-pituitary adrenal (HPA) axis, such as adrenocorticotropic hormone (ACTH) and corticotropin releasing hormone (CRH) may affect immune functions, i.e., ACTH is known to influence immune cell functions *in vitro* (Audhya et al., 1991).

There is a reciprocal communication between the HPA axis and the immune system. Increased glucocorticoid production is associated with the activation of the immune system by numerous antigens (Besedovsky et al., 1991). Moreover, activation of the immune system not only stimulates the activity of HPA axis as part of a regulatory response, but there is evidence that immune cells themselves may produce endocrine products such as, ACTH (Smith and Blalock, 1981), which acts locally as an immunoregulator. The precise mechanism by which the immune system affects HPA axis is unknown, but it probably involves the release of diverse cytokines from activated immune cells (Besedovsky and del Rey, 1996).

Glucocorticoids are released under stressful situations, such as unfavorable environmental conditions, and can compromise cellular and humoral immunity (Levi et al., 1988). Also, circadian variations in several immune parameters are negatively correlated with adrenal hormones secretion (Angeli et al., 1992). The HPA axis may be involved in the mediation of melatonin action on the immune system. Melatonin treatment can ameliorate the immunocompromising effect of glucocorticoids (Aoyama et al., 1987; Haldar et al., 2004), however, the immunostimulatory action of melatonin also varies with the steroid milieu (Persengiev et al., 1991), suggesting a bi-directional interaction between the two. For example, cortisol treatment of ducklings reduced the number of thymic melatonin receptors (Poon et al., 1994) and melatonin treatment decreased the density of thymic glucocorticoid and progestin receptors in rats (Persengiev et al., 1991).

With the demonstration of a steroid-dependent modulation of the rat thymic glucocorticoid receptors by melatonin, and modulation of the duck thymic $2[^{125}I]$ iodomelatonin binding sites, the thymus has been suggested to be the target site for the immunomodulatory interactions between pineal melatonin and adrenal steroids (Persengiev et al., 1991; Poon et al., 1994). Melatonin may completely counteract thymus involution and immunological depression induced by stress or glucocorticoid treatment (Maestroni and Conti, 1990).

The results of the authors' experiments on Indian ground squirrels showed that long-term (60 days) treatment with dexamethasone, which can be compared to long-term natural stress, causes a significant decrease of immune status and plasma melatonin level in these animals (Haldar et al., 2004). However, significant restoration of immunity (i.e., antibody production) and plasma melatonin level was found in simultaneous dexamethasone and melatonin treated animals. On the other hand, it was observed that in Siberian hamster, short-day (prolonged melatonin synthesis) caused a significant increase in plasma cortisol level, which, in turn, might have affected the activity of the immune system of these animals (Pawlak et al., 2005). Taken together, these results suggest that melatonin may act on two different levels – as a signaling molecule transducing the information about the shortening day length, in which case it would inhibit the immune activity by stimulating the release of stress hormones, or directly on immune cells as an immunoenhancer, probably through the increased synthesis of MIIO by immune cells. This explanation is further supported by the fact that the administration of exogenous melatonin had a similar effect as exposure of hamsters to short-day and resulted in the increased aggression in these animals (Jasnow et al., 2002).

(b) Hormones of the Hypothalamo-Pituitary-Thyroid (HPT) Axis

Thyroid hormones, like glucocorticoids, are important for animal adaptation to changing environmental conditions, especially temperature. Thyroid activity of non-hibernating mammals increases during winter compared to summer, in contrast to hibernating mammals, where it decreases during winter (Tomasi et al., 1998). Generally, thyroid hormones enhance immune function, for example, by promoting thymocyte maturation and differentiation (Gala 1991; Fabris et al., 1995). This effect is probably modulated by the thymic peptide – thymulin, as thyroidectomy reduces the thymulin concentration and thyroid hormone replacement restores it to control values (Fabris and Mocchegiani, 1985). Several human disorders characterized by reduction in thyroid hormones are also associated with reduced thymulin concentrations accompanied by immune deficiency (Fabris et al., 1995). In healthy human populations, however, T_3 appears to have a direct stimulatory effect on B-cell differentiation (Paavonen, 1982). Also, upstream components of the HPT axis influence immune function. Receptors for thyrotropin (TSH), possessing many similarities to TSH-binding sites on thyroid cells, have been found on lymphocytes (Pekonen and Weintraub, 1978). In addition, TSH enhances the immune response to specific antigens: it stimulated, for example, the antibody response to T-cell dependent and T-cell independent antigens *in vitro* (Kruger and Blalock, 1986). However, TSH alone does not enhance proliferation independent on antigen stimulation (Provinciali et al., 1992). HPT axis also seems to be responsive to seasonal fluctuations in melatonin levels and might be involved in mediating seasonal changes in immune functions. Melatonin is known, for example, to stimulate the release of hypothalamic thyrotropin-releasing hormone (TRH), which possesses immuno-reconstituting and antiviral activity (Pierpaoli and Yi, 1990). It may also be speculated that the effects of thyroid hormones are secondary to those evoked by steroid hormones or melatonin. It has been demonstrated that the thyroid gland and HPA axis can modulate each other's activity, restoring to some extent their proper functions during different abnormalities. Thus, thyroid hormones may affect immune function by reducing HPA axis activity, which in turn suppresses the immune function.

The thyroid is probably involved in the regulation of immune function/ status in Indian palm squirrels, as thyroidectomy in *F. pennanti* decreased weight of thymus and spleen, TLC, percentage LC of blood and bone marrow, and percentage stimulation ratio of thymocytes and splenocytes, and these parameters were restored to control levels after treatment with melatonin. Moreover, an interesting feature of the functional relationship between thyroxin and melatonin has been noted during two different reproductive phases (i.e., active and inactive), which supports the authors' idea of "Trade off" hypothesis between these two hormones in the control of the immune status. During the reproductively active phase, peripheral melatonin level was low while thyroxin was high, hence thyroxin might have acted as mitogen for lymphoid tissues. On the other hand, during the reproductively inactive phase, when thyroxin level was low and melatonin was high, probably melatonin acted as an immuno-enhancer. Therefore, a complete "Trade off" interrelationship between thyroxin and melatonin was proposed as a mechanism maintaining high immune activity during both phases for the benefit of survival of animals (Fig. 2; Rai et al., 2005).

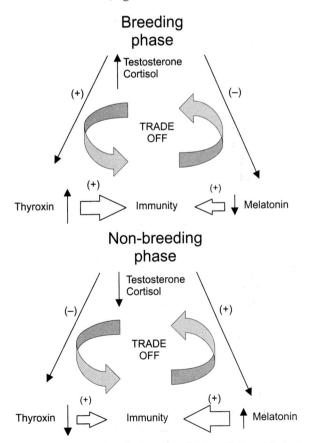

Fig. 2. Diagram showing hypothetical trade-off interrelationship between steroids, thyroxin and melatonin during breeding and non-breeding phase.

(c) Gonadal Steroids and Immunomodulation

The effect of gonadal steroids on lymphoid tissues is well known (Hammar, 1929). Even before the importance of the thymus in immune function had been recognized, researchers noted thymic hypertrophy in response to gonadectomy, particularly in female animals (Eidinger and Garrett, 1972). Consequently, receptors for estrogens and androgens have been found on lymphoid tissue, suggesting that the gonadal steroids can act directly on immune organs (Barr et al., 1984). Most species exhibit a seasonal pattern in reproductive behavior and physiology, including humans (Bronson, 1995), which can be correlated with seasonal variations in immune status (Bilbo and Nelson, 2003). In anticipation of winter or in laboratory-simulated winter conditions, most rodent species exhibit a dramatic decline in gonadotropin and gonadal steroid concentration, eventually leading to the regression of the reproductive system (Bronson and Heideman, 1994; Bronson, 1995).

As with the HPA axis, numerous interactions exist between the HPG axis and the immune system. Importantly, sex steroid hormones are likely mediators of seasonal patterns of immune function (Fig.1). Testosterone generally suppresses the immune function (Singh and Haldar, 2005). Castration of adult male rodents results in increased humoral and cell mediated immunity, as well as increased lymphatic organ size, including thymic, splenic and lymph nodal masses (Schuurs and Verheul, 1990). Treatment of adult castrated males with physiological doses of testosterone compromised their immune function to pre-castrated levels. During short photoperiod, blood androgen level decreases; hence short photoperiodic treatment is similar to a functional castration and simultaneously increases peripheral melatonin level (Schuurs and Verheul, 1990). Thus, enhancement of immune function could be due to both, the removal of the immunosuppressive effects of the androgens and immunoenhancement by melatonin.

The effects of physiological doses of estrogens appear to enhance immune function. Blood estrogen levels are low in short-day females with high circulatory levels of melatonin (Haldar and Singh, 1995). Also, in Siberian hamsters kept in short-day, a significant decrease in estradiol level in females was observed, which can be responsible for a winter decline in the immune activity noted in these animals (Pawlak et al., 2005). On the other hand, the enhancement of winter immune function observed in squirrels is unlikely to involve photoperiod-mediated changes in estrogen levels, again suggesting the major role of melatonin as an immunoenhancer in this species.

CONCLUSION

Immune responses had been studied only in laboratory animal model where the involvement of season and other endocrine organs were scarcely discussed. The study of immune status of wild seasonally breeding species is highly important to save some endangered fauna leading to a new area of research "Immunoecology". The hormones of adrenals (glucocorticoids) and thyroid (L-thyroxine) have some direct modulatory effects on immune status. Not only the gonadal steroids influence immune function, but the products of immune system also modulate the HPG axis. This reciprocal cross talk allows the maintenance of both the immune system and endocrine systems to be regulated within narrow limits. This provides a condition of "Trade off" interrelationship between immune factors and the HPG axis. This hypothesis may reply to some of the clinical problems of patients with chronic immune disorders experiencing gonadal abnormalities or dysfunction. In certain cases of immune disorders, melatonin can act as an immunoenhancer, hence may be of high clinical value.

ACKNOWLEDGMENTS

Research of C. Haldar, S.S. Singh and S. Rai was supported by University Grants Commission (F.3/202/2001) and Council of Scientific and Industrial Research (37(1125)/03/EMR-II), New Delhi, India. Research of J. Pawlak and and K. Skwarlo-Sonta was partly supported by the Polish State Committee for Scientific Research Grant No. 2P04C 09027.

REFERENCES

Angeli, A., Gatti, G. and Sartori, M.L. (1992). Chronobiological aspects of neuroendocrine-immune network: regulation of human natural killer cell activity as a model. *Chronobiologia* **19**: 93.

Aoyama, H., Mori, W. and Mori, N. (1987). Anti-glucocorticoid effects of melatonin in adult rats. *Acta Pathol. Jpn.* **37**: 1143-1148.

Audhya, T., Jain, R. and Hollander, C.S. (1991). Receptor-mediated immunomodulation by corticotropin-releasing factor. *Cell Immunol.* **134**: 77-84.

Barr, I.G., Pyke, K.W., Pearce, P. and Funder, J.W. (1984). Thymic sensitivity to sex hormones develops post-natally. *J. Immunol.* **132**: 1095-1099.

Besedovsky, H.O., del Rey, A., Klusman, I., Furrukawa, H. and Monnge Arditi, G. (1991). Cytokines a modulator of the hypothalamus-pituitary-adrenal axis. *J. Steroid Biochem. Mol. Biol.* **40**: 613-618.

Besedovsky, H.O. and Del Rey, A. (1996). Immuno-neuro-endocrine interactions: facts and hypotheses. *Endocrine Rev.* **17**: 64-102.

Bilbo, S.D., Drazen, D.L., Quan, N., He, L. and Nelson, R.J. (2002). Short day lengths attenuate the symptoms of infection in Siberian hamsters. *Proc. Biol. Sci.* **269**: 447-454.

Bilbo, S.D. and Nelson, R.J. (2003). Sex differences in photoperiodic and stress-induced enhancement of immune function in Siberian hamsters. *Brain Beh. Immun.* **17**: 462-472.

Black, P.H. (1994). Central nervous system – immune system interactions: Psychoneuroendocrinology of stress and its immune consequences. *Antimicrob. Agents Chemother.* **38**: 1-6.

Brainard, G.C., Knobler, R.L., Podolin, P.L., Lavasa, M. and Lubin, F.D. (1987). Neuroimmunology: Modulation of the hamster immune system by photoperiod. *Life Sci.* **40**: 1319-1326.

Bronson, F.H. and Heidman, P.D. (1994). Seasonal regulation of reproduction in mammals. In: *Physiology of Reproduction*, Vol 2. (Knobil, E., Neill J.D., eds.) Raven Press, New York, 541-584.

Bronson, F.H. (1995). Seasonal variation in human reproduction: Environmental factors. *Q. Rev. Biol.* **70**: 141-164.

Calvo, J.R., Rafii-El-ldrissi, M., Pozo, D. and Guerrero, G.M. (1995). Immunomodulatory role of melatonin: specific binding sites in human and rodent lymphoid cells. *J. Pineal Res.* **18**: 119-126.

Carillo-Vico, A., Guerrero, J.M., Laronde, P.J. and Reiter, R.J. (2005). A review of the multiple actions of melatonin on the immune system. *Endocrine* **27**: 189-200.

Caroleo, M.C., Frasca, D., Nistico, G. and Doria, G. (1992). Melatonin as immunomodulator in immunodeficient mice. *Immunopharmacol.* **23**: 81-89.

Csaba, C. and Barath, P. (1975). Morphological changes of thymus and thyroid gland after postnatal extirpation of pineal body. *Endocrinol. Exp.* **9**: 59-67.

Del Gobbo, V., Libri, V., Villani, N., Calio, R. and Nistico, G. (1989). Pinealectomy inhibits interleukin-2 production and natural killer cell activity in mice. *Int. J. Immunopharmacol.* **11**: 567-573.

del Rey, A., Besedovsky, H.O. and Sorkin, E. (1984). Endogenous blood levels of corticosterone control the immunologic cell mass and B cell activity in mice. *J. Immunol.* **133**: 572-575.

Demas, G.E., Klein, S.L and Nelson, R.J. (1996). Reproductive and immune responses to photoperiod and melatonin are linked in *Peromyscus* subspecies. *J. Comp. Physiol. A.* **179**: 819-825.

Dubocovich, M.L. and Markowska, M. (2005). Functional MTl and MT2 melatonin receptors in mammals. *Endocrine* **27**: 101-110.

Eidinger, D. and Garrett, T.J. (1972). Studies of the regulatory effect of the sex hormone on antibody formation and stem cells differentiation. *J. Exp. Med.* **136**: 1098-1116.

Fabris, N. and Mocchegiani, E. (1985). Endocrine control of thymic serum factor in young adult and old mice. *Cell Immunol.* **91**: 325-335.

Fabris, N., Mocchegiani, E. and Provianciali, M. (1995). Pituitary-thyroid axis and immune system: A reciprocal neuroendocrine-immune interaction. *Harm. Res.* **43**: 29-38.

Gala, R.R. (1991). Prolactin and growth hormone in the regulation of immune system. *Proc. Soc. Exp. Bial. Med.* **198**: 513-527.

Garcia-Maurino, S., Pozo, D., Calvo, J.R. and Guerrero, J.M. (2000). Correlation between nuclear melatonin receptor expression and enhanced cytokine production in human lymphocytic and monocytic cell lines. *J. Pineal Res.* **29**: 120-137.

Garidou-Boof, M.L., Sicard, B., Bothorel, B., Pitorsky, B., Ribelayga, C., Simonneaux, V., Pevet, P. and Vivien-Roels, B. (2005). Environmental control and adrenergic regulation of pineal activity in the diurnal tropical rodent, *Arvicanthis ansorgei*. *J. Pineal Res.* **38**: 189-197.

Gitto, E., Tan, D.X., Reiter, R.J., Karbownik, M., Manchester, L.C., Cuzzocrea, S., Fulia, F. and Barberi, I. (2001). Individual and synergistic antioxidative actions of melatonin: studies with vitamin E, vitamin C, glutathione and desferrioxamine (desferoxamine) in rat liver homogenates. *J. Pharm Pharmacol.* **53**: 1393-1401.

Guerrero, J.M and Reiter, R.J. (2002). Melatonin immune system relationship. *Current Topics in Medicinal Chemistry* **2**: 162-169.

Haldar, C., Haussler, D. and Gupta, D. (1991). Effect of pineal gland on rhythmicity of colony forming units for granulocyte-macrophages (CFU-GM) from rat bone marrow cell culture. *Int. J. Biochem.* **23**: 1483-1485.

Haldar, C., Haussler, D. and Gupta, D. (1992a). Effects of opioids (met-enkephalin and β-endorphin, epithalamine, melatonin, thymosin α-1 and noradranalin on colony forming units for granulocyte-macrophages (CFU-GM)) in bone marrow cell cultures from intact and pinealectomized rats. *Biogenic Amine* **911**: 1-6.

Haldar, C., Haussler, D. and Gupta, D. (1992b). Response of CFU-GM (colony forming units for granulocytes and macrophages) from intact and pinealectomized rat bone marrow to murine recombinant interleukin-3 (rIl-3), recombinant granulocyte-macrophage colony stimulating factor (rGM-CSF) and human recombinant erythropoietin (rEPO). *Progress in Brain Res.* **91**: 323-325.

Haldar, C., Haussler, D. and Gupta, D. (1992c). Effect of the pineal gland on circadian rhythmicity of colony forming units for granulocytes and macrophages (CFU-GM) from rat bone marrow cell cultures. *J. Pineal. Res.* **12**: 79-83.

Haldar, C. and Singh, S. (1995). Modulation of ovarian function by various indolearmines during the sexually active phase in the Indian palm squirrel *F pennanti*. *Canadian J. Zool.* **73**: 266-269.

Haldar, C. and Singh, S.S. (2001). Seasonal changes in melatonin and immunological adaptation in birds. *J. Endocrinol. Reprod.* **5**: 13-24.

Haldar, C., Singh, R. and Guchhait, P. (2001). Relationship between the annual rhythm in melatonin and immune system in the tropical palm squirrel *F. pennanti*. *Chronobiol. Internat.* **18**: 61-69.

Haldar, C., Rai, S. and Singh, R. (2004). Melatonin blocks dexamethasone-induced immunosuppression in a seasonally breeding rodent Indian palm squirrel, *Funambulus pennanti*. *Steroids* **69**: 367-377.

Haldar, C., Sharma, S. and Singh, S.S. (2006). Reproductive phase dependent circadian variations of plasma melatonin, testosterone, thyroxine and corticosterone in Indian palm squirrel, *Funambulus pennanti*. *Biol. Rhythm Res.* **37**: 1-10.

Hammar, J.A. (1929). Die Menschenthymus in Gesundheit und Krankheit. Teil II Das organ culture unter anomalen korperverhaltnissen. *Zeitung Mikroskopanatomie Forschung.* **16**: 1-49.

Homo-Delarche, F. and Durant, S. (1994). Hormones neurotransmitters and neuropeptides as modulators of lymphocyte functions. In: *Immunopharmacology of Lymphocytes.* (Rola-Pleszczynski, M., ed.), Academic Press Limited, London, 169-240.

Jasnow, A.M., Huhman, K.L., Bartness, T.J. and Demas, G.E. (2002). Short days and exogenous melatonin increase aggression of male Syrian hamsters (*Mesocricetus auratus*). *Horm. Behav.* **42**: 13-20.

Kruger, T.E. and Blalock, J.E. (1986). Cellular requirements for thyrotropin enhancement of *in vitro* antibody production. *J. Immunol.* **137**: 197-200.

Levi, F.A., Canon, C., Touttou, Y., Sulon, T.J. and Mechkouri, M. (1988). Circadian rhythms in circulating lymphocytes subtypes and plasma testosterone, total and free cortisol in five healthy men. *Clin. Exp. Immunol.* **71**: 329-335.

Lochmiller, R.L. and Ditchkoff, S.S. (1999). Environmental influences on mass dynamics of the cotton rat (*Sigmodon hispidus*) thymus gland. *Biol. Rhythm Res.* **29**: 206-212.

Lopez-Gonzalez, M.A., Calvo, J.R., Osuna, C., Sequra, J.J. and Gurrero, J.M. (1993). Characterization of melatonin binding sites in human peripheral blood neutrophils. *Biotechnol. Ther.* **4**: 253-262.

Maestroni, G.J.M., Conti, A. and Pierpaoli, W. (1987). Role of pineal gland in immunity. II. Melatonin enhances the antibody response via an opiatergic mechanism. *Clin. Exp. Immunol.* **68**: 384-391.

Maestroni, G.J.M. and Conti, A. (1990). The pineal neurohormone melatonin stimulate activated CD4+, Thy-1 + cells to release opioid agonist(s) with immunoenhancing and anti-stress properties. *J. Neuroimmunol.* **28**: 167-176.

Maestroni, G.J.M. (1995). T-helper-2 lymphocytes as a peripheral target of melatonin. *J. Pineal Res.* **18**: 84-89.

Maestroni, G.J.M., Zammaretti, F. and Pedrinis, E. (1999). Hematopoietic effect of melatonin. Involvement of k-1 opioid receptor on bone marrow macrophages and interleukin-1. *J. Pineal Res.* **27**: 145-153.

Mahmoud, I., Salman, S.S. and Al-Khateeb, A. (1994). Continuous darkness and continuous light induced structural changes in the rat thymus. *J. Anal.* **185**: 143-149.

Majewski, P., Dziwnski, T., Pawlak, J., Waloch, M. and Skwarlo-Sonta, K. (2005). Anti-inflammatory and opioid-mediated effects of melatonin on experimental peritonitis in chickens. *Life Sci.* **76**: 1907-1920.

McNaulty, J.A., Relfson, L.M., Fox, L.M., Kus, L., Handa, R.J. and Schneider, G.B. (1990). Circadian analysis of mononuclear cells in the rats following pinealectomy and superior cervical ganglionectomy. *Brain. Behav. Immunol.* **4**: 292-307.

Molinero, P., Soutto, M., Benot, S., Hmadcha, A. and Guerrero, J.M. (2000). Melatonin is responsible for the nocturnal increase observed in serum and thymus of thymosin αl and thymulin concentrations: observations in rats and humans. *J. Neuroimmunol.* **103**: 180-188.

Munk, A. and Guyre, P.M. (1986). Glucocorticoid physiology, pharmacology and stress. *Adv. Exp. Med. Biol.* **196**: 81-96.

Nelson, R.J. and Blom, J.M. (1994). Photoperiodic effects on tumor development and immune function. *J. Biol. Rhythm* **9**: 233-249.

Nelson, R.J., Demas, G.E., Klein, S.L. and Kriegsfeld, L.J. (1995). The influence of season, photoperiod and pineal melatonin on immune function. *J. Pineal Res.* **19**: 149-165.

Nelson, R.J., Demas, G.E. and Klein, S.L. (1998). Photoperiodic mediation of seasonal breeding and immune function in rodents: A multi-factorial approach. *Amer. Zool.* **38**: 226-237.

Nelson, R.J. and Drazen, D.L. (1999). Melatonin mediates seasonal adjustment in immune function. *Reprod. Nutr. Dev.* **39**: 383-398.

Paavonen, T. (1982). Enhancement of human B lymphocyte differentiation *in vitro* by thyroid hormone. Scand. *J. Immunol.* **15**: 211-215.

Pawlak, J., Majewski, P., Maarkowska, M. and Skwarlo-Sonta, K. (2005). Season- and gender-dependent changes in the immune function of Siberian hamsters (*Phodopus sungorus*). *Neuroendocrinol. Lett.* **26**: 55-60.

Pekonen, F. and Weintraub, B.D. (1978). Thyrotropin bimding to cultured lymphocytes and thyroid cells. *Endocrinol.* **103**: 1668-1677.

Persengiev, S., Patchev, V. and Velev, B. (1991). Melatonin effects on thymus steroid receptors in the course of primary antibody responses: significance of circulating glucocorticoid level. *Int. J. Biochem.* **23**: 1487-1489.

Pierpaoli, W. and Yi, C. (1990). The involvement of pineal gland and melatonin in immunity and aging, thymus mediated immuno-reconstituting and antiviral activity of thyrotropin releasing hormone. *J Neuroimmunol.* **27**: 99-109.

Poon, A.M., Liu, Z.M., Tang, F. and Pang, S.F. (1994). Evidence for a direct action of melatonin on the immune system. *Biol. Signals* **3**: 107-117.

Provinciali, M., Di Stefano, G. and Fabris, N. (1992). Improvement in the proliferative capacity and natural killer cell activity of murine spleen lymphocytes by thyrotropin. *Int. J. Immunopharmacol.* **14**: 865-870.

Rafii-El-Idrissi, M., Calvo, J.R., Pozo, D., Harmouch, A. and Guerrero, J.G.M. (1995). Specific binding of 2[^{125}I] iodomelatonin by rat splenocytes: Characterization and its role on regulation of cyclic AMP production. *J. Neuroimmunol.* **57**: 171-178.

Rafii-El-Idrissi, M., Calvo, J.R. and Harmouch, A. (1998). Specific binding of melatonin by purified cell nuclei from spleen and thymus of the rat. *J. Neuroimmunol.* **86**: 190-197.

Rai, S. and Haldar, C. (2003). Pineal control of immune status and hematological changes in blood and bone marrow of male squirrels (*Funambulus pennanti*) during their reproductively active phase. *Comp. Biochem. Physiol.* (C). **136**: 319-328.

Rai, S. (2004). Immunomodulatory role of melatonin and thyroxin in seasonally breeding vertebrate, *Funambulus pennanti*. Ph.D. thesis, BHU, Varanasi, India.

Rai, S., Haldar, C. and Singh, S.S. (2005). Trade-off between L-thyroxin and melatonin in immune regulation of the Indian palm squirrel, *Funambulus pennanti* during the reproductively inactive phase. *J. Neuroendocrinol.* **82**: 103-110.

Reiter, R.J., Tan, D.X., Osuna, C. and Gitto, E. (2000a). Actions of melatonin in the reduction of oxidative stress. A review. *J. Biomed. Sci.* **7**: 444-458.

Reiter, R.J., Calvo, J.R., Karbownik, M. and Qi Wand Tan, D.X. (2000b). Melatonin and its relation to the immune system and inflammation. *Ann. NY Acad. Sci.* **917**: 376-386.

Schuurs, A.H. and Verheul, H.A.M. (1990). Effects of gender and sex steroids on the immune response. *J. Steroid Biochem.* **35**: 157-172.

Singh, S.S. (2003). Avian pineal organ: Role in maintenance of immunity in Indian jungle bush quail, *Perdicula asiatica*. Ph.D. thesis, BHU, Varanasi, India.

Singh, S.S. and Haldar, C. (2005). Effect of exogenous dexamethasone and melatonin on the immune status of Indian jungle bush quail, *Perdicula asiatica. Proc. XX Symp. Reprod. Biol. Comp. Endocrinol.*, Tiruchi, *Gen. Comp. Endocrinol.* **141**: 226-232.

Skwarlo-Sonta, K. (2002), Melatonin in immunity: a comparative aspect. *Neuroendocrinol. Lett.* **23**(1): 67-72.

Skwarlo-Sonta, K., Majewski, P., Markowska, M., Oblap, R. and Olszanska, B. (2003). Bidirectional communication between the pineal gland and immune system. *Can. J. Physiol. Pharmacol.* **81**: 342-348.

Smith, E.M. and Blalock, J.E. (1981). Human lymphocyte production of corticotropin and endorphin like substances: association with leukocyte interferon. *Proc. Natl. Acad. Sci. USA* **78**: 7530-7534.

Tomasi, T.E., Hellgren, E.C. and Tucker, T.J. (1998). Thyroid hormone concentration in black bear (*Ursus americanus*): Hibernation and pregnancy effects. *Gen. Comp. Endocrinol.* **109**: 192-199.

Treeck, O., Haldar, C. and Ortmann, O. (2006). Antiestrogens modulate MT1 melatonin receptor expression in breast and ovarian cancer cell lines. *Oncol. Rep.* **15**: 231-235.

Vermeulen, M., Palerno, M. and Giordano, M. (1993). Neonatal pinealectomy impairs murine antibody dependent cellular toxicity. *J. Neuroimmunol.* **43**: 97-101.

Werb, Z., Foley, R. and Munk, A. (1978). Interaction of glucocorticoids with macrophages: Identification of glucocorticoids in monocytes and macrophages. *J. Exp. Med.* **147**: 1684-1694.

Winczyk, K., Pawlikowski, M. and Karasek, M. (2000). Effects of somatostatin analogs octreotide and lanreotide on the proliferation and apoptosis in colon 38 tumor: interaction with 5-fluorouracil. *Neuroendocrinol. Lett.* **21**: 137-142.

Wurtman, R.J. and Weisel, J. (1969). Environmental lighting and neuroendocrine function: Relationship between spectrum of light source and gonadal growth. *Endocrinol.* **85**: 218-1221.

Yellon, S.M., Teasley, L.A. and Fagoaga, O.R. (1999). Role of photoperiod and pineal gland in T cell dependent humoral immune reactivity in the Siberian hamsters. *J. Pineal Res.* **27**: 243-248.

Index